Erhard Godehardt

Graphs
as
Structural Models

Advances in System Analysis
Editor: Dietmar P. F. Möller

Manuscripts submitted to *Advances in System Analysis* must be original, pointing out the advancement of the contribution with respect to the actual a-priori knowledge.

Manuscripts or exposés should be sent to the Editor of the Series:
Dietmar P. F. Möller, Johannes Gutenberg Universität Mainz, Physiologisches Institut, Saarstr. 21, D-6500 Mainz 1, W.-Germany.

Erhard Godehardt

Graphs as
Structural Models

The Application of Graphs
and Multigraphs in Cluster Analysis

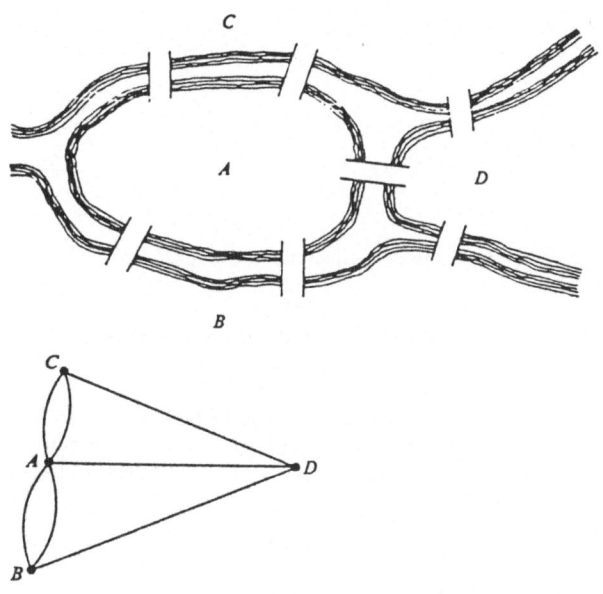

Friedr. Vieweg & Sohn   Braunschweig / Wiesbaden

CIP-Titelaufnahme der Deutschen Bibliothek

Godehardt, Erhard:
Graphs as structural models: the application
of graphs and multigraphs in cluster analysis/
Erhard Godehardt. — Braunschweig; Wiesbaden:
Vieweg, 1988
   (Advances in system analysis; Vol. 4)
   ISBN 978-3-528-06312-2   ISBN 978-3-322-96310-9 (eBook)
   DOI 10.1007/978-3-322-96310-9

NE: GT

Vieweg is a subsidiary company of the Bertelsmann Publishing Group.

Printed and bound by Lengericher Handelsdruckerei, Lengerich

ISBN 978-3-528-06312-2

ISSN   0932-593X

PREFACE

The advent of the high-speed computer with its enormous storage capabilities enabled statisticians as well as researchers from the different topics of life sciences to apply multivariate statistical procedures to large data sets to explore their structures. More and more, methods of graphical representation and data analysis are used for investigations. These methods belong to a topic of growing popularity, known as "exploratory data analysis" or EDA.

In many applications, there is reason to believe that a set of objects can be clustered into subgroups that differ in meaningful ways. Extensive data sets, for example, are stored in clinical cancer registers. In large data sets like these, nobody would expect the objects to be homogeneous. The most commonly used terms for the class of procedures that seek to separate the component data into groups are "cluster analysis" or "numerical taxonomy". The origins of cluster analysis can be found in biology and anthropology at the beginning of the century. The first systematic investigations in cluster analysis are those of K. Pearson in 1894. The search for classifications or typologies of objects or persons, however, is indigenous not only to biology but to a wide variety of disciplines. Thus, in recent years, a growing interest in classification and related areas has taken place. Today, we see applications of cluster analysis not only to biology but also to such diverse areas as psychology, regional analysis, marketing research, chemistry, archaeology and medicine.

These applications indicate not only the importance of the existing classification tools. They also stress the need for further development and investigation into classification theory as a mathematical topic. This progress in the development of mathematical procedures and models is not only stimulated indirectly by the advances in electronic data processing (since lots of data are waiting to be evaluated), but also directly inspired directly by the improvements in electronic computers which allow thorough study and simulation of more complex models. The spectrum of mathematical models for classification reaches from multivariate analysis to the theory of graphs and random graphs.

The roots of graph theory are obscure. The famous eighteenth-century Swiss mathematician Leonard Euler was perhaps the first to solve a problem using graphs when he was asked to consider the problem of the Königsberg bridges (in the 1730s). Problems in (finite) graph theory are often enumeration problems, and thus can become rather intricate to solve. However, in the late 1950s and early 1960s the Hungarian mathematicians Paul Erdős and Alfred Rényi founded the theory of random graphs and used probabilistic methods (limit theorems) to by-pass enumeration problems. (These problems also became secondary with the emergence of powerful computers.) Thus, perhaps no topic in mathematics has enjoyed such explosive growth in recent years as graph theory. This stepchild of combinatorics and topology has emerged as a fascinating topic for research in its own right. Moreover, during the last two decades, calculus of graph theory has proved to be a valuable tool in applied mathematics and life sciences as well. Using graph-theoretic concepts, scientists study properties of real systems by modelling and simulation. The aim of graph-theoretic investigations is, in fact, the simplest topological structure after that of isolated points: The structure of a graph is that of "points" or "vertices", and "edges" or "lines". A graph can be conveniently pictured as a diagram where the vertices appear as small circular dots and the edges are indicated with line segments joining two appropriate dots (or arcs instead of lines if the direction is of relevance).

These graphs can be used to model many real systems. This usually works if the vertices can be identified as physical objects, and the edges can be identified as relations or equations between the objects. In fact, every binary structural relation can be described by a graph. Thus it was a natural consequence to use graphs in classification theory; here, the similarity between the objects defines the relation. Structures in data sets can also be depicted very simply and impressively by dendrograms, a special kind of graph. The advantage of using graphs when describing and modelling structures is the fact that the analyst can rely on a plethora of mathematical theorems and results to gain insight into the structure of a real system. Graph theory provides researchers of different topics with a single (mathematical) language. Thus, it promotes the exchange of ideas to a degree which probably would be impossible if the scientists relied only on the technical terms of their own subject. This is reflected in the fact that a lot of theoretical results in graph theory were not found by pure mathematicians but by scientists working in the field of applied mathematics. Like every mathematical model, graph theory is only useful for applications so long as it describes our scientific observations with a certain amount of accuracy. Its immediate popularity, however, shows that it provides the researcher with a lot of good ideas — at least within the limitations of our actual scientific knowledge.

In many topics in the natural and life sciences, the interpretation of quantitative results from data is tied to mathematical ideas and models. This, however, implies an obligation for the true biometrician: In contrast to pure mathematics and statistics, mathematical concepts in data analysis not only have to be "true" on a theoretical basis, but also must fit into different environments like physiology and biology. And to be applicable, they must stay as simple as possible. This is impossible without constant discussions between the applied mathematicians and the researchers from life sciences. The classification model proposed in this book has been discussed with many physicians and biologists, among others with Professor K.-E. Richard (Department of Neural Surgery of the University of Köln), Priv.-Doz. Dr. H. Borberg (Department of Internal Medicine I of the University of Köln), Priv.-Doz. Dr. H. Feltkamp (Department of Medical Research of Bayer Pharmaceutics, Wuppertal) and Dr. E. Rehse (Institute of Medical Documentation and Statistics of the University of Köln).

The idea of applying results from the theory of random multigraphs rose in 1980 from my dissertation in pure mathematics on probability models for random multigraphs (a generalization of the concepts of P. Erdős and A. Rényi). I am grateful to Professor H. Klinger from the Institute of Mathematical Statistics and Documentation of the University of Düsseldorf, where I wrote this thesis. He and Professor J. Steinebach (now at the Statistical Institute of the University of Hannover) not only discussed the mathematical proofs with me, but, together with Professors J. Krauth (Psychological Institute of the University of Düsseldorf) and O. Richter (now at the Division of Statistics of the Agricultural Institute of the University of Bonn) encouraged me to apply my theoretical results in practice. I also wish to thank Professors V. Weidtman and P. Bauer from the Institute of Medical Documentation and Statistics of the University of Köln where I wrote my habilitation thesis in biometrics on the application of graph-theoretic models in exploratory statistics, and where I was employed until 1986 after I moved from Düsseldorf to Köln. We had very interesting and stimulating discussions on multivariate data analysis and on the problems of the application of graph-theoretic models and concepts to medicine.

It is a great pleasure to acknowledge the generous help of Professors D. Matula (Department of Computer Science, Southern Methodist University, Dallas) and J. Wierman (Mathematical Sciences Department, Johns Hopkins University, Baltimore). I am especially greatful to Professors J.W. Kennedy and L.V. Quintas (Dyson College, Pace University New York) and to Professor M. Karoński, Dr. Z. Palka and Dr. J. Jaworski (Mathematical Institute of the University of Poznań), who not only gave valuable hints for my work, but also invited me for research work to New York and to Poznań. (In Poznań, 1987, the work on this book started: I began to translate my habilitation thesis into English.)

In this monograph, I review the principles and properties of numerical classification. The different chapters deal with its possibilities as well as with its limitations. I emphasize the application of graph-theoretic concepts as tools in the structural analysis of data sets which are composed of "mixed data" as they often occur in medicine. I further propose a graph-theoretic classification model which I developed on the basis of multigraphs.

Chapter 0 deals with the mathematical symbols and notations which are necessary for the understanding of this book. In Chapter 1, a short review of the elementary ideas of mathematical modelling, graph theory, and exploratory statistics are given. They are followed by a short description of the basic ideas of cluster analysis. In Chapter 2, I give a survey of the current methods and different algorithms of cluster analysis. Chapter 3 deals with graph-theoretic concepts in classification theory, especially with the single-linkage and k-linkage procedures. An algorithm for the uncovering of clusters is proposed which has been developed on the basis of my multigraph model.

Up to this point, judging the "relevance" of a classification of a data set is not considered. The following two chapters are devoted to probabilistic models for evaluating the validity of the suggested clusters. In Chapter 4, investigations in the development of different statistical tests are reported. Again, I focus on graph-theoretic based probability models, which are discussed in detail. I propose a conditional statistical test for the homogeneity of a data set which is derived from random multigraphs. In Chapter 5, a probability model for random multigraphs is derived. This part is written as an independent section in this book. It also can be read as a mathematical theory on its own right. The main test principles reported in the previous chapter, however, are founded upon this theory.

The last chapter then deals with three examples, all from medicine. The first two examples are the pharmacokinetics of Urapidil and Lidocaine. In these medical trials, our interest is focussed on studying whether differences in the kinetics of these drugs can be explained by certain external criteria, namely by the occurrence of additional kidney or liver impairments. That is, we want to find whether the a priori groups (according to diagnoses) are reflected by the a posteriori groups (according to the kinetic data). The third example is from the habilitation thesis of Priv.-Doz. Dr. H. Feltkamp. It is part of a long-term study of the significance of pregnancy-induced hypertensive disorders as a prognostic index for a manifestation of an essential hypertension lateron.

It was intended originally, to include only the data from the studies on the pharmacokinetics of Urapidil and Lidocaine as purely numerical examples of the application of the multigraph method. However, since the interpretation of cluster analyses makes little sense without a knowledge of the scientific background to the data, it was decided to include a much fuller exposition of these studies. More details of the compartmental models — together with some modifications — are given in my habilitation thesis. These models are the result of the discussions with Dr. R. Haerlin and Dr. V. Steinijans (Byk Gulden Lomberg Chemicals, Konstanz). The complete data can be found in the doctorial dissertations of R. Dworatzek and W. Heitz (both in the Department of Internal Medicine II of the University of Köln).

The idea of developing interactive programs for numerical classification on the Prime computer of the Institute of Medical Documentation and Statistics would not have been realized without the enormous work of J. Kunert and H. Herrmann. The program package is composed of two independent parts. Part 1 is used to find clusters (it is described in Chapter 3). With the second part, the hypothesis of homogeneity of a data set, i.e., the hypothesis of randomness of clusters, can be tested using finite or asymptotic test statistics from the theory of random multigraphs (see Chapter 5). The algorithms for the finite tests are based on our doctorial dissertation. At the present, H. Herrmann, J. Kunert and I are re-writing these programs at the Department of Thoracic and Cardiovascular Surgery of the University of Düsseldorf, where I have been employed since summer 1986, and at the Institute of Medical Documentation and Statistics — and at home, of course — so that they will be user-friendly and will run on Prime computers and personal computers.

I want to thank all those persons who encouraged and supported me in writing this book. H. Herrmann showed me how to use the TEX scientific text system and wrote most of the macros for the preparation of the manuscript. Professor J. Wierman kindly assisted my first attempts in translating my thesis in Posnań, and Professor A.D. Barbour from the Institute of Applied Mathematics of the University of Zürich immediately agreed to read the final manuscript and made suggestions to improve it.

Finally, I must thank the editor of this series of monographs, Dr. D.P.F. Möller, and the Vieweg Verlag for their patience and encouragement. I also thank Byk Gulden Lomberg Chemicals, Konstanz, and Prime Computers, Wiesbaden, for their financial support.

Neuss, Spring 1988   Erhard Godehardt

TABLE OF CONTENTS

CHAPTER 0

MATHEMATICAL SYMBOLS AND NOTATION

$\mathcal{A}, \mathcal{B}, \mathcal{C}, \ldots$	Sets.
$\lvert \mathcal{A} \rvert$	The number of elements in the set $\mathcal{A}$.
$\emptyset$	The set with no elements, the empty set; $\lvert \emptyset \rvert = 0$.
$\{a, b, \ldots\}$	The set with the elements a, b, ... In a set, the order of the elements is not important, and all elements are mutually different.
$\{a, b\}, \{a_1, a_2\}$	Two-element sets with a and b, or a_1 and a_2 as elements. By definition, $a \neq b$, $a_1 \neq a_2$, $\{a, b\} = \{b, a\}$ and $\{a_1, a_2\} = \{a_2, a_1\}$ all hold true.
$(a_n)_{n \in \mathcal{N}}$	The sequence (ordered set) of the elements (numbers) $a_1, a_2, \ldots$. The order (ranking) of the elements is important in sequences. Elements on different ranks maybe equal.
$(a, b), (a_1, a_2)$	Ordered pair (or tuple) with a and b, or a_1 and a_2 as elements. By definition, $(a, b) \neq (b, a)$ and $(a_1, a_2) \neq (a_2, a_1)$. Furtheron, in ordered pairs we may have $a = b$ or $a_1 = a_2$.
$\mathcal{N}$	The set of natural numbers exclusive the number 0.
$\mathcal{N}_0$	The set of natural numbers inclusive the number 0.
$\mathcal{R}$	The set of real numbers. Its elements usually are denoted by a, b, $c, \ldots$, or r, s, $t, \ldots$.
$\mathcal{R}^t$	The set of t-dimensional real numbers.
$a \in \mathcal{A}$	The set $\mathcal{A}$ contains the element a.
$a \notin \mathcal{A}$	The set $\mathcal{A}$ does not contain the element a.
$\mathcal{A} - \mathcal{B}$	The difference of the two sets $\mathcal{A}$ and $\mathcal{B}$; $\mathcal{A} - \mathcal{B} := \{a : a \in \mathcal{A}, a \notin \mathcal{B}\}$.
$\mathcal{A} \subseteq \mathcal{B}$	$\mathcal{A}$ is a subset of $\mathcal{B}$ (including the case $\mathcal{A} = \mathcal{B}$).
$\mathcal{A} \subset \mathcal{B}$	$\mathcal{A}$ is a proper subset of $\mathcal{B}$ (i.e., $\lvert \mathcal{A} \rvert < \lvert \mathcal{B} \rvert$).
$\mathcal{P}(\mathcal{A})$	The power set of the set $\mathcal{A}$.
$\mathcal{A} \cup \mathcal{B}$	The union of the sets $\mathcal{A}$ and $\mathcal{B}$.
$\mathcal{A} \cap \mathcal{B}$	The cut of the sets $\mathcal{A}$ and $\mathcal{B}$.
$\overline{\mathcal{A}}$	The complement of the set $\mathcal{A}$.
$\mathcal{A} \times \mathcal{B}$	The Cartesian product of the sets $\mathcal{A}$ and $\mathcal{B}$; $\mathcal{A} \times \mathcal{B} := \{(a, b) : a \in \mathcal{A}, b \in \mathcal{B}\}$.
$\min_n a_n$ or $\min\{a_1, a_2, \ldots\}$	Minimum of the numbers $a_1, a_2, \ldots$.

$\max\limits_n a_n$ or $\max\{a_1, a_2, \ldots\}$ — Maximum of the numbers $a_1, a_2, \ldots$.

$\vec{x}, \vec{x}_{*l}$ — Column vectors; for $x_1, \ldots, x_n \in \mathcal{R}$ and $x_{1l}, \ldots, x_{nl} \in \mathcal{R}$,
$$\vec{x} = \begin{pmatrix} x_1 \\ \vdots \\ x_k \end{pmatrix}, \quad \vec{x}_{*l} = \begin{pmatrix} x_{1l} \\ \vdots \\ x_{nl} \end{pmatrix}.$$

$\vec{x}^T, \vec{x}_{i*}$ — Row vectors; for $x_1, \ldots, x_t \in \mathcal{R}$ and $x_{i1}, \ldots, x_{it} \in \mathcal{R}$,
$\vec{x}^T = (x_1, \ldots, x_t)$, $\vec{x}_{i*} = (x_{i1}, \ldots, x_{it})$.

$\mathbf{X}, \mathbf{S}, \mathbf{\Sigma}$ — Matrices; $\mathbf{X} = (x_{ij})$, $\mathbf{S} = (s_{ij})$, $\mathbf{\Sigma} = (\sigma_{ij})$, etc.

$\mathbf{X}^T, \mathbf{S}^T, \mathbf{\Sigma}^T$ — Transposed matrices of $\mathbf{X}, \mathbf{S}, \mathbf{\Sigma}$.

$\mathbf{X}^{-1}, \mathbf{S}^{-1}, \mathbf{\Sigma}^{-1}$ — Inverse matrices of $\mathbf{X}, \mathbf{S}, \mathbf{\Sigma}$.

$\mathbf{E}$ — Unity matrix; matrix with diagonal elements 1, and 0 at all other places.

$a := b$ — The expressions a and b are equal by definition.

$\mathcal{A} :\Leftrightarrow \mathcal{B}$ — The properties (sets) $\mathcal{A}$ and $\mathcal{B}$ are equivalent by definition.

$\to (\uparrow, \downarrow)$ — Convergence (monotonic increasing, monotonic decreasing) to a limit.

$\lim\limits_{n \to \infty} a_n$ — Limit (limes) of the sequence $(a_n)_{n \to \infty}$.

$\limsup\limits_{n \to \infty} a_n$ — Upper limit (limes superior) of the sequence $(a_n)_{n \to \infty}$.

$\liminf\limits_{n \to \infty} a_n$ — Lower limit (limes inferior) of the sequence $(a_n)_{n \to \infty}$.

$o(f(n))$ — Landau symbol; for $f(n) > 0$, $n \in \mathcal{N}$: $g(n) = o(f(n)) :\Leftrightarrow$ $\lim\limits_{n \to \infty} \dfrac{g(n)}{f(n)} = 0$.

$O(f(n))$ — Landau symbol: for $f(n) > 0$, $n \in \mathcal{N}$: $g(n) = O(f(n)) :\Leftrightarrow \dfrac{g(n)}{f(n)}$ is bounded.

$\sim$ — Asymptotical equal; for $f(n) > 0$, $n \in \mathcal{N}$: $g(n) \sim f(n) :\Leftrightarrow$ $\lim\limits_{n \to \infty} \dfrac{g(n)}{f(n)} = 1$.

$\equiv$ — Identical; $g(n) \equiv f(n) :\Leftrightarrow g(n) = f(n)$ for all n.

$\sum\limits_{\nu=a}^{b}, \prod\limits_{\nu=a}^{b}$ — Sum or product sign; for $a > b$, the whole sum is defined as 0, and the product is defined as 1.

$\lfloor x \rfloor$ — The integer part of a real number x, i.e., the largest integer which is smaller or equals x.

$\lceil x \rceil$ — The smallest integer which is greater or equals x, where x is real.

$e^x, \exp(x)$ — Exponential function of x.

$\log x$ — Natural logarithm of x.

$n!$ — Special type of product: n factorial. For $n \in \mathcal{N}$ it is defined as $n! := 1 \cdot 2 \cdots n = \prod\limits_{\nu=1}^{n} \nu$. For $n = 0$, we put $0! := 1$.

$n_{(m)}$ — Special type of product: m-th factorial of n. For $n, m \in \mathcal{N}_0$ and $n \geq m$, it is defined as $n_{(m)} := \dfrac{n!}{(n-m)!}$. For all other cases, we put $n_{(m)} := 0$.

$\binom{n}{m}$ Binomial coefficient n choose m. For $n, m \in \mathcal{N}_0$ and $n \geq m$, it is defined as $\binom{n}{m} := \frac{n_{(m)}}{m!} = \frac{n!}{m!(n-m)!}$. For all other cases, we put $\binom{n}{m} := 0$.

$\binom{n}{n_1, \ldots, n_k}$ Multinomial coefficient n choose n_1 to n_k. For $n_1, \ldots, n_k \in \mathcal{N}_0$ with $n_1 + \cdots + n_k = n$, it is defined as $\binom{n}{n_1, \ldots, n_k} := \frac{n!}{n_1! \cdots n_k!}$. For all other cases, we put $\binom{n}{n_1, \ldots, n_k} := 0$. As a special case, we get $\binom{n}{m, n-m} = \binom{n}{m}$.

$(\Omega, \mathcal{A}, P)$ Probability space with the event set Ω, the σ-algebra $\mathcal{A}$ and the probability measure P.

$T, U, \ldots, Z$ Random variables defined on $(\Omega, \mathcal{A}, P)$.

$EX, E(X)$ Expectation of the random variable X.

$E(X|y)$ Conditional expectation of the random variable X under the condition y.

$EE(X|Y)$ Expectation of the conditional expectation of X under the condition Y.

$Cov(X, Y)$ Covariance of the random variables X and Y (a matrix if X and Y are vector-valued).

$Var\, X, Var(X)$ Variance of the random variable X.

$P(\lambda)$ Poisson distribution with parameter λ. A random variable X is $P(\lambda)$-distributed $:\Leftrightarrow P(X = k) = e^{-\lambda} \frac{\lambda^k}{k!}$. The main parameters are $EX = Var\, X = \lambda$.

$B(n, p)$ Binominal distribution with parameters n and p. A random variable X is $B(n, p)$-distributed $:\Leftrightarrow P(x = k) = \binom{n}{k} p^k (1 - p)^{n-k}$. The main parameters are $EX = np$, $Var\, X = np(1 - p)$.

$H(N, M, n)$ Hypergeometric distribution with parameters N, M and n (N, the number of elements; M, the number of "good" elements; n, the number of elements drawn). A random variable X is hypergeometric distributed $:\Leftrightarrow P(X = k) = \binom{M}{k} \binom{N-M}{n-k} \binom{N}{n}^{-1}$. The main parameters are $EX = n\frac{M}{N}$, $Var\, X = n\frac{M(N-M)(N-n)}{N^2(N-1)}$.

$N(\mu, \sigma^2)$ Normal distribution with parameters μ and σ^2 (univariate). A normally distributed random variable X has the density $f(x) = \frac{1}{\sqrt{2\pi\sigma^2}} e^{(x-\mu)^2/(2\sigma^2)}$. The main parameters are $EX = \mu$, $Var\, X = \sigma^2$.

$N(\vec{\mu}, \boldsymbol{\Sigma})$ Multivariate normal distribution with vector of expectations $\vec{\mu}$ and covariance matrix $\boldsymbol{\Sigma}$. For $N(0,1)$-distributed random variables $Z_1, \ldots, Z_m$, be $\vec{Z} := (Z_1, \ldots, Z_m)^T$, $\vec{\mu}$ a n-dimensional vector and $\mathbf{A}$ an (n, m)-matrix, then the vector-valued random variable $\vec{X} = \mathbf{A}\vec{Z} + \vec{\mu}$ is $N(\vec{\mu}, \boldsymbol{\Sigma})$-distributed with $\boldsymbol{\Sigma} = \mathbf{A}\mathbf{A}^T$.

For a better and easier reading, we generally omit commas to separate indices (with the exception that we will explicitly separate indices to prevent misinterpretations, e.g., if numbers occur together with letters). Thus, Γ_{tnN} and $\Gamma_{t,n,N}$ are the same as are $T_{ij.}$ and $T_{i,j,.}$. However, more easy can be $\Gamma_{t,n,N(n)}$ read. Also is $\Gamma_{t,n,\lfloor EV \rfloor}$ easier to be read than $\Gamma_{tnN(n)}$ or $\Gamma_{tn\lfloor EV \rfloor}$. Thus, it is better to use commas here. If the indices are clear from the context then we omit them completely. We, e.g., write $\Gamma_{n,\lfloor EV \rfloor}$ instead of $\Gamma_{n,\lfloor E_{tnN}V_s \rfloor}$. The dot index in $T_{ij.}$ denotes summation over that index, i.e., $T_{ij.} := \sum_{l=1}^{t} T_{ijl}$.

The following monographs deal with the terms and notations which we introduced here. Thus, these books are a good basis for a better understanding of this work. We especially recommend the first and third book of this list because they provide the reader with the concepts of probability theory and graph theory. In our opinion, they are excellent introductory texts, complete, compact, and very readable.

Behzad, M., Chartrand, G., Lesniak-Foster, L.: *Graphs and Digraphs*. Prindle, Weber & Schmidt, Boston 1979

Cavalli-Sforza, L.: *Biometrie — Grundzüge biologisch-medizinischer Statistik*. Gustav Fischer Verlag, Stuttgart – New York 1980

Grimmett, G., Welsh, D.: *Probability — An Introduction*. Oxford University Press, Oxford – New York – Toronto 1986

Hartung, J., Elpelt, B.: *Multivariate Statistik*. Oldenbourg, München – Wien 1984

Hartung, J., Elpelt, B., Klösener, K.-H.: *Statistik*. Oldenbourg, München – Wien 1984

Lorenz, R.J.: *Grundbegriffe der Biometrie*. Gustav Fischer Verlag, Stuttgart – New York 1984

Mott, J.L., Kandel, A., Baker, T.P.: *Discrete Mathematics for Computer Scientists and Mathematicians*. Prentice-Hall, Englewood Cliffs, N.J., 1986

Nöbauer, W., Timischl, W.: *Mathematische Modelle in der Biologie*. Vieweg, Braunschweig – Wiesbaden 1979

Peil, J.: *Grundlagen der Biomathematik*. Verlag Harri Deutsch, Thun – Frankfurt 1985

Weber, E.: *Grundriß der biologischen Statistik*. Gustav Fischer Verlag, Stuttgart – New York 1980

CHAPTER 1

INTRODUCTION, BASIC CONCEPTS

Ideas about and hypotheses concerning our environment are formulated as "models". These models should express the essentials of the environment in a simplified and easy-to-understand manner. Models are images of reality or of our understanding of reality. Every scientific perception bears this model character. Scientific research is composed of two complementary areas: The **generation** of hypotheses and their **validation**. In 1934, K.L. Popper stated in his monograph "LOGIK DER FORSCHUNG" (see [316]):

> "Die Tätigkeit des wissenschaftlichen Forschers besteht darin, Sätze oder Systeme von Sätzen aufzustellen und systematisch zu überprüfen; in den empirischen Wissenschaften sind es insbesondere die Hypothesen, Theoriesysteme, die aufgestellt und an der Erfahrung, durch Beobachtung und Experiment überprüft werden."

The generation of hypotheses as well as their testing are always based on **data**. Often, scientific ideas rise from data of casual or fortuitous observations. Postulation and validation of scientific hypotheses, however, is usually performed by planned and controlled experiments. Here, sampling is done according to elaborate plans, and the data are analysed by use of (multivariate) exploratory or confirmatory statistical procedures. A serious disadvantage in most planned experiments is, however, that only few of the possible factors can be varied, contrary to reality. Thus, laws in natural or life sciences result from (and are verified by use of) simplified experiments or small random samples. For this purpose, abstract (mental) models are created which should behave under experimental conditions in the same way as in real life.

1.1 MODELLING IN MEDICINE AND BIOLOGY

In biological and medical research, abstract models are used mainly to prove scientific hypotheses and theories empirically. Another purpose of modelling is to try out new relations or to forecast responses to certain actions. The major difficulty in systems analysis and in the generation of scientific hypotheses is that our knowledge of the true relations between influencing and responding entities in biology is rather rudimentary (in contrast to our knowledge of technical systems which are far more simple than biological systems). Therefore, we are forced to develop rather crude models by simply omitting the bulk of the influencing factors. As a consequence, we can create different models as answers to the same scientific problem; and all of them will "fit" the data in some respect (illuminating different aspects of the problem). Usually, from these possible models the simplest is chosen which just suffices to explain the observed phenomena (**minimal model**). We can say:

(a) Scientific modelling is an art;
(b) all models are wrong;
(c) some models are better than other ones;
(d) our task is to find the best ones.

To prove (or disprove) scientific hypotheses, we usually compare the logical results of mental experiments with real data (Figure 1-1). Hypotheses usually are formulated as plausible and simplified models (for example, as equations or differential equations in classical physics, as systems of usually linear differential equations in pharmacokinetics, and as systems of partial or nonlinear differential equations in physiology). In our understanding of science, we have variables which act as factors, and variables which show the responses or effects.. The systems of mathematical equations above obey this principle of cause and effect.

In life sciences, however, real systems are not as simple as our models of them. Biological systems show a rich entanglement with the environment. Most often, the true behaviour of some system components is unknown as are the relations between them. Therefore, the cause-effect principle is replaced by probabilistic models. Here, the (unknown) influence of so-called "nuisance factors" or "disturbance factors" on the response variables is interpreted as a **random error** or **random noise** in the response variables. We speak of **biological variation**. Hypotheses are postulated in form of probability models on fictive, fixed **populations**. These **abstract models** then are proved or disproved by **statistical tests**. For this purpose, random samples are drawn (which can be interpreted as **real models**, see Figure 1-1). Estimates of the parameters for the whole population are computed from these samples. The behaviour of these estimates serves as basis for making predictions on the behaviour of the model parameters of the whole population. Statistical tests usually serve as a tool which enables us to decide between different models (**null hypothesis** H_0 and **alternative** H_1). The risk of misjudgement is controlled by error probabilities. Furthermore, we can minimize the risk of misjudgements by careful design of experiments ([190], [238], [335]).

This procedure of deciding between hypotheses already given is known as **confirmatory statistics** (also **testing, analytical** or **inference statistics**). Stochastic models are chosen since deterministic ones are often far too complicated. The cause-effect principle is abandoned because stochastic models yield results which can be understood and interpreted with less expense. It is, however, a philosophical problem whether such a stochastic model — or other simplified models — mirrors reality. (An interesting alternative to probabilistic models, by which one can model images of complex dynamical systems and which can produce non-predictible events on a purely deterministic basis, is offered by the **theory of fractals**, originally founded by B.B. Mandelbrot, see [310].) Confirmatory statistics are used to decide between different hypotheses, that is, to **test** hypotheses. By use of methods of confirmatory statistics, we hope to pick that hypothesis which is the best of those existing at the moment. The methods of **exploratory statistics**, on the other hand, are used to **generate** models.

Every — mathematical — model is an **abstraction** and **idealization** of a real

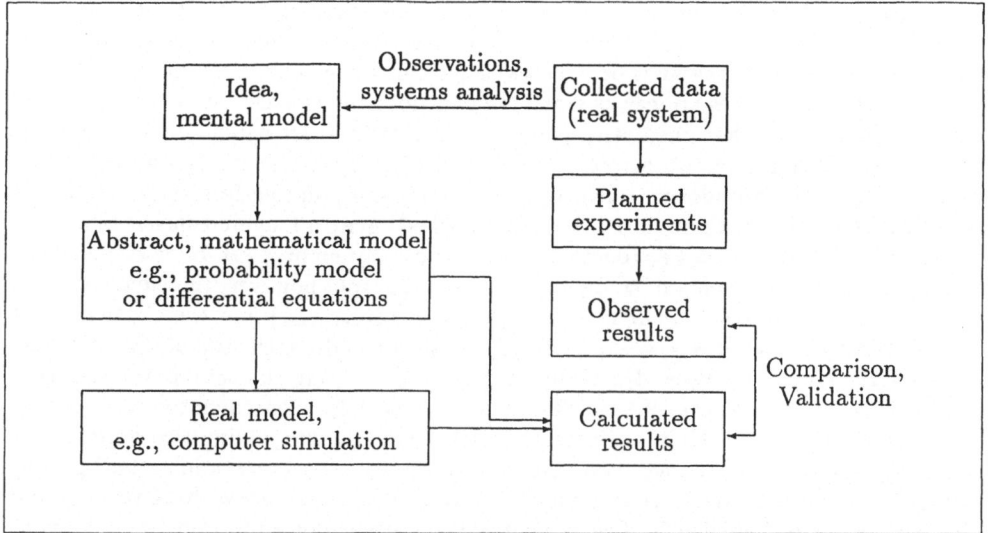

Figure 1-1 *The process of scientific research for the improvement of the knowledge about our environment; the computer simulation in the real model can be a Monte Carlo simulation (for probability models) or the numerical solution of a system of differential equations.*

system. During the process of abstraction, certain details are dropped to keep the model controllable. This often results in rather crude models. (In classification problems, when using discriminant analysis, we propose that all variables which are relevant for the grouping have been observed. Furthermore, we propose that the different groups can be separated by hyperspheres; that means that all relations between the variables are linear.) We thus gain a first impression of the system's behaviour. We often improve such crude, first models later on by adding previously neglected elements and by reshaping the relations between the variables.

Moreover, during the process of data collection we are faced with inaccuracies and errors in measurement; and we have to take biological variation into account (in stochastic models). Thus, agreement between real data and an abstract model cannot be perfect. Compliance is possible within certain tolerance limits only. It becomes important to prove whether or not a theoretical model describes the behaviour of the real system sufficiently, that is, whether or not it can be fitted sufficiently to the data (**model validation**). Two procedures of validation are possible: The procedure of **verification** and that of **falsification**. By verifying a model we want to show by a sequence of experiments that there is — sufficient — agreement between our abstract model and the real system. In contrast, falsifying a model means looking for a single example which proves disagreement between our model computations and the data from an experiment (see [346], [347]).

1.2 Graphs as Tools in Mathematical Modelling

Graph theory emerged from combinatorics and topology. A **directed graph** (or **di-**

graph) $\Delta = (\mathcal{G}, \mathcal{H})$ is composed of two sets: A finite nonempty set $\mathcal{G}$ of **points** or **vertices** together with a (possibly empty) set $\mathcal{H}$ of two-element ordered subsets of distinct elements of $\mathcal{G}$, the **arcs** or **arrows**. Digraphs are conveniently pictured as diagrams where the vertices appear as small circular dots and the arcs are drawn pointing from one dot to another one. Any pair of points of $\mathcal{G}$ — let us call them ξ_i and ξ_j — can be connected by two arrows: One called (ξ_j, ξ_i) (or shorter δ_{ji}) and pointing from ξ_i to ξ_j, and another one called (ξ_i, ξ_j) (or δ_{ij}) and pointing from ξ_j to ξ_i (an arc is a **directed pair**, cf. Chapter 0). An (**undirected**) **graph** Γ again consists of a finite nonempty set $\mathcal{G}$ of vertices ξ_i together with a (possibly empty) set $\mathcal{H}$ of two-element subsets of distinct elements of $\mathcal{G}$. In contrast to a digraph, however, the elements of $\mathcal{H}$ are **nonordered pairs** of elements of $\mathcal{G}$. (Let $\mathcal{G} = \{\xi_1, \ldots, \xi_n\}$ and the elements of $\mathcal{G}$ be denoted by κ_{ij} then $\kappa_{ij} = \kappa_{ji} = \{\xi_i, \xi_j\}$.) In an undirected graph, the elements of $\mathcal{H}$ are the **lines** or **edges**. No vertex is preferred as being the source or cause (the origin of the arrow), or the sink or effect (the arrowhead). Consequently, in diagrams of (undirected) graphs, edges are pictured with line segments joining two appropriate dots; only one edge can connect two vertices. Figure 1-2 shows two (undirected) graphs (one with 58 vertices and 58 edges, which is "unlabelled", and one with 14 vertices and 20 edges, which is "labelled"). The geometric structure is not important, we only want to know which vertives are connected by lines and which are not. Thus, the structure of a graph is the simplest structure after that of isolated points.

Because of this simple structure, calculus of graph theory has proved to be a valuable tool in applied mathematics and life sciences as well. Using graph-theoretic concepts, scientists study properties of real systems by modelling and simulation. This usually works if the vertices can be identified as physical objects, and the edges can be identified as relations or equations between the objects. In fact, every binary structural relation can be described by a graph. In [405], G. Uhlenbeck published his investigations in statistical mechanics. Molecules are interpreted as vertices of a graph. Two molecules are connected by an edge if they are close enough for physical reciprocation.

In digraphs and graphs, no arc or edge can connect a vertex to itself. Thus, if $\mathcal{G}$ contains n vertices then we can say the following about the number N of elements in $\mathcal{H}$: $0 \leq N = |\mathcal{H}| \leq n(n-1)$ in a digraph Δ, and $0 \leq N = |\mathcal{H}| \leq \binom{n}{2} = \frac{1}{2}n(n-1)$ in a graph Γ. If we admit **loops** (that are edges or arcs which connect a vertex with itself) then we get **pseudographs**. In **multigraphs**, more than one edge can join two different vertices or more than one arc can point from a vertex to another one. Graphs, di-, pseudo- or multigraphs, respectively, are **labelled** if all vertices and edges (or arcs) are labelled uniquely. If in graphs, di- and pseudographs, the vertices are denoted by $\xi_1, \ldots, \xi_n$ then the edges usually are labelled as pairs $\{\xi_i, \xi_j\}$ or (ξ_i, ξ_j), or by double indices (like κ_{ij} or δ_{ij}). In undirected graphs, the lower index is put in front of the higher one without any restriction (see Chapter 0). In Figure 1-2, two undirected graphs are pictured. The graph in the upper part is not labelled, while the graph in the lower part is labelled (not all labels of the edges are attached here, we only want to explain the principle of labelling). Often it is useful to attach additional numbers as **weights** to the edges. For example, we can define the Euclidean distance d_{ij} of two vertices ξ_i and ξ_j as the weight of the edge κ_{ij} (if the vertices are points of the $\mathcal{R}^t$). We thus get **edge weighted graphs** (or simply **weighted graphs**).

A graph Γ of size n is called **complete** if it contains all $\binom{n}{2}$ edges, that is, if its edge set is complete. It is **connected** if every pair (ξ_{i_1}, ξ_{i_m}) of vertices in $\mathcal{G}$ is connected by an alternating sequence $\xi_{i_1}, \kappa_{i_1, i_2}, \xi_{i_2}, \ldots, \xi_{i_{m-1}}, \kappa_{i_{m-1}, i_m}, \xi_{i_m}$ of vertices and edges

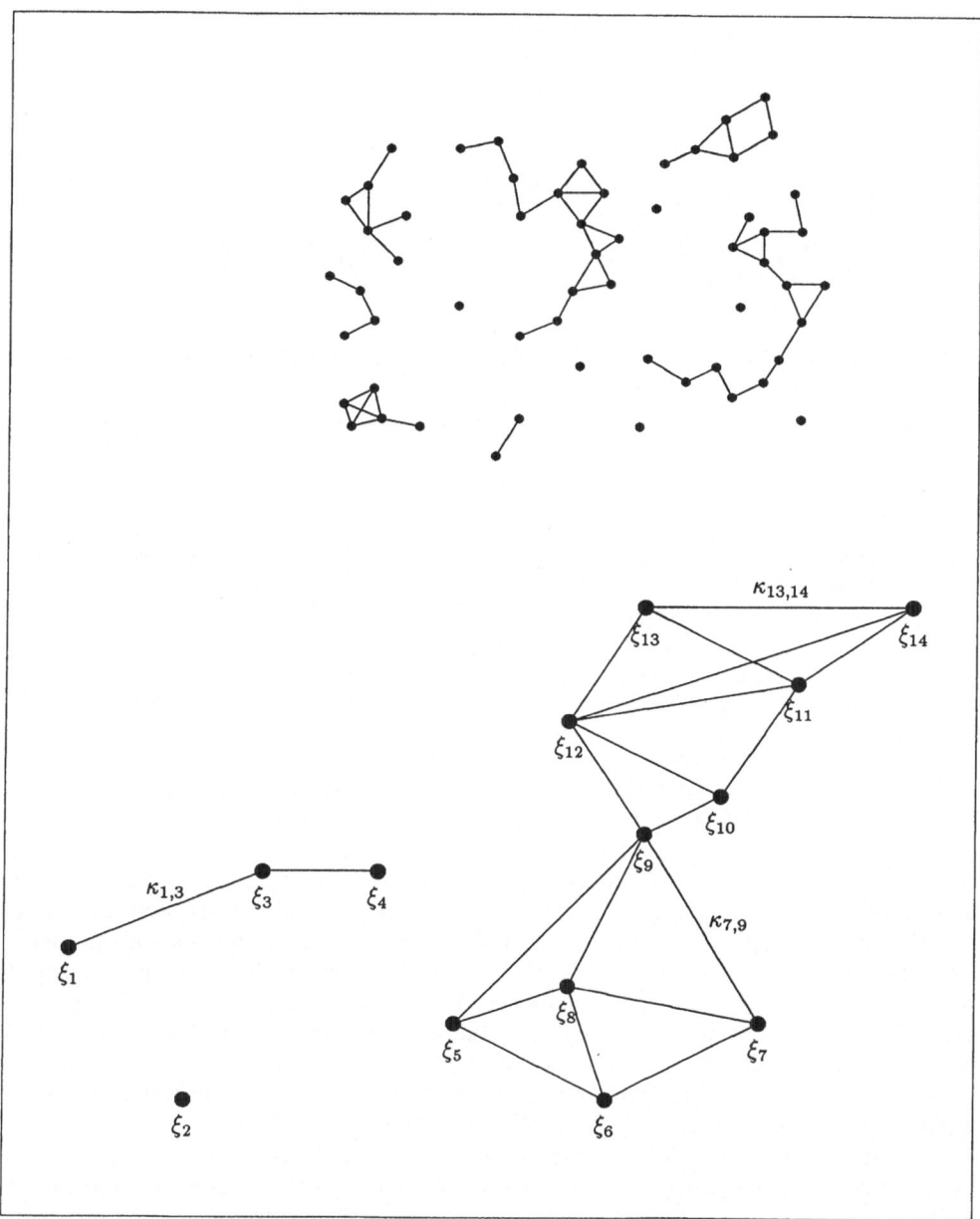

Figure 1-2 *Two graphs, the vertices are pictured as dots, lines connecting pairs of dots indicate the edges. The graph in the upper part of the picture is not labelled since no labels are attached to the different vertices and edges; the graph in the lower part of the picture is labelled, the labels of the vertices are $\xi_1, \ldots, \xi_{14}$, the edges therefore can be labelled uniquely as described in the text.*

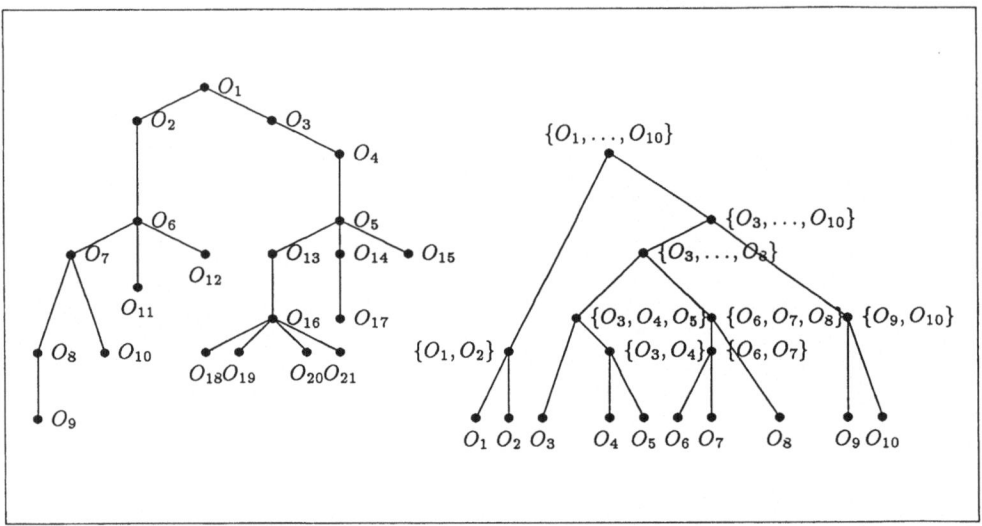

Figure 1-3 *Two different kinds of dendrograms (pictured as rooted trees with the "root" at the top of the tree):*

(a) The vertices are single objects (like in phylogenetic trees);

(b) only the vertices at the bottom indicate single objects, all other vertices are subsets of the set $\{O_1, \ldots, O_{10}\}$, which is the root;

note that both rooted trees are labelled (in two different ways).

in Γ. This alternating sequence is called a **path**. Either the number of edges in a path or the sum of their weights are denoted as its length.

By removing edges from a graph Γ we get a subgraph Γ'; by removing some vertices together with all those edges having one of these vertices as an endpoint, we obtain a subgraph $\Gamma^\star$ of Γ. The graph in Figure 1-2 is obviously not connected. It is split into disjoint subgraphs $\Gamma^\star$, however, which are connected but which lose this property when further vertices (together with the related edges) are added to them. To say it more precisely: A subgraph $\Gamma^\star$ is **connected** if every pair (ξ_{i_1}, ξ_{i_m}) of its vertices is connected by a path lying completely in $\Gamma^\star$ (all vertices and edges of that path are part of this subgraph). It is called a **component** of Γ if it is maximal connected, (if it is maximal with respect to the fact that any two of its vertices are interconnected by an s-path). The **size** m of a component is the number of its vertices.

As we can see from Figure 1-2, the degree of connectedness of the vertices varies in the different components: Some vertices are linked together by only one path, some by two paths and so on. Thus it makes sense to define a degree of connectedness or compactness of subgraphs by certain modifications of the term "component". A component, in which every two vertices are connected by one single path only, is called a **tree**. (In a tree of size m are only $m-1$ edges linking the m vertices together; isolated vertices are trees of size 1.). A **spanning tree** Γ' of a connected graph $\Gamma = (\mathcal{G}, \mathcal{H})$ consists of all n vertices in $\mathcal{G}$ and $n-1$ edges from $\mathcal{H}$ such that Γ' is connected, too. A special kind of trees are the **rooted trees** (or **dendrograms, hierarchies**), where one distinct vertex serves as a **root** (see Figure 1-3 for two examples).

By a **weak component of degree** k of a graph Γ, we denote a connected subgraph Γ^* which is maximal with respect to the fact that every vertex is directly connected with at least k other vertices of Γ^*. That is, in such a component the degree of every vertex is at least k. A **strong component of degree** k (or shorter, a k-**component**) of Γ is a connected subgraph, which is maximal with respect to the fact that every two of its vertices are connected by at least k different paths which mutually share no edges (edge disjoint paths). A connected subgraph Γ^* is a k-**block** if it has at least $k+1$ vertices and is maximal with respect to the fact that at least k vertex disjoint paths in Γ^* connect every two vertices of this subgraph ("vertex disjoint" means that the paths have no vertex in common with the exception of the endpoints which they connect). A **clique** (or: maximal complete subgraph) is a subgraph, which is maximal regarding the fact that every two of its vertices are directly linked together by an edge. While different components, weak or strong components are vertex disjoint, different k-blocks or cliques can have vertices in common.

If digraphs are used to model biological systems, an arrow pointing from a vertex to another one, indicates that the vertex (symbolizing an object of the system) from which the arrow starts, controls the vertex at which the arrowhead points (that is, the second vertex is a function of the first one). Digraphs are used to describe and solve systems of coupled linear differential equations. Complex systems can be analysed and reduced to simpler ones — which then can provide a basis for numerical simulations — by using digraphs (see [151], [159], [191], [192], [209], [254]). Further applications of digraphs are mentioned in [75] (biological networks, food webs), in [318], [381], [382] (networks), and in [68] (sociology).

The theory of random graphs — a topic of probabilistic combinatorics — provides the researcher with solutions to applied problems in probability theory and statistics ([261], [390], [420]). For example, in classification problems like those considered in the following chapters, the objects of a sample $\mathcal{S}$ can be considered as vertices of an (undirected) graph. Here, two vertices are linked together by an edge if and only if the related objects satisfy a particular similarity relation (if, for example, the Euclidean distance of their variables is smaller than a given threshold). Single-linkage clusters then can be defined as the components of a graph. We know about the probabilities of peculiar graphs under "random conditions". From these results, we can derive assumptions on the structure of a data set. Thus, graph theory is not only useful in exploratory statistics (as in finding clusters in a possibly heterogeneous data set). Beyond this, it provides the scientist with tools for confirmatory statistics. In the language of random graphs, testing the homogeneity of a data set (and thus the "randomnes" of the outlined clusters) means testing whether the components of the related graph can be thought of as found by randomly drawing edges between the different objects (which means random attachment of similarities to pairs of objects).

Like every mathematical model, graph theory is only useful for applications so long as it describes our scientific observations with a certain amount of accuracy. Its immediate popularity, however, shows that it provides the researcher with a lot of good ideas — at least within the limitations of our actual scientific knowledge. The advantage of using graphs when describing and modelling structures is based on the fact that the analyst can rely on a plethora of mathematical theorems and results to gain insight into the structure of a real system. For the remainder of this monograph, we always think of undirected graphs or multigraphs if we simply speak of graphs or multigraphs. That is, an edge connecting two vertices indicates a reciprocation or an alternating effect between the vertices.

1.3 THE SCOPE OF EXPLORATORY DATA ANALYSIS

In almost all topics of scientific research, a lot of data is simply collected and stored without specific a priori hypotheses to be tested. Extensive data sets are stored, for example, in clinical cancer registers. Certain **properties** or **attributes** (like age, cancer stage or another stage of illness) of **objects** or **individuals** (patients, for example) are recorded. The possible structures of such large, multidimensional data sets cannot be seen and interpreted directly. It becomes necessary to prepare the data by means of statistical procedures to make it possible to analyse their structure. This type of data analysis — which is used to generate hypotheses and not to test them — is called **exploratory statistics** (another popular term is "data snooping").

Methods of exploratory statistics are usually applied to multivariate, high-dimensional data. Experimental design, data collection and recording in medicine can be a very cumbersome task. Thus, after finishing a medical trial to test a given a priori hypothesis, the recorded data may be used for further inquiries. New correlations between different variables (attributes) or structural relations between the objects may be uncovered by these retrospective analyses. The aim of exploratory statistics thus is (see [43])

(a) the characterisation of the distribution of the data (**descriptive statistics**);
(b) the uncovering of mutually dependent attributes (**regression, correlation**);
(c) the elimination of redundant information, that is, the elimination of unnecessary and unimportant data which may be highly correlated to other variables (**reduction of dimensionality**);
(d) the uncovering of heterogeneity or homogeneity of the objects of a data set (**classification**),
(e) the reconstruction of spatial structures and chronological processes and their mutual correlations (**seriation**).

Thus, a biometrician's task is not only the validation and testing of hypotheses but the **generation** of hypotheses as well. Most procedures of multivariate statistics are very complex. This especially holds true for those procedures used for the exploration of a data set, since here the biometrician cannot rely on a special a priori model for the data. Therefore, these procedures presume the availability of electronic data processing and good numerical programs. Storage space and computation time are the critical requirements that limit both the number n of objects and the number t of attributes or dimensions for many statistical inquiries. Thus, until recently, biometricians who wanted to explore a data set carefully were restricted to simple univariate descriptive methods or regression and correlation analysis. Although theoretical concepts of exploratory statistics have been developed as early as those of confirmatory statistics, the practical application of such complex procedures like cluster analysis was made possible only by the advent of the high-speed computer with its enormous storage capabilities.

In exploratory data analysis, we usually discriminate between **dependent** statistical methods and **interdependent** ones. Dependent methods are applied if the attributes of a data set can be divided into two groups: The variables of the first group are the factors, influencing the variables of the second group which mirror the responses (effects). If we cannot divide the variables into these classes then we are bound to interdependent statistical procedures where all variables are considered as being equivalent..

Of the dependent procedures, the methods of **multiple linear regression** are well known and often applied. Since the evolution of powerful computers, biometricians have also become familiar with **nonlinear regression**, for example, in compartmental analysis in pharmacokinetics where simple mathematical models can be assumed (usually in form of coupled systems of ordinary or stochastic differential equations, see [4], [7], [11], [23], [144], [145], [209], [282], [296], [324], [325], [326]). Less known are the principles of **canonical analysis** and of **path analysis**. The principles of canonical analysis were derived and published in 1936 by H. Hotelling ([199]). Although not popular today, canonical analysis is of significant theoretical importance since multiple linear regression and discriminant analysis both are topics of this theory (see also [190], [255]). It is used to uncover correlations between two distinct groups of variables (the variables of one group being interpreted as causes and the attributes of the other one as effects). With methods of path analysis, again cause-effect models are analysed. These models are related to those of canonical analysis. They, however, are based on the calculus of digraphs (see [96], [97], [151], [159], [254]).

Methods of interdependent statistical analysis to be used for the exploration of high dimensional data are, among others,

(a) **factor analysis** and **principal component analysis** to be used to reduce dimensionality and to uncover and describe correlations between continuous variables like "blood pressure" and "body weight";

(b) the classification methods of **correspondence analysis, analysis of multivariate contingency tables** and the related **analysis of configuration frequency** to outline correlations between qualitative data like "blood group" and "smoker, yes or no";

(c) the classification methods of **discriminant analysis** and **cluster analysis** to answer the question whether a data set is homogeneous or composed of different, separate subgroups, and to split the sample into these clusters.

All these methods, although meanwhile based on mathematical theories, originally rose from practical problems in biology, psychology and sociology. Those classification procedures named in (b) and (c), are common to researchers in medicine and in the life sciences by now (see [6], [25], [38], [42], [48], [66], [76], [77], [93], [102], [116], [117], [127], [183], [190], [210], [216], [242], [243], [250], [255], [258], [335], [341], [373], [385], [406], [413], [415]).

Factor analysis and principal component analysis are developments of linear correlation analysis. Both techniques are based on the calculus of matrix algebra and need heavy computations when applied to multivariate data sets. They virtually cannot be performed without computer programs (see [10], [181], [190], [243], [402], [404], [416]).

The beginnings of classical factor analysis are connected to research in human intelligence. C. Spearman analysed the correlation coefficients between different intelligence tests. He published the results of his investigations in 1904 in a paper on the theory of intelligence ([389]). More than 40 years later, L.L. Thurstone elaborated and generalized Spearman's theory in [402] using the concepts of matrix algebra. Factor analysis is used to combine highly correlated attributes and to trace them back to few essential influencing **factors**. It is hoped by factor analysis to uncover those causal variables which cannot be observed directly. (Psychologists think that the many scores in psychological tests can be traced back to a small number of hidden factors like "intelligence" or "memory".) Methods of factor analysis are often applied in psychology and sociology. They are, however, not commonly used in other topics of natural and life sciences since they postulate a rather special **linear model**: The relations between different variables are proposed to be of the form of linear equations. Thus, in many cases the factors uncovered cannot be interpreted sufficiently or even seem to be purely artificial.

Principal component analysis is connected with the names of K. Pearson and H. Hotelling ([199], [307]). It bears strong relations to factor analysis. The mutually dependent variables are transformed into a set of orthogonal and uncorrelated components. Mathematically, the computation of this transformation is an eigenvalue problem. The transformation is uniquely determined if the eigenvalues of the empirical correlation matrix all are different. Methods of principal component analysis are used to gain insight into the structure of the data.

As with factor analysis, techniques of principal component analysis are applied to a large extent in psychology and sociology only. They are restricted to linear models, too. Thus, their application is problematical. Moreover, because of the many transformations and computing operations needed, a critical analysis and interpretation of the results is often rather difficult. The danger of an uncritical acceptance of possibly artificial results at least must be taken into consideration.

Correspondence analysis was founded by H.D. Hirschfeld and R.A. Fisher ([124], [196]). By this method and by the so-called "analysis of contingency tables", correlations between qualitative or classified variables can be uncovered, and types or syndromes (factors) can be found. Their main purpose, however, is not the detection of factors themselves but the attachment of individuals to these factors by using measurements of qualitative attributes. The same holds for the nonparametric analysis of configuration frequencies ([242]).

At the beginning of this century, procedures had been asked for in anthropoloy, by which archaeological discoveries could be attached uniquely to one of several groups or populations (knowledge about the number and kind of the groups was presupposed). The optimal splitting of a sample of objects into the different groups as well as — later on — the attachment of a new single object to one of these groups was to be performed according to objective measurements only. The first methods which were proposed for this purpose, were tailored very closely to anthropology and to the particular questions to be answered. At that time, it seemed impossible to generalize these methods to other sciences. In 1936, however, in his paper "THE USE OF MULTIPLE MEASUREMENTS IN TAXONOMIC PROBLEMS" R.A. Fisher proposed a method which is known today as "discriminant analysis". This method allows one to attach objects to different groups using only measurements (continuous variables only) which have been taken on them. It is not restricted to a particular science ([123]). Nowadays, under discriminant analysis we find a collection of multivariate procedures, which perform an attachment of objects of a sample to different groups which are given a priori. The optimal partition of the objects into the different groups is achieved by calculation of the discriminant functions which contain those measurements which maximise a criterion for the quality of the partition (or which minimise the number of incorrectly grouped objects from a training data set). These discriminant functions usually are linear or quadratic.

None of the methods described above are direct classification methods. They propose an a priori knowledge about the different groups like the number of different classes and their shape (external criteria). For example, the discriminant functions are calculated using training data samples. The portion of incorrectly classified objects then provides a measure of goodness for different sets of discriminant functions. It can also be used to decide whether we can separate objects (on the basis of the measurements we took) with sufficient success into the different groups or not. New objects can then be attached to the different known groups using these discriminant functions. In medicine, such procedures are used as tools for prognosis and decision-making.

Again in anthropology, another problem for a researcher was (and still is) not only to attach new objects to well-known groups (which already are defined by parameters like a mean vector and a covariance matrix) but to **find** and define different classes on the basis of the measurements of a sample of objects found at a site. Using techniques of cluster analysis, he can examine the structures of previously ungrouped data, and can split the objects into classes. By "cluster analysis", on one hand, we denote procedures which can be used to answer the question whether a sample is homogeneous. On the other hand, procedures which partition the objects of a (possibly heterogeneous) sample into clusters are described as "cluster analysis". Another traditional name for these **cluster-detecting algorithms** is "(numerical) taxonomy". Unter this name, they are widely used in biology.

The principle of cluster analysis procedures — when detecting and outlining groups — is that of optimization. The attributes of each individual object are replaced by attributes of "average objects" which are typical only for the whole cluster (class or group). The many objects are replaced by only few classes (or their representatives, respectively), which means both a loss of individual information and the benefits of economy and a better insight into the structure of the data. A balance between the loss of precision and information and the benefit of clearness and economy is sought. This is the main purpose of performing a cluster analysis for practical or operational reasons. In most cases we then are interested in compact, round groups. Ready-made clothes are an often cited example of a classification for pure operational or commercial reasons. The population does not "naturally" divide into groups coinciding with the different measures of ready-made clothes. These measures are chosen arbitrarily according to a criterion based only on a commercial standpoint. All the people in a group are then represented by a "typical" average representative. Very often, however, scientists want to uncover "natural clusters" which are contained in a sample of biological, anthropological or medical data. Whatever the case, the clusters should be chosen so that the objects within a cluster are "mutually similar" (**homogeneity within the classes**) while the representatives of distinct clusters are "mutually dissimilar" (**heterogeneity between the classes**). Thus, the task of cluster-detecting algorithms is either to reconstruct the unknown membership of different objects in natural clusters or to find an optimal partition.

The first systematic investigations in cluster analysis had been done already in 1894. In that year, K. Pearson published his paper "CONTRIBUTIONS TO THE MATHEMATICAL THEORY OF EVOLUTION" ([306]). Probably the first algorithm for automatic classification which was not heuristic and intuitive but mathematically based appeared in 1951. It was published by K. Florek, J. Perkal and co-authors. In this algorithm, graph-theoretic concepts together with similarities and dissimilarities between the pairs of objects instead of the geometric structure of the data are used to detect clusters ([128], [312]). The use of distances for outlining groups is extensively discussed by L.S. Penrose in [311]. Many of the cluster-detecting methods and algorithms known today, are based on similarities between the objects. Other methods rely on estimations of the point density.

It is not only the objects of a sample that can be clustered. Another application of classification procedures is the grouping of attributes. If an empirical correlation matrix for the attributes in a sample has been computed, this also defines similarities for the different items. They thus can be grouped into different classes. As opposed to factor analysis, however, we are not now interested in uncovering factors and attaching different attributes to these factors; we are instead interested only in the mutual relations of the attributes. Thus, the difference between cluster analysis and factor analysis is **not** that by the first method we classify objects while by the second method we group attributes. The difference is that the well-defined (and restrictive) linear model of factor analysis has no equivalent in most classification procedures.

Such linear models make it easier to derive statistically testable hypotheses on the data structure. Nevertheless, linear model can rarely be assumed to hold in samples where little is known about their structure. Thus, they are of no importance in the classification of objects (see [117], [126], [127]).

Methods of exploratory data analysis are designed to support researchers in uncovering new phenomena. The essential problem in the interpretation of the results of such an exploratory analysis lies in the fact that we are tempted to generalize those models or hypotheses, which have been derived from one special sample, to a whole population. This, however, is admissible only if models from exploratory studies have been validated with methods of confirmatory statistics and with new data. Model validation on the basis of exploratory methods alone is impossible. The purpose of confirmatory statistics (together with careful experimental design), on the other hand, is to validate phenomena and hypotheses from investigations which have been previously performed. Its aim is at least to keep the probability for wrong decisions as low as possible (if not to prevent researchers from premature and wrong decisions). This confirmation is necessary. At the same time, pure confirmation alone is not sufficient for progress. For this reason, both principles of statistics have their own importance in scientific research, and exploratory methods are indispensable for the advance of scientific research.

1.4 THE BASIC CONCEPTS OF CLUSTER ANALYSIS

In this section, we want to explain the basic ideas of cluster analysis. We are interested in techniques of uncovering structures in a heterogeneous data set, and we want to outline natural (or real) groups. Since the outcome of a cluster analysis (the different groups as well as their number and shape) depends on the objects of the sample as well as on the measurements taken, a cluster analysis does not begin with the application of a specific cluster-detecting procedure. It already starts with the drawing of the sample. In their monograph "PRINCIPLES OF NUMERICAL TAXONOMY", the microbiologists R.R. Sokal and P.A.H. Sneath define the different passes of a cluster analysis as follows (see Figure 1-4 and [373]):

(a) The sample is drawn.
(b) The researcher decides on the attributes to be measured and (if necessary) on the similarity measure to be used for the classification.
(c) The measurements are taken; data recording and preparation is performed.
(d) The computations of pairwise similarities and of the values of group-specific representatives are done.

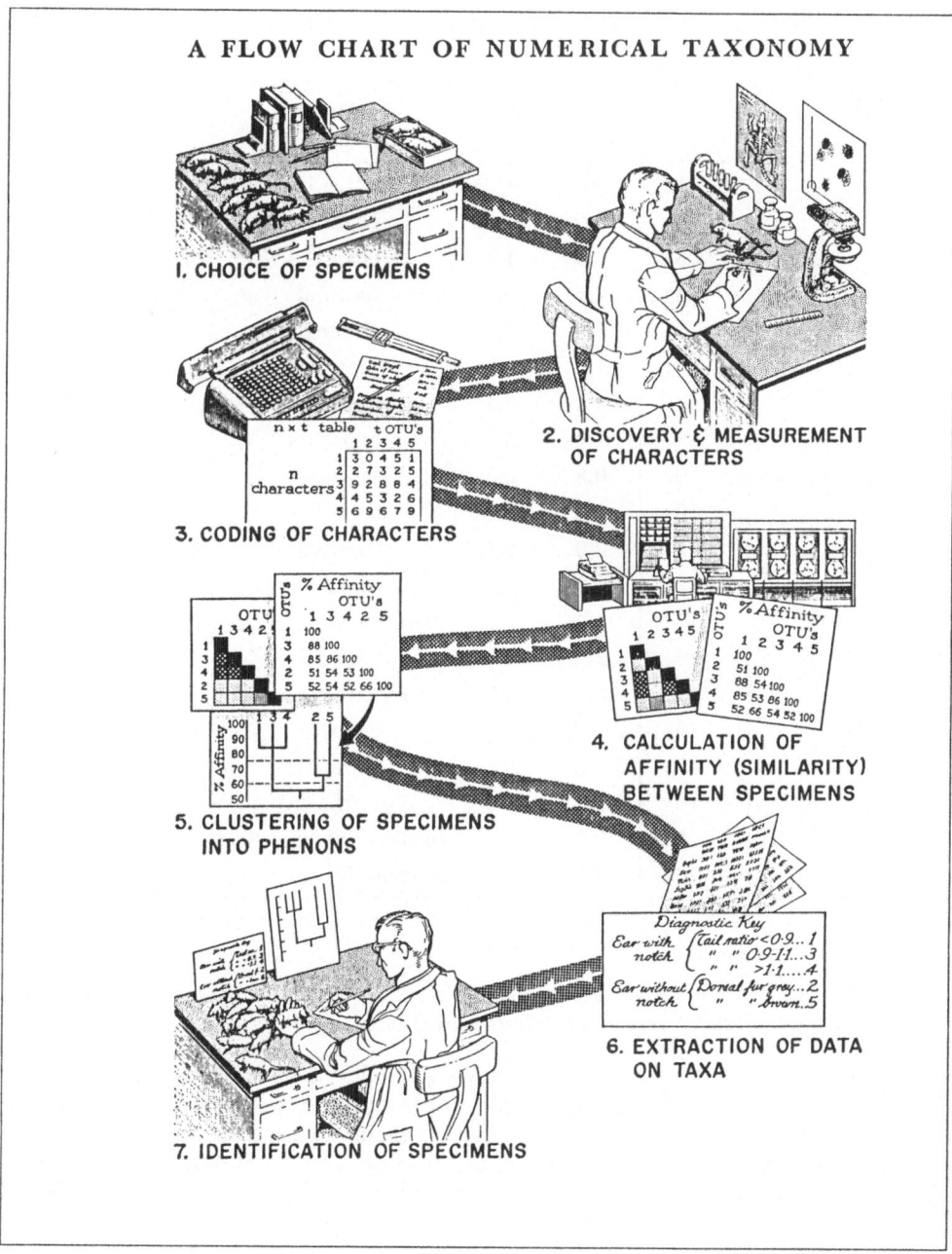

Figure 1-4 *A flow chart of numerical taxonomy (from [373]).*

Table 1-1 *Table of the serum concentrations of alkaline phosphatasis (AP) and iron (Fe) in 20 subjects together with the diagnoses (cited from [90]).*

Subject	AP (U/l)	Fe (mg/l)	Diagnosis
1	4.0	1.0	obstructive jaundice
2	3.0	1.7	cirrhosis
3	2.6	1.8	cirrhosis
4	1.5	0.7	normal
5	2.5	2.2	cirrhosis
6	1.1	1.0	normal
7	2.8	3.1	hepatitis
8	1.7	3.2	hepatitis
9	0.8	0.5	normal
10	2.1	3.0	hepatitis
11	2.0	0.7	normal
12	1.2	0.5	normal
13	4.5	0.7	obstructive jaundice
14	2.5	3.0	hepatitis
15	3.5	0.7	obstructive jaundice
16	2.2	3.2	hepatitis
17	2.1	3.5	hepatitis
18	2.1	2.0	cirrhosis
19	3.5	1.2	obstructive jaundice
20	3.2	1.1	obstructive jaundice

(e) Every object of the sample is attached to a cluster. Usually, an object is attached to that cluster to the representative of which it is closest.

Using a simple example from medicine, we will follow these steps. (This example is cited from [90]). In Table 1-1, the blood serum concentrations of alkaline phosphatasis and iron in 20 subjects are given (together with a diagnosis which we are not interested in at this moment). Since we have measurements of two attributes only, we can picture a scattergram of the data (see Figure 1-5). At a first glance, the sample can be split into three or four groups.

This example illustrates the typical classification problem. Given is a sample of individual observations, a sample of possibly heterogeneous data. Wanted is a classification of these objects into groups and, beyond this, a statement as to whether this classification mirrors the "natural" structure of the data. If we now look at the row with the diagnoses in Table 1-1, we see that all the measurements in the upper part of Figure 1-5 have been taken from six patients with hepatitis. Below them, we find the values of four patients with liver cirrhosis. At the lower right part of the scattergram,

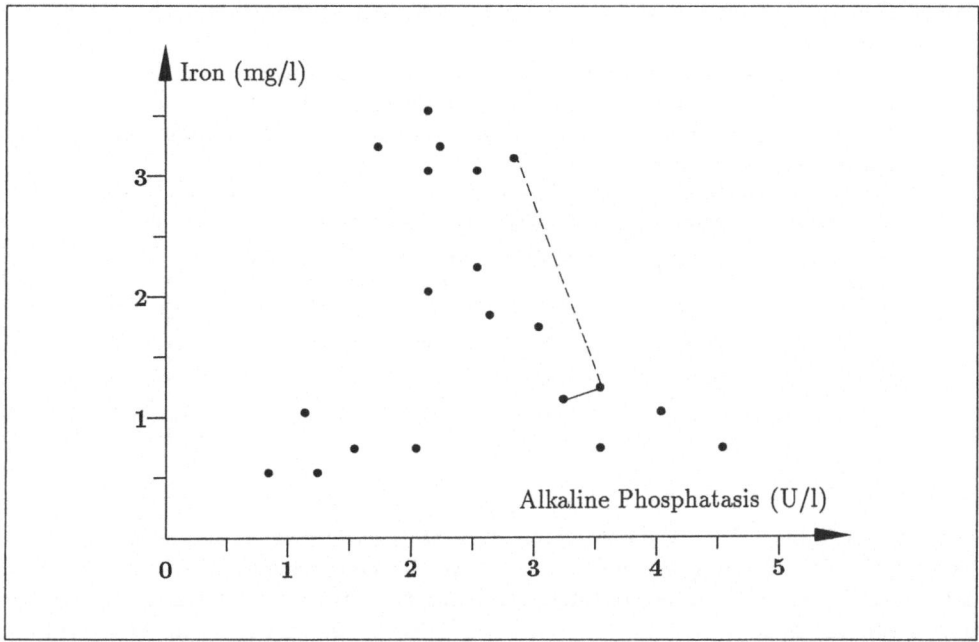

Figure 1-5 *Scattergram of the serum concentrations of alkaline phosphatasis and iron in 20 subjects; a pair of similar subjects (straight line) and a pair of dissimilar subjects (dashed line). (cited from [90]).*

the values of five patients with obstructive jaundice can be seen, while the values of five normal controls are close to the origin. (Thus, the a posteriori classification of 20 subjects based on two of their physiological parameters coincides with the a priori groups of the diagnoses.)

On what basis should classification be performed? It seems to be obvious that within a group, the objects should be as similar as possible; while between the objects of different groups, the similarity should be rather low. However, what does **similarity** mean? Moreover, how can we define this term, which dominates classification theory, precisely or mathematically; when are two objects to be called similar? This seems to be not too difficult for objects with continuous attributes. We can postulate that two objects are the more similar the lower their Euclidean distance is. They become dissimilar if their mutual distance passes beyond a given threshold. In Figure 1-5, we indicate two mutually similar subjects by connecting them by a straight line (they belong to the same diagnosis group as we have seen). The patients connected by a dashed line in Figure 1-5, are dissimilar.

Defining similarities between objects is not very difficult for continuous data: We can choose distance measures, the properties of which are well-known from mathematical calculus. However, when dealing with qualitative data or even with dichotomous data, we have to choose other measures of similarity. In Table 1-2, answers of four subjects from a hypertension study are reported. Among others, the persons had been asked whether their parents had hypertensive disorders, and whether they smoked cigarettes. Then their blood pressure was measured at rest as well as during bicycle

Table 1-2 *Table of qualitative data from a hypertension study (Hypert., hypertension; Ex., exercise):*

(a) *Four coincidences in the answers of subjects nos. 2 and 3;*

(b) *three coincidences in the answers of subjects nos. 1 and 2;*

(c) *two coincidences in the answers of subjects nos. 1 and 3, 1 and 4, and 2 and 4;*

(d) *one coincidence in the answers of subjects nos. 3 and 4.*

| Subject | Anamnesis: Hypert. | | Hypertension | | Smoker |
	in Father	in Mother	at Rest	during Ex.	
1	yes	yes	no	yes	no
2	no	yes	yes	yes	no
3	no	yes	yes	yes	yes
4	no	no	no	no	no

exercise. In Table 1-2, only dichotomous data are reported since we see only the answers "yes" or "no". Subjects nos. 3 and 2 have the same responses in four of the five attributes. Thus, they can be considered as being mutually more similar than subjects nos. 3 and 1, which coincide in only two attributes. Similarity now can be measured by the number attributes which are equally answered by the subjects.

After having defined the term "similarity", how will we perform a classification of the data? The answer is rather easily given if the data are normally distributed with different means for the different groups, and if the number of different groups is known in advance. Figure 1-6 gives an example of a sample composed of objects from two different groups with normally distributed measurements. The task of a cluster-detecting algorithm is to split the sample into two disjoint clusters. This task is optimally solved (and the within-groups similarities are high) if the sum of the within-groups variances is low. And this is the case when we draw a straight line from top to bottom right across Figure 1-6, separating the two ellipsoidal collections of points. Such classification procedures, which are based on the minimization of the variability (of the sum of within-groups variances, for example), are called "variance-based procedures".

The situation changes abruptly if the assumption of normally distributed data is wrong. Everybody sees immediately how to split the objects of Figure 1-7 into two disjoint clusters: A curve should be drawn separating the clusters where the point density is low. Variance-based procedures, however, are linear in the sense that they again try to separate the data into two groups by drawing a straight line as boundary (see also Figures 2-13a and 2-13b). In two-dimensional data, this misbehaviour can be easily checked. Cluster analyses, however, are performed very often with high-dimensional data. Bad or wrong classification then cannot be seen by pure visual inspection. Another reason not to choose procedures which are based on variances is that these procedures need not work correctly when the number of real groups is unknown. Therefore, if natural groups within a data set are to be uncovered, it is better to rely on classification methods which do not depend on the distribution of the variables or assume any knowledge of the true number of the groups.

One such method is the **single-linkage** or **nearest-neighbour procedure**. Let us again consider the data of Table 1-1. With these data, the single-linkage method

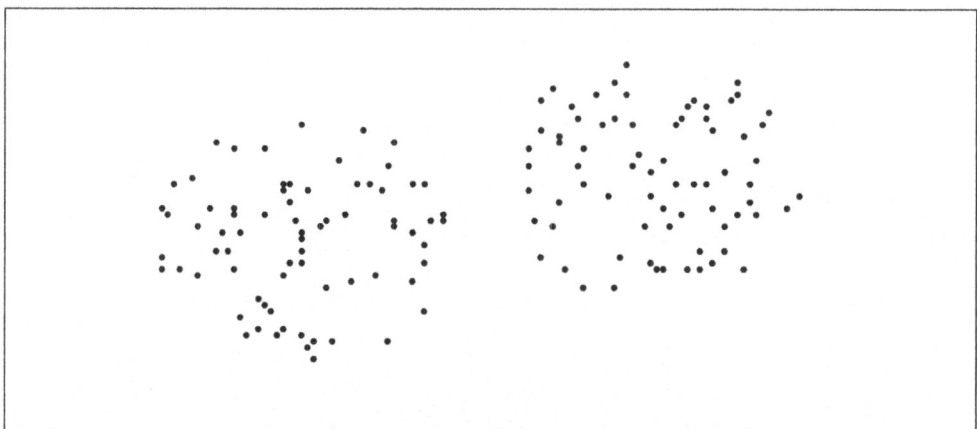

Figure 1-6 *Mixture of two normal distributions: Disjoint compact groups in the $\mathcal{R}^2$.*

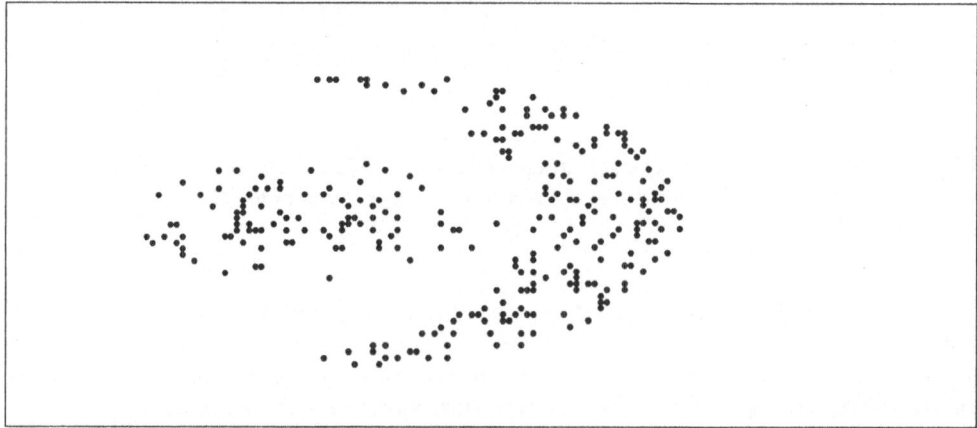

Figure 1-7 *Disjoint sickle-shaped groups in the $\mathcal{R}^2$.*

works as follows. We define a threshold *d* and call two patients similar if the distance between their measurements does not exceed this value of *d*. The points of each pair of mutually similar subjects are joined by line segments. No line is drawn between points associated with dissimilar persons. Those points which are connected either directly by a line or by an alternating sequence of lines and other points, belong to the same cluster. (This procedure defines a graph, see Section 1.2. We therefore speak of **graph-theoretic cluster analysis**.) We speak of "single-linkage clusters", since clusters are joined by the single shortest link (that is, by the mutually closest elements). A single object is attached to an already existing group if its distance to only one element of that group is not larger than a threshold *d*. Two clusters are joined if their mutually closest elements have a distance at most *d*. For this procedure, no knowledge about the number of groups is required. It also can outline twisted or sickle-shaped groups like those of Figure 1-7.

The classification of the data of Table 1-1 is given in Figure 1-8 for a special threshold *d*. We see that, in this example, the partition of the 20 objects agrees completely

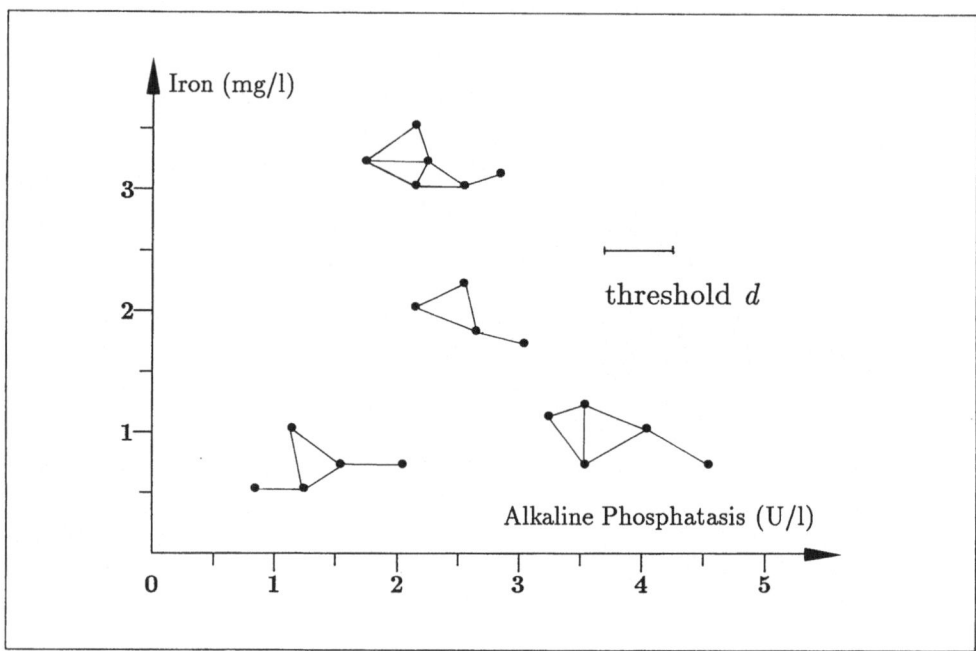

Figure 1-8 *Scattergram of the serum concentrations of alkaline phosphatasis and iron in 20 subjects; a threshold d and the single-linkage clusters for this threshold (cited from [90]).*

with the a priori diagnosis groups. As opposed to methods which are based on variance criteria, single-linkage procedures can also be applied to qualitative data, since they use similarities for the classification process instead of the geometric distribution of the points. Single-linkage procedures, however, also have some disadvantages. The main disappointment is that the threshold d must be chosen "appropriately", but how can we do that without proper information? If d is too small then we get too many clusters with few elements only. This is indicated at the upper left corner of in Figure 1-9: Only one line is pictured, connecting the objects no. 1 and 2, and we get tree clusters. If d is too large, however, the whole sample is joined to one single cluster (see the upper right part of Figure 1-9).

This problem can be by-passed if we don't specify a single threshold d. We let d increase from 0 to ∞ and get **dendrograms**. The lower part of Figure 1-9 shows the dendrogram of the single-linkage procedure applied to the four points above. We have four one-element clusters as long as $d < 0.5$ holds true. At $d = 0.5$, objects nos. 1 and 2 are merged into a cluster. If d is increased to 0.75 then objects nos. 3 and 4 form a second cluster which is joined with the first cluster into a single one at $d = 1.25$. Obviously, dendrograms give a deeper insight into the structure of a data set than classifications at a particular level d.

Another displeasing property of single-linkage procedures is an effect called "bridging" or "chaining": Otherwise well-separated clusters may be linked together by a single path as is shown in Figure 1-10 (the dashed lines indicate the bridge). This disappointing property lead to different modifications of the graph-theoretic definition of a cluster.

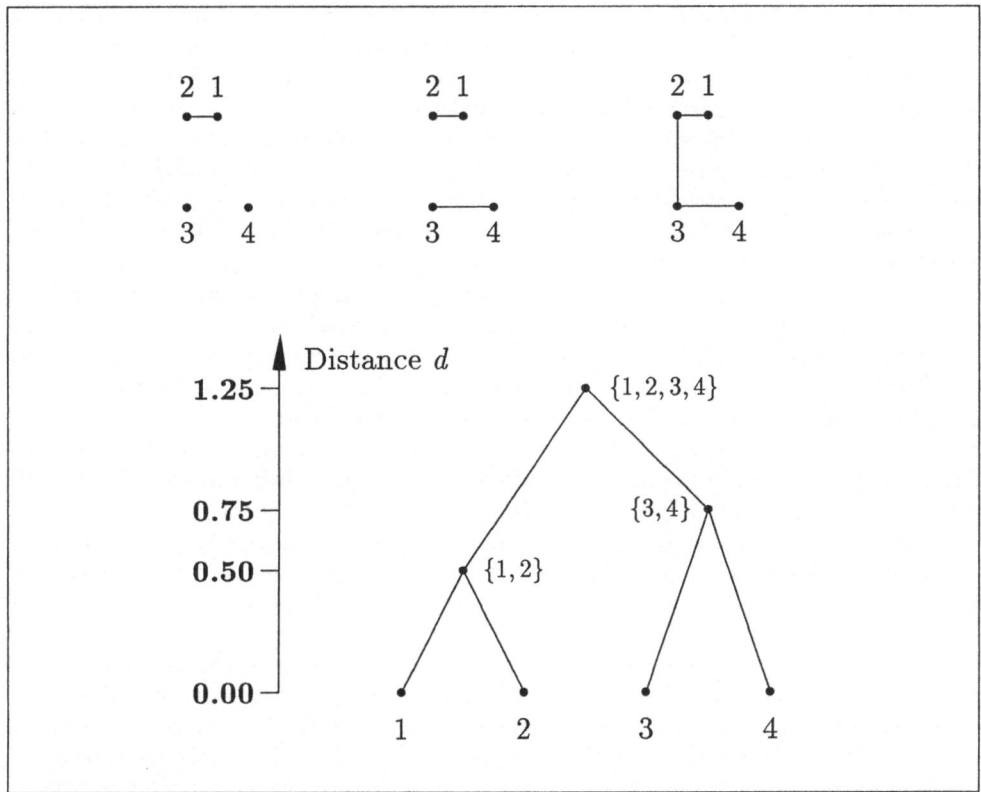

Figure 1-9 *Different thresholds and dendrogram for the single-linkage clustering of a sample of four data.*

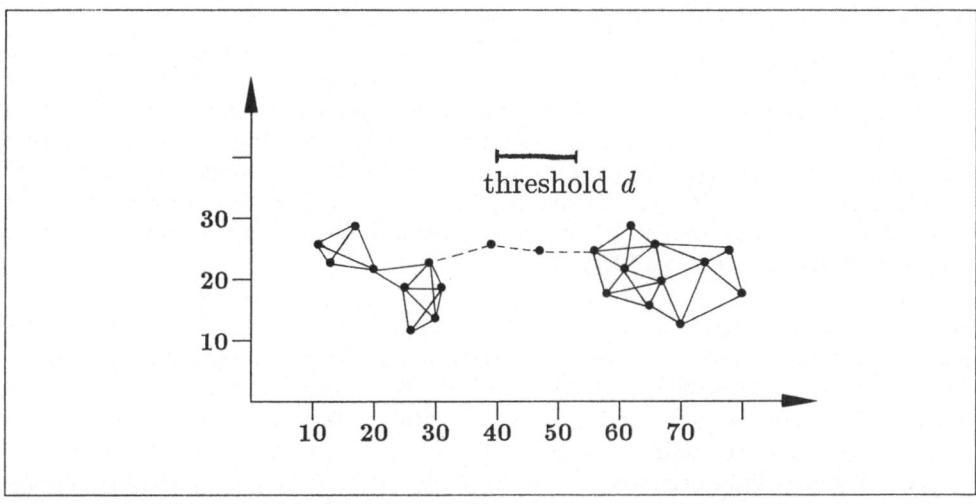

Figure 1-10 *Bridges or chains in graph-theoretic clustering procedures; the 3-linkage cluster to break such chains.*

Instead of "single linkage" we postulate "k linkage": Every object must be similar to at least k objects of an existing cluster before it is merged to this cluster. Each of the two objects of the bridge in Figure 1-10 is similar to only two other objects. For $k = 3$, we therefore get two separated **3-linkage clusters**, the points of the bridge belonging to none of them. A side effect of this modification of the term "cluster" is that the shape of the clusters to be uncovered can be controlled by k. For small values of k, the clusters can be stretched and twisted. For large values of k, however, they must be compact and ellipsoidal. Furthermore, for the same threshold d, a $(k + 1)$-linkage cluster always must be a subset of a k-linkage cluster.

Up to this point attention has been focussed on samples of data units with attributes of the same type. In our first example, the data all are continuous. The data of the hypertension study are dichotomous. Many data sets from the life sciences or from medicine, however, involve a mixture of data from different scale levels. If, in the hypertension study, the original values of blood pressure at rest and during bicycle exercise had been listed, we would have had continuous data together with qualitative data. We speak of **mixed data**. Mixed data as well as **missing values** and **outliers** cause problems in defining similarities or distances between objects.

Like all methods of exploratory data analysis, cluster analysis helps to "explore" and uncover structures within a data set. Methods of cluster analysis should primarily be used to formulate hypotheses (mathematical models which are well-fitted to the data). Every researcher, however, must note that cluster analyses are very subjective even if we use "objective" mathematical methods to outline the different groups. This holds since the resulting clusters depend not only on the computational procedure, but also on the choice of attributes to be measured. And since the researcher (alone or together with a biometrician) decides on the basis of his or her personal knowledge which attributes and objects should be drawn for a sample, this choice may be biased. Therefore, the results of a cluster analysis are chiefly valid for the specific sample only, and we cannot generalize them to a larger population without careful inspection. Moreover, every application of a clustering algorithm to a set of data results in a classification of objects, whether or not the data exhibit a true or "natural" grouping structure. (This can happen if in single-linkage classification the threshold d has not been chosen appropriately.) This is irrelevant if clustering is done to obtain a practical stratification of the given set of objects for organisational purposes. In exploratory data analysis however, the interest is in detecting an unknown clustering structure from the data. Here, the result of a clustering procedure should reflect the real structure (real or natural clusters). From the group structure of the objects of a sample we usually make inference about the whole population. Thus, an artificial clustering is not acceptable. The classes resulting from the algorithm therefore must be investigated for their **relevance** and their **validity**.

While the researcher may judge the relevance from a possibly more qualitative point of view, the statistician has to rely on appropriate significance tests or other methods of data analysis. Using graph-theoretic models, we can develop test procedures for testing the validity of the clusters outlined by a classification procedure. One model is as follows. We interpret homogeneity within a data set as random attachment of the distances to the pairs of objects. In an n-element sample where the first N smallest distances have been drawn, let $Y = k$ objects be in the union of single-linkage clusters (and the remaining $X = n - k$ objects form as many one-element groups). For small values of N, we expect that the N smallest distances are widely spread

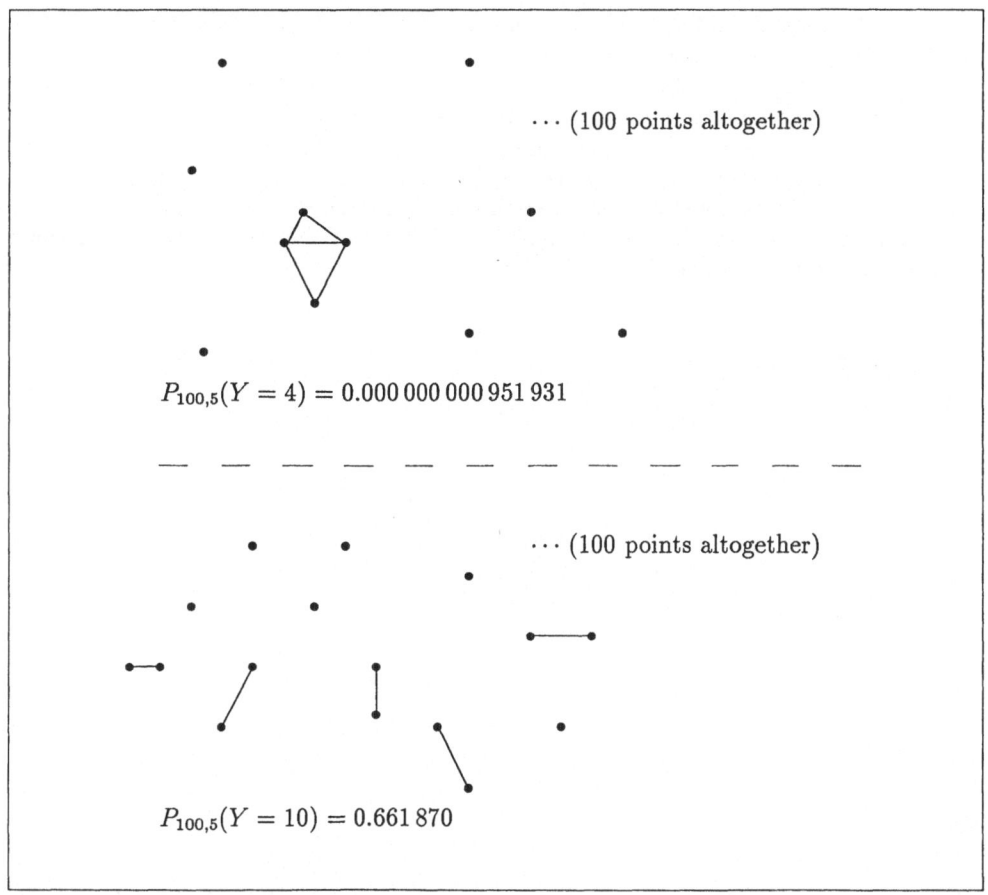

$P_{100,5}(Y = 4) = 0.000\,000\,000\,951\,931$

$P_{100,5}(Y = 10) = 0.661\,870$

Figure 1-11 *Graph-theoretic test for the detection of real clusters, $P_{100,5}(Y = k)$ is the probability that in a graph with 100 vertices and 5 edges drawn "at random", k vertices are not isolated. We get*

(a) $P_{100,5}(Y = 4)$: The first five edges connect four vertices (one single "cluster" immediately at the beginning of the edge-drawing process);

(b) $P_{100,5}(Y = 10)$: The first five edges connect ten vertices to five pairs (five "clusters" immediately at the beginning of the edge-drawing process).

among the data, that is, that they most probably connect $2N$ objects to N pairs. If now $P_{nN}(Y \leq k) \leq \alpha$ holds for a given error probability α of the first kind, then the null hypothesis of a random attachment of distances to pairs of objects can be rejected at this level α. In a sample of 100 objects, attaching the five smallest distances at random as weights to the $\binom{100}{2} = 4\,950$ pairs of objects most probably results in 90 isolated objects and five two-element "groups" (see the lower part of Figure 1-11). The probability for this event is about 2/3. On the other hand, we do not expect that the five smallest distances mutually connect four objects within homogeneous data, that is, under random conditions. The probability for this event is smaller than 10^{-9} (see the upper part of Figure 1-11). If now this event occurs within a data set, we can reject the null hypothesis of homogeneity. The four points connected by the five smallest distances can be considered as the kernel of a real group.

These developments show that, today, we can use methods of cluster analysis not only for pure exploration of samples but also to test statistical hypotheses. This presupposes a well-defined mathematical classification model like that of single-linkage clusters. The development of test procedures to test the homogeneity of a sample (and thus to infer from a sample to a larger population) as sketched above, is a first step from pure exploration to confirmatory statistics. It illustrates the strong intercorrelations between exploratory and confirmatory statistics.

CHAPTER 2

CURRENT METHODS OF CLUSTER ANALYSIS: AN OVERVIEW

For the most part, statisticians have been concerned with uncovering the interrelationships that exist among variables. The objects upon which the measurements were taken were assumed to be homogeneous. Rarely was interest in the possibility that a given set of objects could be grouped into subsets that displayed systematic differences. In many applications, however, there is reason to believe that a set of objects can be clustered into subgroups that differ in meaningful ways. In large data sets like those stored in clinical cancer registers, nobody would expect the objects to be homogeneous. The most commonly used term for the class of procedures that seek to separate the component data into groups is **cluster analysis**. Other names which are used synonymously are **automatic classification, numerical classification, numerical taxonomy**, and sometimes **pattern recognition**.

Cluster analytical methods gained distinctive importance while being used as tools to derive dendrograms of animals or plants in biology. Personal intuition was the basis of many classification procedures which have been proposed for biological or anthropological problems. Though initially primitive, the field of taxonomy gradually grew to rely on the more objective techniques of numerical taxonomy. The origins of mathematical classification can be found in two famous papers of K. Pearson at the end of the previous and the beginning of this century ([306], [307]). In these papers and in [308] and [309], however, K. Pearson defined only the **procedures** for detecting clusters on a statistical basis. Until the early fifties, the term **cluster** itself was not defined mathematically. Then, the first mathematical definitions of cluster came out together with the deduction of theoretical models and concepts of classification. These definitions were based on a mathematical discipline known as "graph theory" (see [128], [129], [312], [383]).

The search for classifications or typologies of objects or persons is indigenous to a wide variety of disciplines. At the same time, the evolution of more complex cluster-analytical methods was made possible by the advent of the high-speed computers with their enormous storage capabilities. The rapid progress in the computer sciences and industry makes it possible to write more complex programs, which perform more arithmetic operations and are applied to larger sets of high-dimensional data. Thus, over the last years a growing interest in classification and related areas has been noticed. To-day, we see applications of cluster analysis not only to biology but also to such diverse areas as psychology (classifying individuals into personality types), regional analysis (classifying cities into typologies based on demographic and fiscal variables), marketing research (classifying customers into segments on the basis of psychographic factors and product use), chemistry (classification of compounds based on their performance properties), archaeology (classifying ceramics by means of their colours and paintings into epochs), and medicine (classifying patients into groups of different cancer stages). Papers on these topics are, for example, [5], [51], [52], [77], [79], [87], [89], [114], [125], [131], [140], [179], [205], [206], [207], [219], [252], [253], [285], [300], [301], [305], [321], [363], [367], [383], [378], [386], [392], [419], [423], [440]. More papers can be found in [49] and [139], respectively.

Because of the many fields in which classifications of objects are performed, and because of the different types of data to which classification procedures are applied, the number of new clustering procedures increased rapidly during the last two decades. (Some of these algorithms had been developed just for one special type of data, or just for the problem under consideration. They cannot be applied to other problems.) The direction for this evolution was shown by the monograph "PRINCIPLES OF NUMERICAL TAXONOMY" by R.R. Sokal and P.A.H. Sneath (see [373], [380]). In this book, different methods for cluster analysis have been derived by pure empirical intuition. In addition to discussing various clustering methods, this book emphasizes an empirical approach to classification eschewing reliance on theoretical approaches. This advocacy of empiricism is widely spread in the general literature of cluster analysis; it caused scientists to propose lots of different methods of clustering. Usually, new cluster procedures are introduced by applying them to data sets like R.A. Fisher's famous iris data, for which the right classification is already known ([123]), or they are applied to data generated by Monte Carlo simulation. That means, instead of a correct model validation it is only asked whether a well-known classification of a data set can be reproduced fairly well by the new procedure.

In contrast, new methods are rarely introduced by deducing them from a theoretical model. Thus, comparatively few articles and monographs exist enunciating theories of classification, that is, dealing with the classification theory as a mathematical discipline. Only few authors compare the different procedures methodologically trying to trace the different heuristic ideas back to statistical models which make sense for classification theory or which allow the testing of the "realness" of clusters which have been found as the result of a cluster detecting procedure (see [1], [15], [16], [21], [37], [83], [114], [133], [135], [148], [178], [186], [212], [213], [214], [215], [222], [232], [233], [237], [259], [261], [272], [288], [350], [359], [400], [437], and [439]).

The applications of classification procedures to different topics like anthropology, archaeology, astronomy, biology, business, chemistry, computer science, economics, engineering, geography, geology, information and library science, linguistics, marketing, medicine, political science, psychology and sociology and soil sciences indicate not only the importance of existing classification tools, but also stress the need for further development and investigations in classification theory as a mathematical topic. This holds even more, since, for a researcher, the availability of so many different procedures is, at the same time, both an advantage and a real disadvantage: Only few scientists who want to perform a cluster analysis of their data, are familiar with classification theory, such that they can choose the best (or even the right) program for their special problem. Moreover, some critical points must be known by the possible user to prevent him from being lead to misinterpretations of the result of a cluster analysis.

Therefore, the relations of classification theory to recent developments in computer science, operations research, combinatorics, probability theory and multivariate statistics become important for scientists. Moreover, beyond pure classification problems the environment of classification, the general topic of data analysis or information science, is increasingly a subject of consideration and investigation. This situation has recently led to the foundation of the "INTERNATIONAL FEDERATION OF CLASSIFICATION SOCIETIES (IFCS)" which is intended to promote activities in this realm.

In this chapter, we want to give an overview of the different ideas and models, on which cluster procedures are founded. We also want to explain the procedure of performing a cluster analysis. Thereby, we shall describe different methods without going too deep into mathematical theories. Books providing more detailed descriptions are, for example, [6], [25], [33], [38], [42], [77], [102], [106], [115], [119], [147], [183], [210], [216], [243], [255], [258], [362], [373], [374], [385], [386], [387], [388], [394] and [443]. These books also contain enough helpful hints to enable the scientist to apply the different cluster procedures to his data. In the series of monographs "STUDIEN ZUR KLASSIFIKATION" (edited by H.H. Bock, I. Dahlberg and P. Ihm) and "FALLSTUDIEN CLUSTER-ANALYSE" (edited by H. Späth), mathematical cluster models are presented and examples of applications are discussed with their pros and contras. The same holds for the "JOURNAL OF CLASSIFICATION". The "PROCEEDINGS OF THE FIRST CONFERENCE OF THE INTERNATIONAL FEDERATION OF CLASSIFICATION SOCIETIES (IFCS)", edited by H.H. Bock ([49]), and the "PROCEEDINGS OF THE 9TH ANNUAL MEETING OF THE GERMAN CLASSIFICATION SOCIETY", edited by W. Gaul and M. Schader ([139]) present a mixture of research papers which represent the state of the art both in applied classification and foundations of classification theory.

A short abstract of cluster analysis can be found in [93], [190] and [335]. The monograph [90] is helpful for the research worker in applied sciences. Different models of cluster analysis are described and explained in that booklet by means of carefully chosen examples and without using many mathematical formulas.

2.1 THE AIM OF CLUSTER ANALYSIS

Cluster analyses — as methods of exploratory data analyses — are especially performed in cases where no a priori null hypotheses exist. The aim of a cluster analysis procedure is to classify n objects or individuals, upon which t measurements have been taken, into m **clusters** (or: **classes**, **groups**) $\mathcal{C}_1, \ldots, \mathcal{C}_m$. The partition $\mathcal{C} = \{\mathcal{C}_1, \ldots, \mathcal{C}_m\}$ is called a **classification**. "Cluster analysis" is a generic name for a series of mathematical, statistical procedures which perform a classification of the objects of a data set under different assumptions about the data and by use of different theoretical concepts. Such objective procedures rely only on informations contained in the measurements taken on each of the objects. They more and more replace the intuitive classification procedures of applied natural scientists, which have been influenced for the most part by personal experiences only. Those intuitive procedures required small samples and low dimensional data, since they relied on visual impression and hand calculations. The process of "mathematization"in cluster analysis guarantees objectivity in that the same results are reproduced independently of the experimenter if the same data set and the same procedures are used.

We must caution against performing cluster analyses on small data sets even if we do this in some examples for the purpose of demonstration. The possibility of producing artificial or random clusters is fairly high in this case. However, when analysing larger data sets, one can hope to avoid mistakes in the interpretation of the results of a cluster analysis which else may be caused especially by artefacts or outliers. Analyses of large samples, however, require the development of rather complex mathematical algorithms. They only have been made possible by electronic data processing: Because of the relatively huge amount of calculations, cluster procedures can only be performed on computers, even if the data sets are of medium size. This is not only due to the size n and the dimension t of the data set but also due to the complexity of the formulas which must be used for multivariate data analses. (Similarly, many results and methods from the topics of multifactorial analysis of variance, discriminant analysis, factor analysis, or nonlinear regression analysis, respectively, can be used in practice only on high speed computers with their storage capacities. The formulas and procedures, which are needed for performing these analyses, have been known for a long time before they really could be applied in practice.)

It should be pointed out that the natural scientist influence the outcome of a cluster analysis in many ways. He selects the attributes, which he wants to measure, he chooses the cluster model (on which the type of clusters to be outlined depends), and he selects the ultimate algorithm or procedure. (Here, a **stable** algorithm should be chosen if possible. Stability means that the shapes of the resulting clusters are unaffected or only little affected when new data are added to the sample, and that they are robust against isolated objects and cannot be influenced by permutations of the objects in the data set. Naturally, this kind of stability depends to a certain degree on the sample size: Adding some objects is more likely to yield a different classification in a small sample than in a larger one.) And last but not least does the researcher's personal experience influence the outcome in the final step of the analysis: He interprets the resulting clusters.

Classification problems occur in many fields of medical research. The clinician is interested in classifying patients based on different aspects of diseases. Patients from different groups then would be treated according to different regimens. Classification procedures are applied to patients with a certain disease which all are treated according to the same regimen. Here, the aim of the classification procedure is to find groups based on the therapeutic effect ([5], [253]). If especially a subgroup of patients exists who don't show any therapeutic effect then, in a second step, the scientist wants to find out the attributes which mainly resulted into that classification. In psychiatry and social medicine, cluster analysis is a tool which is often used ([305]). The development of systems of diagnostic keys and tumour keys was — and is — based on the methods of numerical classification (see [51], [52], [179], [219]). Medical institutes of environmental hygienics use classification procerdures to locate clusters of regions with high concentrations of dangerous components. (Some procedures are known here as "random clumping".) In epidemiology, cluster analysis is used to describe the spatial and chronological spread of infectious diseases.

Generally, the classification of objects is done according to the principles of optimization: A balance between the loss of precision and information and the benefits of clearness and economy is sought. The individual attributes of each object are replaced by average attributes which are typical only for the whole class, and the many objects are replaced by only few classes. This means both a loss of individual information and the benefits of economy and a better insight into the structure of the data. In his survey paper [42], H.H. Bock emphasizes the following topics, which illustrate the usefulness of classification methods tools of exploratory data analysis: Abstraction and compression of information since few classes and their typical representatives replace the whole sample; detection of possibly nonlinear structures and generation of hypotheses; detection of natural classes; data preparation before application of linear procedures; reduction of the variety of data for reasons of operations research or economical reasons. With H.H. Bock's own words ([42]):

> "(a) Bei umfangreichem Datenmaterial vermittelt das System der (i.a. wenigen) Klassen sowie geeignet gewählter Klassenrepräsentanten eine kurze, geordnete und übersichtliche Beschreibung der Daten (Datenreduktion, Informationsverdichtung).
>
> (b) Die Zusammensetzung der Klassen sowie deren klassentypische Eigenschaften lassen unbekannte (auch nichtlineare) Strukturen innerhalb der Objekt- oder Merkmalsmengen erkennen. Insofern dient die Klassifikation zur Erzeugung von Hypothesen bzgl. der Objektmenge.
>
> (c) Gut separierte, "natürliche" Klassen lassen sich als "Objekt-Typen" interpretieren und tragen so zur Bildung von Begriffen bei (Abstraktion). Sie weisen auf die Nützlichkeit oder Notwendigkeit klassenspezifischer Behandlungs-, Forschungs- oder Entwicklungsmethoden hin. Isolierte Objekte können als Ausreißer erkannt werden.
>
> (d) Während eine globale (d.h. alle Objekte erfassende) lineare Datenanalyse (z.B. multiple lineare Regression) bei einer heterogenen Datenmenge oft sinnlos ist, kann dieselbe Methode, jedoch lokal (d.h. klassenweise) angewandt, eine adäquate Analyse der Daten liefern (lokaler Aspekt der Klassifikation).
>
> (e) Im Operations-Research-Bereich wird die Bildung homogener Objektklassen häufig aus praktisch-organisatorischen Gründen notwendig sein, auch ohne daß die Objektmenge eine "natürliche" Klassenstruktur aufweist."

Point (d) gives the logical sequence of the performance of an analysis of multivariate data: At first, a cluster analysis of the objects should be performed, followed by a factor analysis of the attributes within every group. This would prevent the applied scientist to perform a factor analysis of the ungrouped, possibly inhomogeneous data where the global correlation matrix is significantly different from the correlation matrices within every group. On the other hand, clusters found by a classification procedure can be used as a base for a discriminant analysis by which objects, upon which measurements will be taken later on, can be attached to one of these groups.

2.2 THE DIFFERENT STEPS OF A CLUSTER ANALYSIS

Figure 2-1 provides an overview of the procedure of cluster analysis with its different steps (according to [39], modified). Depending on the type of data and the intentions of the scientist, the type of classification is determined. Then, bases for a valuation of the classification are chosen. These may be criteria measuring the homogeneity or separation of different classes or global optimality criteria for the whole classification. Finally, a clustering algorithm is applied to the data to perform the actual cluster

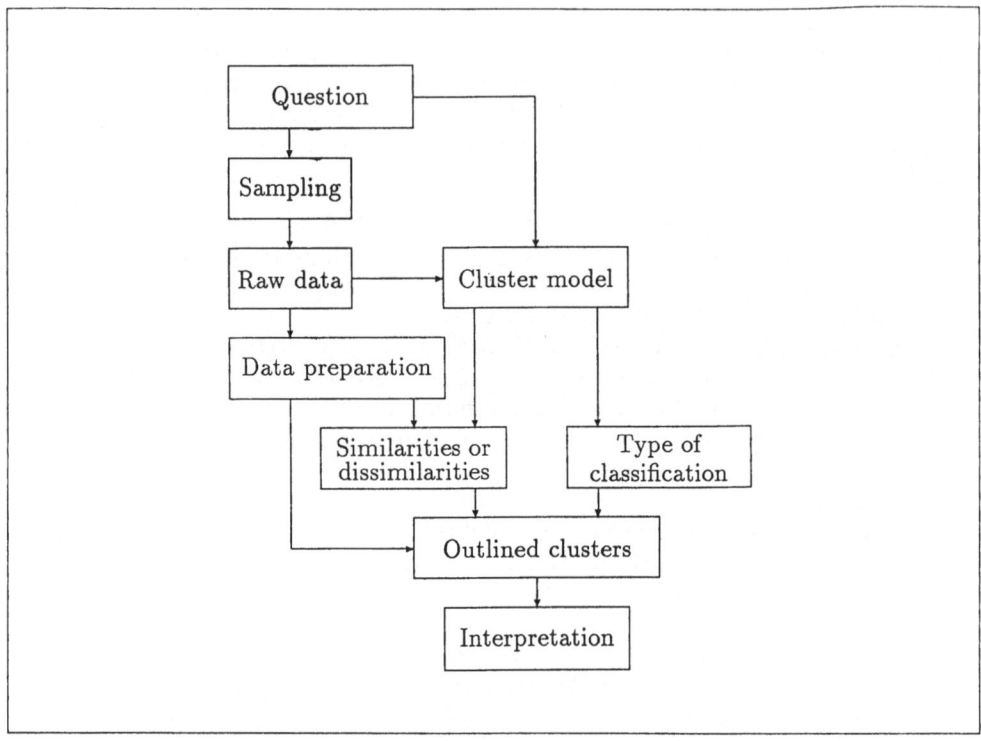

Figure 2-1 *The different steps of a cluster analysis.*

analysis. In the following, the different steps of this procedure are illustrated. Moreover, several of the most important methods are explained.

2.2.1 DATA SAMPLING AND PREPARATION

The objects of data analyses can be persons, animals, or other such entities. The mechanics of a cluster detecting procedure are performed on a sample of these entities, each member being denoted as a **data unit**, **subject**, **observation**, **case**, **element**, **object**, or **event**. From a methodological point of view, we must differ between two situations in the choice of data units. In the first instance a sample is the complete object of data analysis. The purpose is to discover a classification scheme for the given set of observations. It is not intended to generalize the results found for this data set to any additional data units outside the original sample. In the second and more frequent situation, the sample is a portion of a much larger population which is the true object of interest. In this case the statistical principles of random and independent selection of objects should be carefully obeyed even if cluster analyses generally are not involved with hypothesis testing.

The data units must be consistently described in terms of their characteristics, **properties**, **attributes**, or **items**. These descriptors of the objects are the **variables** of the classification problem (in statistical terms, the attributes are the random variables and the measurements taken are their realizations). Thus, at the beginning of a cluster analysis, the researcher determines the attributes to be measured. This is an often neglected but very important part. We can say that it is most probably the choice of variables that has the greatest influence on the ultimate result of a cluster analysis.

Variables, which take largely the same values for all data units, have little or no discriminating power whereas those manifesting consistent differences from one subgroup to another can induce strong distinctions. When a relevant discriminating variable is left out of the analysis some clusters may merge into an amorphous mass. On the other hand, inclusion of strong discriminators, which are not particularly relevant to the purpose of the classification to be performed, can mask the clusters sought and lead to wrong interpretations. The analyst also should be cautious not to choose different variables which merely describe the same property (see [6] for a detailed discussion).

After choosing the objects and their descriptors (n objects with t attributes), measurements of the attributes $M_1, \ldots, M_t$ are taken on every object $O_1, \ldots, O_n$. The values of the different attributes for the objects of the sample $\mathcal{S} = \{O_1, \ldots, O_n\}$ may be determined using tests, questionnaires or taking the measurements directly. For every object, these values are coded numerically and stored as (n, t)-matrix $\mathbf{X} = (x_{il})$, the so-called **data matrix**. Every object O_i in this matrix of **raw data** is represented by a **row vector**, $\vec{x}_{i*} = (x_{i1}, \ldots, x_{it})$. The l-th **column vector** $\vec{x}_{*l}$ gives the values of the attribute M_l for the different objects. Thus, in the data matrix $\mathbf{X}$, the element x_{il} is the measurement of the l-th attribute M_l, taken upon the i-th object O_i. Therefore we can write $\mathcal{S} = \{\vec{x}_{1*}, \ldots, \vec{x}_{n*}\}$ instead of $\mathcal{S} = \{O_1, \ldots, O_n\}$. The symbols $\vec{x}_{i*}$ and O_i, respectively, are interchangeable. By $\mathcal{X}_l$ we denote the range of the l-th variable ($l = 1, \ldots, t$). The Cartesian product $\mathcal{X} = \mathcal{X}_1 \times \cdots \times \mathcal{X}_t$ is the **measurement space** (or **event space**). Obviously, $\vec{x}_{i*} \in \mathcal{X}$ holds true for all $i = 1, \ldots, n$.

Based on their **scales of measurement**, the variables can be divided into different classes. First there is a **nominal scale** which merely distinguishes between classes as the variable "blood group" does. One can only decide whether two objects are equal or not in this variable. An **ordinal scale** induces an ordering of the objects. The Karnofsky index or the Apgar index are variables of that type. We can decide whether a given object has a greater value than another object in this variable or not. The magnitude of the difference between two subsequent values, however, bears no special meaning. An **interval scale** like the body temperature assigns a measure of the difference between two objects. The **ratio scale** is an interval scale with meaningful zero point like the time between the date of operation and the detection of a rezidive or the kinetic constants in a pharmacokinetic model. We can say that the half life of a pharmacon for a subject is twice the half life of that pharmacon for another subject. A statement like this makes no sense for the variable "body temperature". At last we have the **absolutely scaled** variables like the variable "number of leukocytes".

Of greater importance than the classification of variables, due to their scale levels, is the global distinction between **quantitative** and **qualitative** (or **categorical**) variables. Variables on nominal scale are referred to as qualitative ones. In contrast, variables on absolute, interval, or ratio scales are then referred to as quantitative variables. Variables on ordinal scale are sometimes referred to as qualitative variables sometimes as quantitative ones (see [74], [190], [245], [374], [385], [403]). For an attribute M_l on absolute scale, $\mathcal{X}_l \subseteq \mathcal{N}_0$ holds true; such an attribute is **quantitative discrete**. For variables on ratio or interval scales, $\mathcal{X}_l \subseteq \mathcal{R}$ holds true; they are **continuous**. If an item M_l is ordinally scaled, then we also have $\mathcal{X}_l \subseteq \mathcal{N}_0$. For qualitative data, the range consists of all possible different values which usually are coded numerically. In many data sets from the life sciences or from medicine, variables on different scale levels oc-

Table 2-1 *Matrix of the raw data of 21 women (rows), upon whom measurements of 15 attributes (columns) were taken. The attributes are in detail:*

M_1 : *Age of the woman (full years),*
M_2 : *Time between last pregnancy and medical examination (months),*
M_3 : *Familiar anamnesis, hypertensive father (1=yes, 0=no),*
M_4 : *Familiar anamnesis, hypertensive mother (1=yes, 0=no),*
M_5 : *Height of woman (cm),*
M_6 : *Weight of woman (kg),*
M_7 : *Total number of pregnancies,*
M_8 : *Reception of an ovulation inhibitor (1=yes, 0=no),*
M_9 : *Systolic blood pressure at supine rest (mm Hg),*
M_{10} : *Diastolic blood pressure at supine rest (mm Hg),*
M_{11} : *Systolic blood pressure during bicycle exercise (mm Hg),*
M_{12} : *Diastolic blood pressure during bicycle exercise (mm Hg),*
M_{13} : *Heart rate during bicycle exercise (beats per min),*
M_{14} : *Average arterial pressure in the second trimenon of pregnancy (mm Hg),*
M_{15} : *Average arterial pressure in the third trimenon of pregnancy (mm Hg).*

	M_1	M_2	M_3	M_4	M_5	M_6	M_7	M_8	M_9	M_{10}	M_{11}	M_{12}	M_{13}	M_{14}	M_{15}
O_1	23	32	1	0	173	52	1	1	115	75	160	100	138	82	87
O_2	29	23	0	0	163	54	1	0	115	70	160	105	145	90	91
O_3	32	29	0	0	168	61	1	0	105	75	145	95	132	–	–
O_4	36	33	0	0	170	63	1	0	120	85	160	95	144	78	96
O_5	35	23	0	0	172	63	2	1	120	75	175	90	136	75	89
O_6	23	32	1	0	156	50	3	0	110	70	140	85	146	91	80
O_7	43	33	0	0	167	67	3	1	115	80	175	90	140	89	90
O_8	41	23	0	0	172	56	2	0	95	65	160	90	137	76	89
O_9	24	31	0	0	160	52	2	1	115	85	155	95	154	70	72
O_{10}	27	28	0	0	160	70	3	0	110	95	160	95	155	100	115
O_{11}	33	30	0	1	175	86	1	0	115	80	190	105	130	98	103
O_{12}	29	29	1	0	170	76	1	0	125	75	185	98	131	101	105
O_{13}	34	22	0	1	168	64	9	1	105	80	175	115	145	–	–
O_{14}	34	31	0	1	169	71	2	1	125	80	195	102	150	92	97
O_{15}	36	30	0	0	170	70	1	0	130	85	210	110	134	–	–
O_{16}	24	28	0	0	168	63	1	1	130	75	175	90	155	100	103
O_{17}	37	21	0	0	–	–	2	0	105	75	165	110	145	92	99
O_{18}	27	29	0	1	162	62	2	1	120	80	175	110	165	105	99
O_{19}	33	28	0	0	169	73	2	0	125	85	170	90	130	102	106
O_{20}	27	30	0	0	176	129	1	1	125	80	180	93	123	107	110
O_{21}	39	32	0	1	176	72	2	0	130	90	180	110	155	83	91

cur simulaneously. We then speak of **mixed data**. Mixed data can cause problems in defining similarities or distances between objects.

Example 2-1 In a clinical study, we wanted to answer the question whether signifi- cant differences in blood pressures and heart rates during bicycle exercise between two groups of women could be proved. The first group consisted of normotensive women after a pregnancy without hypertensive disorders. The second group also consisted of normotensives; but all women of this group had hypertensive pregnancy disorders. This study was part of a long-term study about the significance of pregnancy-induced hypertension as a prognostic index for a manifestation of an essential hypertension later on. Measurements of 15 attributes were taken upon 21 women. The matrix of the raw data is shown in Table 2-1. Some of the attributes are qualitative dichotomous (binary) like the two variables of the familiar anamnesis. Most of the variables are ratio scaled like the different blood pressure values. The item "total number of pregnancies" is absolutely scaled. Dashes in the table indicate missing values. The first nine probands had no hypertensive disorders during their last pregnancy. The last twelve women had pregnancy-induced hypertensions. An exercise-induced hypertension was characterized by a rise in systolic pressure to at least 200 mm Hg or a rise in diastolic pressure to at least 100 mm Hg during bicycle exercise. This is a special way of applying methods of cluster analysis. We knew in advance that the sample can be divided into two groups according to the fact whether the women had hypertensive disorders during pregnancy or not (so-called **external criterion**). We wanted to know whether this classification could be repeated with the measurements only. This could prove pregnancy-induced hypertension to be an indicator of a manifestation of an essential hypertension later on. (Using these variables, we performed a discriminant analysis. The results together with a detailed discussion are given in [120].) •

Difficulties in cluster analyses are often caused by **missing values** or by **outliers**. We speak of "missing data" or "missing values" if in some cases the measurements have not been taken for all the attributes. There are many reasons for missing data. It often occurs that probands do not answer all questions of a questionnaire. In some experiments, not all laboratory tests could be performed for all objects, and so on. "Outliers" are measurements which do not lie within the normal or generally expected range. Outliers can be caused by faulty data entries or transfers. Other reason can be that the personnel of a laboratory work with standards different to other laboratories, or that the sampling was not correctly performed (see [20]). Both types of incomplete or wrong information cause similar problems and thus are treated in similar ways when the data are prepared for a cluster analysis.

For most of the cluster procedures, we must exclude all objects (rows of the data matrix) with missing values or outliers before performing the actual classification step. Some algorithms allow the masking (extracting) of attributes (columns of the data matrix). A few cluster programs contain statistical procedures to estimate missing values. D. Wishart recommends to estimate missing data of an object by those values of the nearest neighbour ([427], [429]). Other authors suggest that missing values should be estimated by the mean of the variable in question. We cannot recommend this last procedure in cluster analysis. Its disadvantage lies in the fact that in case of inhomogeneous data the mean should be calculated only from those object belonging to the same group as that object with the incomplete data. But such a group-specific calculation is impossible since the groups are unknown before the clustering procedure is performed. Therefore, in our opinion, this suggestion is worse than D. Wishart's method. At the moment, there are no subroutines for the a posteriori estimation of missing values in Wishart's program package CLUSTAN and also in the clustering procedures of BMDP, SAS and SPSS, respectively. Objects with incomplete or faulty informations must be eliminated or the attributes in question must be masked. The other way is to estimate missing data "by hand" before the classification step is performed. The analyst also must examine carefully whether an object is an outlier and must be excluded from the sample. (Often, faulty data can be corrected from the original documents.)

However, we want to point out that values, which are missing originally, should be estimated by statistical procedures in those cases only where the absence is caused by random mechanisms ([35], [100]). In questionnaires, missing values might hint to the fact that a proband really did not want to answer a question. In longitudinal or long-term studies, missing data might indicate that observations were censored. This must be regarded when the variables are coded.

Generally, missing data cause problems which — from a methodological point of view — cannot be overcome if the researcher is interested in a natural classification which is induced by the joint distribution of the objects in the measurement space as J.C. Gower states ([165], [166]). However, under more pragmatic prepositions, we can expect that the comparability of the objects (in terms of similarity) changes only slightly if only few data are missing. In this case, we expect that the measurement spaces differ only to a minor degree which can be neglected.

Storage space and computation time are the critical requirements that limit both the number n of objects and the number t of attributes or dimensions for many statistical inquiries. However, statistical considerations also suppose to choose not too many attributes. For larger investigations, the researcher is expected to restrict the data to the most important variables which should be **statistically independent**. Thus, another aspect of classification is to search — after performing a cluster analysis of the data of a pilot study — for those variables with the most discriminating power. The aim is to exclude variables with little or no discriminating power from further inquiries. We also can reverse the classification idea: To find a set of probably independent variables, we can perform a partition of the attributes into classes based on a — usually small — sample of objects instead of looking for clusters of objects based on attributes. It is expected that the correlation between variables from different clusters is not very high. In subsequent studies, the scientist then can choose a representative from every class of variables. By this method the number of variables can be reduced, too.

Other methods of reducing dimensionality in high dimensional data before performing a cluster determinating algorithm are the principal components analysis and the factor analysis (see [10], [243]). Calculation of the principal components allows us to map the original objects $O_1, \ldots, O_n$ onto points $\vec{y}_1, \ldots, \vec{y}_n$ on the straight line (by calculating the first principal component), in the plane (by calculating the first two principal components), or in the 3-space, respectively (by calculating the first three principal components). Homogeneous clusters then often can be identified by pure visual inspection. However, originally well separated groups can be merged by the projection (**superposition**, see Figure 2-2). Thus, using the original data it should be proved for every such cluster whether it must be divided into subclusters ([38], [208]).

Both principal components analysis and factor analysis are heuristic tools only when applied to reduce dimensionality of the data units to be clustered. One reason is that their statistical properties are unknown when the distribution of the data vectors is not known, and especially when the sample is composed of a mixture of different distributions, which is expected in samples, on which a cluster analysis should be performed. Furthermore, both types of analyses cannot be applied to reduce dimensionality of mixed data because they suppose continuous data and linearity. For inhomogeneous data the original group structure can be destroyed by the projection as Figure 2-2 shows. A method for the reduction of dimensionality, which does not assume a linear model, is that of R. Blomer and co-authors. From its basic ideas, however, this

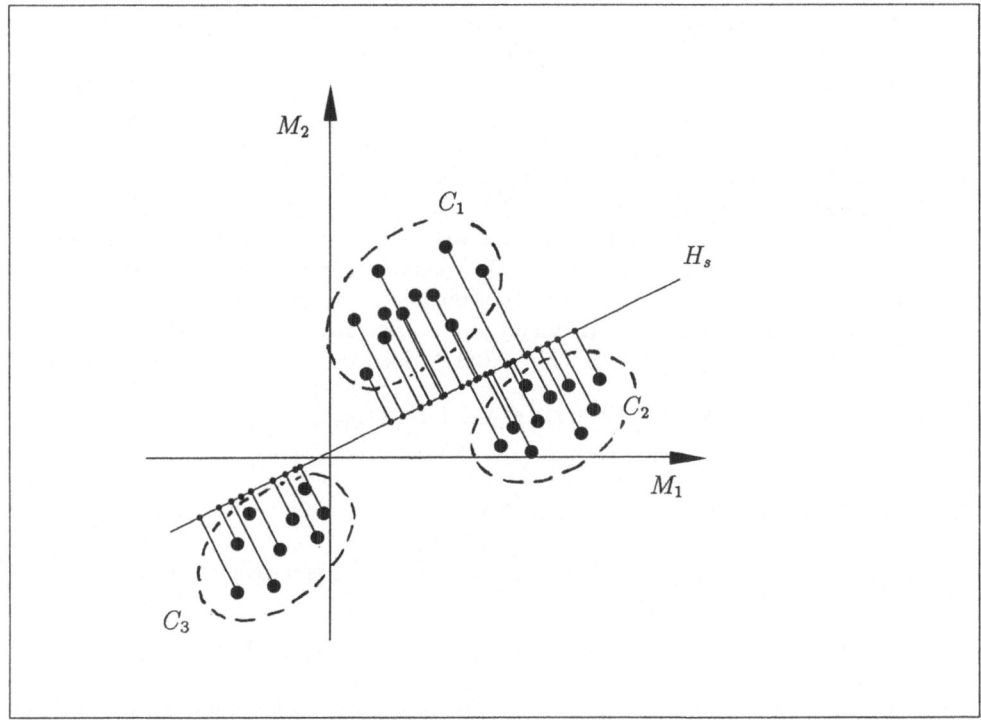

Figure 2-2 *Superposition in principal components analysis: The originally two groups C_1 and C_2 in the $\mathcal{R}^2$ are united to one single group by the projection.*

method supposes data on at least interval scale level (see [5], [36]).

2.2.2 MEASURES OF SIMILARITY OR DISTANCE

Obviously, for an appropriate choice of a classification procedure, the structure of the data to be clustered must be taken into consideration. In some cluster analysis algorithms, the data vectors $\vec{x}_{1*}, \ldots, \vec{x}_{n*}$ are used directly and without further preparations than the elimination of objects with missing data. These methods are based on the geometrical structure of the data as classification criterion. They search for local maxima of the multivariate empirical density function. These maxima then are interpreted as cluster centres. Procedures like these make sense only for continuous data; they cannot be applied to nominally or ordinally scaled data. However, the clustering of variables generally involves correlation or other measures of association between the objects. Thus, the majority of the algorithms is not based on the original raw data. They require measures of **similarity** or **dissimilarity**, defined for every pairwise combination (O_i, O_j) of the entities to be clustered. This presupposes different methods of data preparation: In a first step, the similarities or dissimilarities must be calculated. In this step, the data usually are standardised, and the treatment of missing data and outliers must be defined. The actual classification of the objects follows in a second step and uses the similarities as clustering criterion.

The mutual proximities of individuals are usually expressed as distances. The distance measures interact with cluster analysis criteria so that different measures, applied to a sample, give identical results with one criterion and distinctly different results with another. In fact, the researcher must choose from a series of similarity and dissimilarity measures that one, which is appropriate for the type of data or scale level under consideration. However, in practice this combined choice of variables, similarity measures and cluster algorithms proved to be easier for an experimenter than the choice of the favourable classification algorithm to be applied to the raw data. And generally speaking, the calculation of similarities or distances as cluster criterion makes it possible to choose the same cluster determination algorithm for different types of data (because now no assumptions about the original distribution of the data are needed).

Generally, **similarities** (or **similarity measures**, **similarity relations**, **similarity functions**) are mappings $s : \mathcal{S} \times \mathcal{S} \to [0; 1]$, which satisfy the following conditions:

(2-1a) $\qquad\qquad 0 \leq s(O_i, O_j) \leq s(O_i, O_i) = 1 \qquad$ (normalisation),

(2-1b) $\qquad\qquad s(O_i, O_j) = s(O_j, O_i) \qquad$ (symmetry).

We speak of **metric** similarity functions if additionally

(2-1c) $\qquad\qquad s(O_i, O_j) = 1 \Rightarrow O_i = O_j \qquad$ (uniqueness),

and

(2-1d) $\qquad\qquad \big(s(O_i, O_j) + s(O_j, O_k)\big)\, s(O_i, O_k) \geq s(O_i, O_j)\, s(O_j, O_k),$

hold true. Formula (2-1a) means that every object is most similar to itself. This highest value of similarity is normalised to 1. Often, for qualitative data the claim of uniqueness, (2-1c), is violated: Two different data units can coincide in all attributes. (Even for theoretically continuous data, in practice (2-1c) can be satisfied only in case of sufficient accuracy of measurement.)

A mapping $d : \mathcal{S} \times \mathcal{S} \to [0; \infty)$ is a **disparity function** if it satisfies

(2-2a) $\qquad\qquad d(O_i, O_j) \geq d(O_i, O_i) = 0.$

The mapping is a **dissimilarity fuction** or a **proximity function** if, additionally,

(2-2b) $\qquad\qquad d(O_i, O_j) = d(O_j, O_i) \qquad$ (symmetry)

holds true. The mapping is called a (metric) **distance function** if, together with (2-2a) and (2-2b), the conditions

(2-2c) $\qquad\qquad d(O_i, O_j) = 0 \Rightarrow O_i = O_j \qquad$ (uniqueness),

and

(2-2d) $\qquad\qquad d(O_i, O_j) + d(O_j, O_k) \geq d(O_i, O_k) \qquad$ (triangle inequality)

are satisfied. Again, in practical applications, (2-2c) is often neglected. (Formulas (2-1c) for metric similarity functions, and (2-2c) for distance functions are equivalent as are (2-1d), and (2-2d), respectively.) Especially in psychological or sociological inquiries, sometimes disparity measures must be used (if sympathy or antipathy between persons is to be measured). In most cases, however, at least similarity or proximity measures are used. As common practice, we use s_{ij} as an abbreviation of $s(O_i, O_j)$, and we write d_{ij} instead of $d(O_i, O_j)$. These figures usually are arranged as (n, n)-similarity matrix $\mathbf{S} = (s_{ij})$ or as (n, n)-proximity or distance matrix $\mathbf{D} = (d_{ij})$.

For the remainder of this subsection we name and shortly discuss the most important distance and similarity measures. In case that all variables are ratio or interval scaled, often

(2-3a)
$$d_{ij} = d(O_i, O_j) = d(\vec{x}_{i*}, \vec{x}_{j*})$$
$$:= \|\vec{x}_{i*} - \vec{x}_{j*}\|_p = \left(\sum_{l=1}^{t} |x_{il} - x_{jl}|^p \right)^{1/p} \quad (L^p\text{-distance})$$

is chosen with an appropriate value for p (with $p \geq 1$). Some researchers prefer

(2-3b)
$$d_{ij} := \left((\vec{x}_{i*} - \vec{x}_{j*}) \mathbf{\Sigma}^{-1} (\vec{x}_{i*} - \vec{x}_{j*})^T \right)^{1/2}$$
$$= \left(\sum_{l=1}^{t} \sum_{m=1}^{t} (x_{il} - x_{jl}) \sigma_{lm}^* (x_{im} - x_{jm}) \right)^{1/2} \quad (\text{Mahalanobis distance})$$

instead of the L^p-distance. In (2-3b), $\mathbf{\Sigma}^{-1} = (\sigma_{lm}^*)$ is the inverse of the empirical (t, t)-**covariance matrix** $\mathbf{\Sigma} = (\sigma_{lm})$ **of the variables**. The elements σ_{lm} of $\mathbf{\Sigma}$ are

$$\sigma_{lm} = \frac{1}{n-1} \sum_{i=1}^{n} (x_{il} - \overline{x}_{.l})(x_{im} - \overline{x}_{.m}).$$

For every $l = 1, \ldots, t$, $\overline{x}_{.l}$ is the mean of the l-th variable, taken over all objects, and $\sigma_{ll} =: s_{.l}$ is the empirical standard deviation of that variable (see [266]).

By inserting $p = 1$ into (2-3a), we get the so-called **city-block** or **Manhattan distance**; $p = 2$ gives the **Euclidean distance** which is used more frequently. The limit case $p \to \infty$ gives the **maximum** or **Chebyshev distance**, which can be written as

$$d_{ij} = \max_l |x_{il} - x_{jl}|.$$

The Mahalanobis distance (2-3b) is invariant to translations and scale transformations of the data units. However, the significant disadvantage is the fact that the covariance matrix must be calculated from all objects. It would be more adequate to calculate this matrix for each group separately because we can expect significantly different covariance matrices for the different groups in an inhomogeneous sample — but these groups are unknown and searched. Furthermore, (2-3b) cannot be computed as easy as other distance measures. In contrast, for arbitrary $p > 1$, the L^p-distances are invariant only to translations of the data. The Euclidean distance is invariant to orthogonal transformations of the data. Therefore, before performing the classification step, the t variables are often transformed, so that they have mean 0 and p-th central, absolute moment 1 (they are **normalised** or transformed into **normed random variables**). This is done to allow the direct comparison of different attributes; it is a necessary step before any computation of distances between multivatiate data units can be performed. For example, in case of the Euclidian distance ($p = 2$),

$$z_{il} = (x_{il} - \overline{x}_{.l})/s_{.l}$$

is calculated (in this special case we speak of **standardisation** of the data; see [6], [38], [190], [385]). For other values of p, $s_{.l}$ is replaced by the p-th root of the p-th absolute, central moment. Another method to normalise the data is to map, for $l = 1, \ldots, t$, the n measurements $x_{1l}, \ldots, x_{nl}$ onto the unit interval $[0; 1]$. This usually is done when the city-block distances between the data vectors are calculated. Normalising transformations are used to get rid of the units of measure, thus allowing a direct comparison of the measurements of different attributes. On the other hand, attributes can be overvalued or undervalued by these transformations. Thus, the results of a classification procedure can be quite different depending on the fact whether the data have been normalised or not (and depending even on the normalising method).

For qualitative variables, often metric similarities are preferred instead of distances. For binary data,

$$(2\text{-}4a) \qquad s_{ij} := \frac{0_{ij} + 1_{ij}}{t} = 1 - \frac{\|\vec{x}_{i*} - \vec{x}_{j*}\|_p^p}{t} \qquad \text{(matching coefficient)}$$

can be chosen as similarity measure if the value 1 (presence of the attribute) has the same weight as the value 0 (absence of this attribute). Here, by 1_{ij} we denote the number of attributes which together are present in both objects O_i and O_j; 0_{ij} is the number of attributes which are absent in both data units. In many problems, the presence or absence of attributes cannot be treated in the same way. This especially holds in medical research: For the classification of syndromes the presence of certain symptoms in two data units is of more importance than the absence. In cases like this,

$$(2\text{-}4b) \qquad s_{ij} := \frac{1_{ij}}{t - 0_{ij}} \qquad \text{(S-coefficient of Jaccard and Tanimoto)}$$

can be chosen. Formula (2-4a) easily can be modified if the qualitative attributes are not binary but can take more than two values.

Up to this point, attention is focussed on samples of data units with attributes of the same type. Real data sets, however, often involve a mixture of nominally, ordinally, and interval scaled data. Such mixed variable data sets are troublesome when a clustering of the data units is searched. We have the problem of weighting the contributions of the different variables. Similar problems arise in the simultaneous use of interval and nominally scaled variables. In [6], M.R. Anderberg proposes a reduction of the scale levels of all variables to a uniform type or the partitioning of variables, performing a cluster analysis of the data units for each type of variables. Using scaling methods we also can convert qualitative variables into quantitative ones. Thus, from an (n, t)-matrix of mixed data, a distance matrix can be computed. Other authors suggest separate calculations of similarities for the variables of every scale level. From these similarities, a weighted mean can be computed as an overall similarity for each pair of objects. It is recommended to transform the quantitative variables onto $[0; 1]$. If a data set consists of quantitative and binary data only, then we can use

$$s_{ij} = \frac{1}{t} \left(1_{ij} + 0_{ij} \right) + \frac{1}{t} \sum \left(1 - |x_{il} - x_{jl}|/r_l \right)$$

as a measure of similarity. Here, r_l denotes the range of the l-th variable. The sum is taken over all quantitative variables. Other metric similarity measures are

$$s_{ij} = \frac{u_{ij}}{t} \qquad \text{or} \qquad s_{ij} = \frac{u_{ij}}{t + v_{ij}} \, .$$

Here, u_{ij} is the sum of the number of dichotomous components with the same values in $\vec{x}_{i*}$ and $\vec{x}_{j*}$, and of the number of quantitative components with $|x_{il} - x_{jl}| \leq \delta_1$. By v_{ij} we denote the sum of the number of dichotomous components with different values and of the number of quantitative components with $|x_{il} - x_{jl}| \geq \delta_2$ in both data units. The threshold δ_1 defines the difference which is tolerated as a maximum to call two objects similar in the l-th variable. Similarly, by δ_2 we denote the threshold which must be overcome before two objects are called dissimilar in a continuous variable $(0 < \delta_1 \leq \delta_2)$. The definition of a joint measure of similarity (or distance), however, is problematic for mixed variable data sets. Computing a weighted mean as an overall similarity causes troubles, which are similar to those we get by normalising the variables: The true influence of variables in the classification result may be over- or underestimated.

More problems arise when some values are missing in a data set. Sometimes it is impossible, from a statistical standpoint, to estimate them. And often enough, we cannot exclude all data units $\vec{x}_{i*}$ or all variables $\vec{x}_{*l}$ with missing values. Thus, for data vectors with missing values in some components, it must be defined in each case whether they are similar to other data vectors in those components or not. In case that there are only few missing values in a data matrix $\mathbf{X} = (x_{il})$ we also can modify the similarity measures, or dissimilarity measures, respectively. For data on an interval or ratio scale level, $\delta_{ij} = d_{ij}/t^{1/p}$ can be chosen as a distance measure which is derived from the L^p-distances d_{ij}, defined by (2-3a). If some observations are missing for a pair of objects, then the sum in (2-3a) is taken only over those components of the data vectors where measurements exist. This modified distance d_{ij} then is divided by the p-th root of the number of components with measurements.

Large amounts of memory sometimes are needed for the computation and storage of the s_{ij} or d_{ij}. Here, it is of advantage if metric measures of similarity or distance can be chosen, which result in symmetric (n, n)-matrices. Thus, only the $\binom{n}{2} = n(n-1)/2$ similarities or distances above the main diagonal must be computed. Using

$$(2\text{-}5) \qquad d_{ij} = 1 - s_{ij} \quad \text{or} \quad s_{ij} = 1 - \frac{d_{ij}}{\max_{i,j} d_{ij}},$$

we can always compute dissimilarities from similarities and vice versa. These formulas correspond with our normal — nonmathematical — thinking that two objects must be more similar if their mutual distance becomes smaller.

More measures of proximities, distances, and similarities with detailed discussions of their advantages and disadvantages can be found in [6], [8], [38], [39], [42], [163], [229], [230], [241], [246], [267], [320], [335], [374], [375] and [385]. As has been said at the beginning of this subsection, the measures interact with cluster analysis criteria. Thus, the choice of an appropriate proximity measure should not be underestimated. The choices of variables, transformations, and similarity measures are often made separately in classification problems. They then at least should be reviewed for their composite effect to make sure the result is satisfactory. However, combined choices of the variables, transformations, and measures should be preferred. In [90], the authors use an elementary example to demonstrate that the choice of different distance measures results in distinctly different classifications. For this reason, some classification procedures don't use the distances directly. They use the **ranks** of the distances. This makes the clustering results invariant to monotonic transformations of the elements of the distance matrix. (Measures which can be derived by monotonic transformations from each other, are called **equivalent**.) R.F. Ling's method belongs to this group of classification procedures (see [259], [260], [261], [262]). We will discuss this method in a further chapter.

2.2.3 TYPES OF CLASSIFICATION

Before using a classification algorithm from the many procedures available, the analyst must decide on the type of classification which is appropriate for his data. In many cases, the choice of the clustering procedure depends on this type. Generally, we can distinguish between four classification types: **disjoint classifications** (or **partitions**), **nondisjoint classifications** (or **overlapping clusters, fuzzy partitions**), **hierarchic classifications**, and **quasi-hierarchic classifications**.

(a) **Disjoint classifications** (compare Figure 2-3a): The sample of the n bjects $\{O_1, \ldots, O_n\}$ is split into a number of non-overlapping (or **disjoint**) subsets. Every element O_i of the sample is attached to exactly one cluster (**exhaustive classification**). Sometimes it is tolerated that some objects remain isolated without being attached to any group (**nonexhaustive classification**). Obviously, the different subgroups are the more homogeneous the higher their number is. On the other side, the separation between the different groups will decrease with an increasing number of clusters. Classification procedures using optimization criteria for the detection of disjoint clusters thus need an information about the number of clusters. One of the main difficulties in the application of these procedures is to determine the right or an optimal number of classes. Using cluster analysis as a method of exploratory statistics, the researcher has rarely enough information to determine the appropriate number of classes.

(b) **Nondisjoint classifications** (compare Figure 2-3b): In some classification problems, it is convenient to ask for homogeneous groups and to allow objects to belong to more than one class at the same time. The clusters are "overlapping". In voluminous data bases like cancer registers, or literature data bases, the researcher often is confronted with the problem to attach data units (patients of the cancer register, or books cited in a literature data base, respectively) to more than one class only. Methods from the theory of **fuzzy sets** can be applied when overlapping clusters are searched for. They are described by J.C. Bezdek in [26], [27], and by H.H. Bock in [41]. In [6], M.R. Anderberg recommends a method for finding overlapping clusters in samples of continuous data, which is based on the analysis of the point densitiy and on discriminant analysis. In this procedure the number of mixed densitities (that is, the number of classes) and their parameter are estimated. In [212], [213], [214], [215], N. Jardine and R. Sibson use graph theoretic methods to construct overlapping groups. Their method is a generalization of the so-called "single-linkage" procedure.

(c) **Hierarchic classifications** (compare Figure 2-3c): The objects and groups are arranged and graphically represented in form of a genetic tree (or dendrogram). Every group arises by the fusion of "subgroups" underneath or by the splitting of the superset above it. At the bottom of this hierarchy, the different objects are listed as one-element classes $\{O_1\}, \ldots, \{O_n\}$, the whole sample $\mathcal{S} = \{O_1, \ldots, O_n\}$ of all objects is listed at the top. Embedding this genetic tree in a coordinate system where the y-axis represents an appropriate measure of homogeneity within the groups, we get a **labelled hierarchy** (the actual **dendrogram**, see Figure 2-7b).

Every level in this labelled hierarchy consists of a collection of disjoint clusters of the same homogeneity listed besides them on the scale axis ([42], [46], [67], [84], [86], [182], [213], [218], [251], [281], [331], [379], [414], [426]). We speak of **agglomerative** cluster procedures if the groups in a hierarchy are constructed by amalgamation of subgroups. In contrast, procedures which stepwise split supergroups — beginning with the whole sample as one group — are called **divisive**. Dendrograms can be constructed using methods for generating disjoint clusters. A clustering criterion is chosen (for example, that each of the clusters must have a certain degree of homogeneity). This criterion then is diminished step by step. In [287], [289] and [290], F. Murtagh gives a survey of recent advances in hierarchical clustering algorithms and counting dendrograms. In [288], he derives a probability distribution of **random dendrograms** (a special case of random graphs) which can be used as a probability model to test the randomness of hierachical structures. In [261], R.F. Ling defines an isolation index for clusters based on dendrograms. In the same paper, he derives a probability distribution of this index and uses it to test the randomness of single-linkage clusters.

(d) **Quasi-hierarchic classifications** (compare Figure 2-3d): In quasi-hierarchic classifications, the same holds as in hierarchic classifications with the exception that the clusters at each level may overlap.

Disjoint and nondisjoint classification procedures are **horizontal** procedures, whereas hierarchical and quasi-hierarchcal procedures are **vertical**: In a dendrogram, every level itself shows a disjoint classification of the sample which arises by fusion of clusters of the level below, or by splitting of some clusters of the level above. Thus, for the most part we can confine ourselves to the description of horizontal procedures.

Last but not least, the analyst needs a basis for the valuation of a classification, a **clustering criterion** since he is interested in a "good partition" of the sample into clusters, which should be as "natural" and problem-orientated as possible. In a partition, the groups should be **homogeneous** and mutually **well separated**. Otherwise it would make little sense to speak of different groups and to characterize them by "typical" representatives.

A measure of homogeneity, which is often used for clusters C_k with at least two objects, is the average of all proximities or distances in C_k,

$$(2\text{-}6a) \qquad \hom(C_k) = \frac{1}{2\,n_k\,(n_k-1)} \sum_{O_i \in C_k} \sum_{O_j \in C_k} d_{ij}.$$

Here, n_k is the number of objects in the k-th cluster. For the Euclidean distance as distance measure, another measure of homogeneity is given by the within-group variance. It is obtained by replacing the distances d_{ij} in (2-6a) by the squared distances d_{ij}^2. Other measures of homogeneity for groups with at least two objects are

$$(2\text{-}6b) \qquad \hom(C_k) = \min_{O_i,O_j \in C_k} d_{ij},$$

$$(2\text{-}6c) \qquad \hom(C_k) = \max_{O_i,O_j \in C_k} d_{ij}.$$

The homogeneity of a cluster C is high if its homogeneity index $\hom(C_k)$ is low. Clusters which consist of one element only, naturally must be very homogeneous. Their homogeneity index is defined as 0.

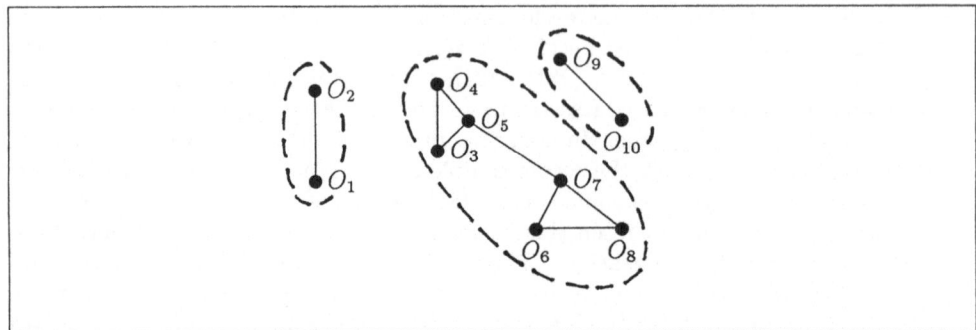

Figure 2-3a *Disjoint classification in $\mathcal{R}^2$.*

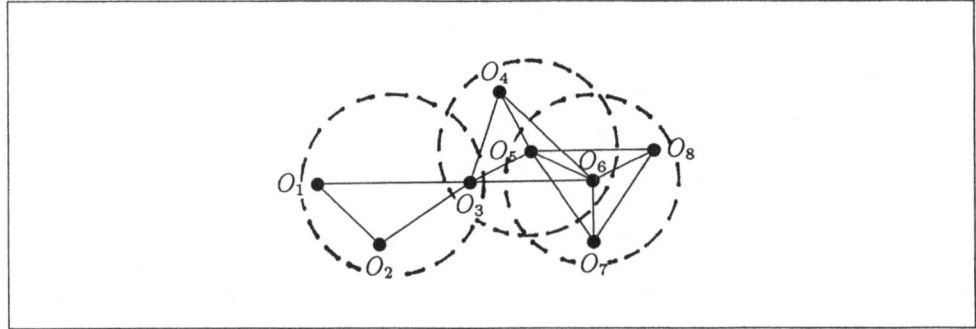

Figure 2-3b *Nondisjoint classification in $\mathcal{R}^2$.*

The separation between two clusters $\mathcal{C}_k$ and $\mathcal{C}_l$ can be measured by the mean of all distances of pairs of objects, one from $\mathcal{C}_k$, and one from $\mathcal{C}_l$,

$$(2\text{-}7a) \qquad \text{sep}(\mathcal{C}_k, \mathcal{C}_l) = \frac{1}{n_k\, n_l} \sum_{O_i \in \mathcal{C}_k} \sum_{O_j \in \mathcal{C}_l} d_{ij}.$$

Sometimes the separation is calculated, using one of the following formulas

$$(2\text{-}7b) \qquad \text{sep}(\mathcal{C}_k, \mathcal{C}_l) = \min_{\substack{O_i \in \mathcal{C}_k \\ O_j \in \mathcal{C}_l}} d_{ij},$$

$$(2\text{-}7c) \qquad \text{sep}(\mathcal{C}_k, \mathcal{C}_l) = \max_{\substack{O_i \in \mathcal{C}_k \\ O_j \in \mathcal{C}_l}} d_{ij}$$

(see [38], [190], [385] for further discussions). The separation between two clusters $\mathcal{C}_k$ and $\mathcal{C}_l$ is good if the separation index is high. The separation of a class $\mathcal{C}_k$ from the remainder of the sample can be simply defined by

$$(2\text{-}8) \qquad \text{sep}(\mathcal{C}_k) := \text{sep}(\mathcal{C}_k, \mathcal{S} - \mathcal{C}_k).$$

The calculations of the separation indices must be slightly modified for overlapping clusters: All objects, which are in both clusters simultaneously, are omitted.

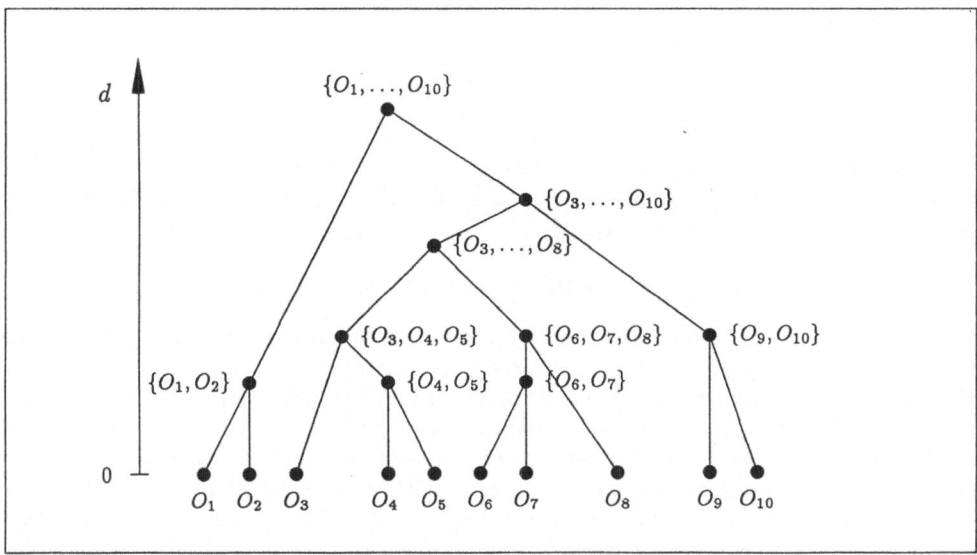

Figure 2-3c *Hierarchical classification of the objects from Figure 2-3a.*

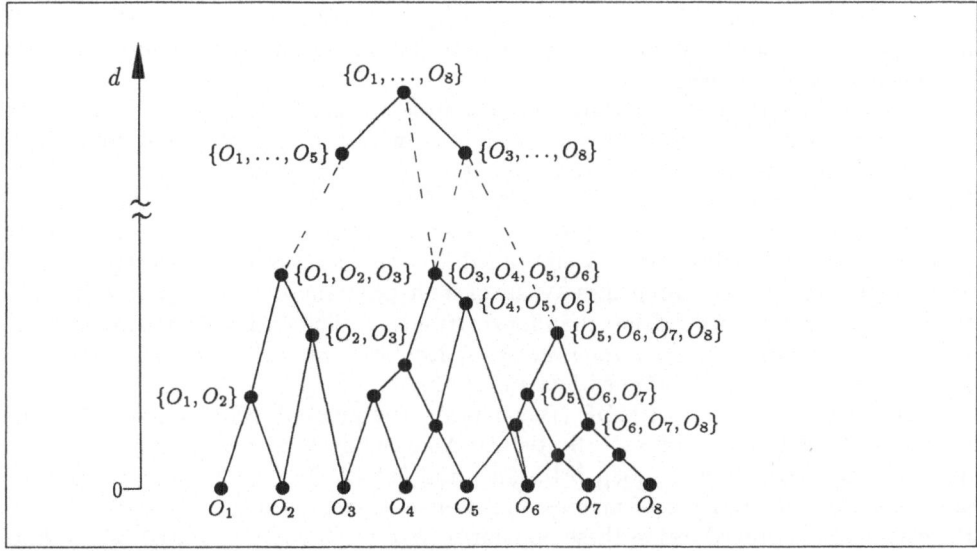

Figure 2-3d *Quasi-hierarchical classification of the objects from Figure 2-3b.*

Formula (2-6c) often is too strong, especially for big groups. In contrast, the condition (2-6b) often is too weak. Small values of (2-6b) may not indicate homogeneous clusters because only two objects may lie a small distance apart. On the other side, (2-7b) is a stronger condition for the separation of two classes than (2-7c). Formulas (2-6a) and (2-7a) are a compromise between these extremities. A measure of the quality of a cluster is the ratio of (2-6a) and (2-7a) which is called the **B-coefficient**.

More generally,

$$(2\text{-}9) \qquad\qquad b(\mathcal{C}_k) = \frac{\hom(\mathcal{C}_k)}{\sep(\mathcal{C}_k)}$$

can be defined as an index denoting the quality of a cluster. From this it is not difficult to derive a criterion to describe the quality of a classification itself. This leads to optimization methods of cluster analysis.

2.2.4 PROCEDURES OF CLASSIFICATION

"Cluster analysis" is a collective name for a series of procedures. Cluster analysis encompasses many diverse techniques for discovering structure within complex bodies of data, some of them deduced under special mathematical assumptions about the data to be clustered, many of them being based on heuristics only. This can be seen most clearly when browsing through lots of different computer algorithms, which are offered today. Programs exist which are especially written for samples consisting of spherical and compact groups as shown in Figure 2-4: They are optimal for this case but they cannot be applied to samples containing elongated and curved clusters like those shown in Figure 2-5. They often presuppose the knowledge of the correct number of groups. Some procedures have been designed to detect small groups besides big ones. Special "bridge-detecting" algorithms can be applied if it is expected that otherwise well-separated clusters are linked by some data units lying between them (these data units being called a **bridge**).

Every researcher needs a certain period of professional adjustment for every cluster-detecting procedure. He must become acquainted with the properties and possibilities of the algorithm he wants to apply to his data. He must know something about the features of data preparation like the handling of missing data and normalisation, and he must choose the appropriate proximity measure if required. Most programs for performing cluster analyses are not embedded in widely used program packages but are available as stand-alone subroutines. Many of these programs originally were written for the authors' personal use, rather than for distribution. The authors' principal concern was to get their own research done, not to write "user-friendly" programs with good documentations and well designed human interfaces.

The program packages BMDP (Biomedical Programs), SPSS (Statistical Package for the Social Sciences) and SAS (Statistical Analysis System) now all contain cluster detecting programs (see [3], [53], [94], [297] and [349]). The latter program package allows the binding and use of macros and user-written procedures. A comparison of the classification procedures in these popular software collections of statistical methods is given in [393]. We especially want to draw the attention to a program package called CLUSTAN. In this package many different classification procedures are collected. Thus, more than one algorithm can be applied selective to a data set. However, difficulties arise when the sample consists of variables of different scale levels. Then the user is involved in intensive and intricate preparations of the data before he can apply a clustering algorithm. In [343] and [428], the use of CLUSTAN is described in detail. At the moment, neither of the packages contains a method to estimate missing values in its data preparation part.

For the remainder of this subsection, we shortly present the principles and backgrounds of some of the most commonly used classification procedures. A discussion of their advantages and disadvatages follows in Section 2.3. Generally speaking, the methods of cluster analysis fall into four categories (this holds for disjoint and nondisjoint classifications as well as for hierarchic and quasi-hierarchic ones).

(a) Optimalization methods;
(b) recursive construction of groups around centres (so-called "kernels");
(c) analysis of the point density;
(d) linkage methods.

For methods from the third class, the entries from the matrix $\mathbf{X} = (x_{il})$ of the raw data are used or those from the matrix $\mathbf{Z} = (z_{il})$ of standardised data. The other methods are not applied to these data directly but to the proximities or similarities defined between every pair of objects.

Besides that, there must be named a fifth class called "other methods". The correspondence analysis, for example, provides a cluster determination method for qualitative data (see [44]). Methods of sequential rank classification which are based on ideas from discriminant analysis are explained in [239]. We also must cite methods which have been derived from the theory of fuzzy sets ([26], [27], [143], [340]). There are methods of clustering by ordering ([400]), and methods which are based on decision theory ([232], [233]). Methods of the first three categories often use either the raw data directly or standardised data. The other methods are applied to the proximities. All these methods are not commonly used. Thus, we don't include them into our further discussion. Let us aim at the four classes named above.

2.2.4.1 OPTIMIZATION METHODS

All optimization methods of cluster analysis naturally need a **clustering criterion** to be optimised. Furthermore, we must define the number of classes. That implies that we need some information about the number of "natural" classes. A commonly used criterion to judge the quality of a classification is the within groups sum of squares. This criterion therefore is called the WGSS criterion. Other names are the L^2- or variance criterion. For an exhaustive partition $\mathcal{C} = \{\mathcal{C}_1, \ldots, \mathcal{C}_m\}$ of $\mathcal{S}$, a quality function $g(\mathcal{C})$ is defined by

$$(2\text{-}10) \qquad g(\mathcal{C}) = \sum_{k=1}^{m} \sum_{O_i \in \mathcal{C}_k} d(O_i, \overline{O}_{\mathcal{C}_k})^2 = \frac{1}{2} \sum_{k=1}^{m} \frac{1}{n_k} \sum_{O_i \in \mathcal{C}_k} \sum_{O_j \in \mathcal{C}_k} d_{ij}^2.$$

Here, by $\overline{O}_{\mathcal{C}_k}$ the "typical" representative of the k-th group is denoted. This fictive object is the mean vector $\overline{\bar{x}}_{\mathcal{C}_k}$. For d, the Euclidean distance is chosen. It is easy to see that (2-10), which is derived from (2-6a), is the sum of the average within-groups sums of squares if the d_{ij}^2 are squared Euclidean distances. Strictly speaking, this criterion postulates metric data on interval or ratio scale level. That partition $\mathcal{C}$ is chosen as being the optimum, for which (2-10) takes its minimum. The subsets $\mathcal{C}_1, \ldots, \mathcal{C}_m$ of $\mathcal{C}$ are the clusters sought.

By (2-10), the quality of a partition is solely defined by the sum of the homogeneity indices of all classes. As well as (2-10), the sum of the separation indices or a function based on (2-9) could be taken as criteria for the quality of a classification. For overlapping clusters, the modifications discussed in the previous subsection must be obeyed. Procedures, which are based on optimization criteria, construct partitions, for which a quality function $g(\mathcal{C})$ takes its optimum. However, every quality function would take its optimum for a classification of the data units into n one-element classes if the number m of clusters would not be fixed before applying a clustering procedure. This holds for (2-10) as clustering criterion as well as for

$$g(\mathcal{C}) = \sum_{k=1}^{m} b(\mathcal{C}_k)$$

with $b(\mathcal{C}_k)$ from (2-9). This is because $g(\mathcal{C}) = 0$ always holds for $\mathcal{C} = \{\{O_1\}, \ldots, \{O_n\}\}$. To prevent cluster-detecting procedures from producing such trivial and useless results, the number of clusters and the maximum number of classes, which may contain only one data unit, must be defined in advance (see [283]).

The most crucial point preventing scientists from applying optimization procedures in practice, is the computation time needed to find an optimum of $g(\mathcal{C})$, since the value of the validity function must be computed for all partitions of n elements into m non-empty classes (see Section 2.3). Thus, iterative methods have been developed to replace the optimal strategies (detection of **sub-optimal** or **quasi-optimal partitions**). For the same reason, hierarchic divisive or agglomerative procedures are used instead of optimal ones.

The numerical detection of sub-optimal partitions starts from an arbitrary classification of the data into m groups which must be performed by the scientist. The groups of this outset classification should be of roughly equal size and as compact as possible. This classification of the objects then is improved stepwise. The elements of each class are attached to another group, and the value of the quality function g for this new classification is computed. This value is compaired with that of the former classification. If it indicates an improvement, then the previous classification is replaced by the new one. This process continues until the clustering criterion reaches its local optimum, that is, cannot be improved by further reclassifications of single elements. Procedures like these are called **hill-climbing-algorithms**. The quality of the classifications found by these algorithms depends to a high degree on the outset classification as does the number of iterations. Some of these procedures are described in [88] and [117].

In divise algorithms, a sample $\mathcal{S} = \{O_1, \ldots, O_n\}$ is not split into m subsets directly. Usually, in the first step of a divisive algorithm, the optimal partition of $\mathcal{S}$ into two non-empty subsets is chosen out of the $2^{n-1} - 1$ possible two-class partitions. In the next step, for every cluster of this optimal partition another optimal partition into two group is computed. The procedure stops if either the partition consists of at least m classes or the clustering criterion reaches a pre-defined threshold. Divisive methods are discussed in [38], [67], [107], [164], and [251].

In [414], J.H. Ward proposes an agglomerative algorithm. Starting with the n objects as n classes, those clusters are merged stepwise, for which the increase in (2-10) is a minimum. This fusion is stopped if $g(\mathcal{C})$ passes a given threshold. Ward's method is commonly used in the social sciences but only to a minor degree in biometric research. Other agglomeratice procedures and different stopping rules are studied in [281].

In most classification problems, the adequate number of classes is not known a priori. For optimization methods, using the WGSS-criterion, M.G. Kendall proposes in [229] the following statistical test to determine the optimal number of classes. For m clusters, let $g_m = g(\mathcal{C})$ be the mean sum of squares from (2-10); for $k < m$ let g_k be the value of that sum. From these values,

(2-11)
$$F(m,k) = \frac{g_k - g_m}{g_m} \left(\frac{n-k}{n-m} \left(\frac{m}{k} \right)^{2/t} - 1 \right)$$

is calculated. Then $F(m, k)$ is compared with the quantile $F_{t(m-k),t(n-m);1-\alpha}$ of the F-distribution with $n_1 = t(m-k)$ and $n_2 = t(n-m)$ degrees of freedom and given level α. The sample is assumed to consist of more than k clusters if $F(m, k)$ is smaller than the given level of significance. This test is based on ideas from [21]. M.G. Kendall points out that it is by no means an exact statistical test. It is just an analogy to methods known from the analysis of variance and thus can serve as a hint to a good choice of the number of clusters only. For further information about optimization methods of cluster analyis, the reader is referred to [268], [269], [289], [379], and [429].

2.2.4.2 RECURSIVE CONSTRUCTION OF GROUPS

Any recursive construction of groups around kernels starts from an arbitrarily chosen element O_i of a sample, which is believed to be a "typical" representative of a group. This object is the so-called **kernel** of the cluster. All elements nearest to the kernel are attached to its group. This process stops if the cluster becomes too heterogeneous. Another element from the remainder of the sample is chosen as the kernel of the next cluster, and the process is repeated. This procedure is repeated until either all objects would be classified (exhaustive classification) or conditions for the homogeneity would be violated by the attachment of further objects to the clusters already existing (non-exhaustive classification). The number of clusters must not be determined in advance.

Two ways of defining new cluster kernels during the iterative process are possible. The simplest one is to leave the object unchanged which was chosen as kernel at the beginning of each process. The second possibility is to compute a new cluster kernel each time after attaching a new object to that cluster. Usually, the geometric centre of a cluster becomes the new kernel. This kind of "adaptive estimation" of a kernel is obviously better than the first method since an object chosen arbitrarily as a cluster kernel at the beginning of such a process can be a great distance apart from the true group centre.

During such iterative processes, it often happens that objects, which already have been attached to a cluster, get closer to the centre of a new group than to the centre of the group they previously have been attached to. These objects must be reclassified. Some procedures allow the amalgamation of groups if their kernels get too close together. Both the definition of the centres of gravity as cluster kernels and the need of reclassification of data units imply heavy computations (see [38]).

2.2.4.3 ANALYSIS OF THE POINT DENSITY

For interval or ratio scaled data units $O_1, \ldots, O_n$, the data vectors $\vec{x}_{1*}, \ldots, \vec{x}_{n*}$ can be plotted in the t-dimensional space $\mathcal{R}^t$ as is shown in Figures 2-4 and 2-5 for the plane. "Clusters" then are connected areas of high point densities. They are determined by the maxima of these densities, the so-called **modes**. Under ideal conditions, distinct clusters are separated by areas of low density, the so-called **valleys** of the distribution mixture.

In [235] and [236], W.L.G. Koontz and K. Fukunaga propose non-parametric **valley seeking techniques** for the separation of the distribution mixtures and the detection of clusters. These methods can be applied to samples containing an unknown number of groups which may be sickle-shaped like those shown in Figure 2-5. The techniques rely on a graph-theoretic interpretation of the clusters (see [237]).

Other authors rely on the detection of modes which they define as cluster centres. These **mode seeking techniques** are always parametric. The multivariate point

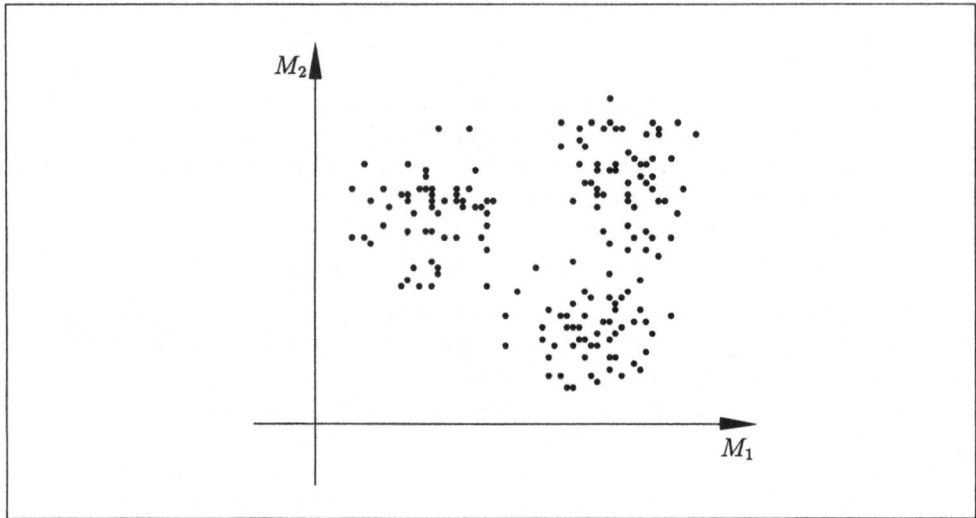

Figure 2-4 *Normal distributions in the $\mathcal{R}^2$.*

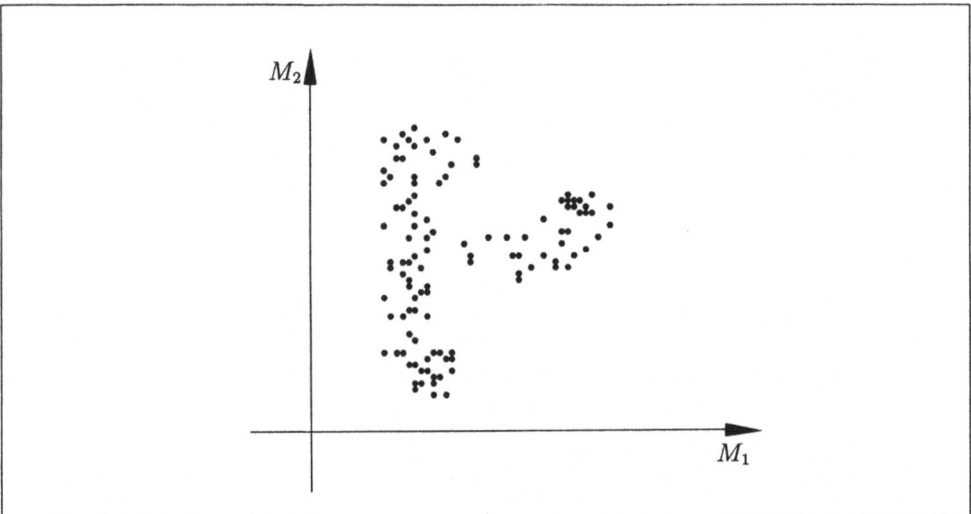

Figure 2-5 *Stretched and bended, "sickle-shaped" clusters in the $\mathcal{R}^2$.*

density $f(\vec{x})$ is assumed to be of the form

$$(2\text{-}12) \qquad f(\vec{x}) = \sum_{k=1}^{m} c_k f_k(\vec{x}).$$

The number m of clusters must be known in advance. By $f_k(\vec{x})$, we denote the probability density of the k-th cluster, and by c_k the **a priori probability** of the k-th cluster, that is, the probability of an object to belong to this group ($c_1 + \cdots + c_m = 1$). The $f_k(\vec{x})$ are assumed to be multivariate normal (see [37], [243]).

The portions $c_1, \ldots, c_m$, the mean vectors and the covariance matrices of the different normal distributions $f_1(\vec{x}), \ldots, f_m(\vec{x})$ in the mixture (2-12) are estimated by **maximum likelihood methods**. Then, using **discriminant rules**, each data unit is attached to one of m clusters. Very heterogeneous clusters are tolerated besides homogeneous ones. Furthermore, because of the assumption of normal distributions, sickle-shaped clusters cannot be detected with mode-seeking techniques ([277]). Moreover, all attributes in the data units must be on an interval or ratio scale, no mixed data are allowed. For these methods, no proximities or distances are needed. They are directly applied to the matrices of the raw or standardised data. However, mixture analysis involves heavy computations even if no calculations of proximities are needed. More about classifications by mixture analysis can be found in [29], [30], [62], [63], [80], [92], [136], [137], [162], [189], [198], [222], [276], [352], [353], [384], [396], [424], [425], [430], [432], and [433], respectively.

To these methods of density analysis, another procedure can be counted, which R.A. Jarvis and E.A. Patrick proposed in [217]. This k-**nearest-neighbours method**, however, needs a similarity relation to classify the objects. For every object, its k nearest neighbours are determined ($k \leq n$). Every two objects are attached to the same group if they have at least k_d of their k nearest neighbours in common ($k_d \leq k$). The authors show by example that this procedure can be used for the detection of elongated and bended clusters. The degree of compactness of the groups is controlled by the numbers k and k_d. The same holds for the number of clusters. Thus, it must not be known in advance but follows from the degree of compactness, which the user desires.

2.2.4.4 Linkage Methods

The use of optimization methods as cluster determination procedures does not presuppose a definition of the term "cluster". The clusters are defined constructively through statement of the criterion and an implementing algorithm: Every subset C_k of the optimal or sub-optimal partition $C = \{C_1, \ldots, C_m\}$ is a cluster. In recursive procedures, the clusters are defined as subsets of S, which don't pass a given threshold of homogeneity. In the nearest-neighbour method, a cluster is defined by the number of neighbours which are common for every pair of its elements. Here, a subset of S is a cluster if and only if it satisfies certain a priori conditions, it is not just the result of a clustering procedure; the sole role of an algorithm now is that of **finding** clusters, not of **defining** them.

This also holds true for the linkage methods in cluster analysis. Starting from a matrix $\mathbf{S}$ of similarities or a matrix $\mathbf{D}$ of proximities or distances between every pair of objects $O_1, \ldots, O_n$, similarities or distances between subsets of a sample are computed. These then are used for the construction of classes. Linkage procedures are studied by many authors, for example, by P.H.A. Sneath and R.R. Sokal in [373], by H.H. Bock in [38], and by B.S. Everitt in [115]. At least twelve different linkage procedures are known from literature. According to R.K. Blashfield, M.S. Aldenderfer and L.C. Morey, the following methods are most commonly used ([32], [33], [34]):

(a) **single-linkage** methods (or nearest-neighbour methods),
(b) **complete-linkage** methods (or furthest-neighbour methods),
(c) **average-linkage** methods,
(d) **centroid** and **median** methods.

All these methods can be used to detect clusters with properties, which have been defined by a researcher in terms of their separation or homogeneity. (We speak of "clusters of level d",) They, moreover, can be applied to construct dendrograms as we also show in this part of the monograph.

For **single-linkage methods**, a proximity between two subsets C_k and C_l of S is defined by

$$(2\text{-}13a) \qquad\qquad d(C_k, C_l) = \min_{\substack{O_i \in C_k \\ O_j \in C_l}} d_{ij}.$$

Thus, the dissimilarity between two subsets C_k and C_l is the dissimilarity between the two mutually closest objects, one from C_k and one from C_l. Usually, L^p-distances according to (2-3a) are used as inter-object dissimilarities d_{ij} (see Figures 2-7a and 2-7b for an example). This method had been proposed, for the first time, in 1951 by K. Florek, J. Perkal and others (in [128], [129], [312]). In 1957, L.L. McQuitty (in [275]) and P.H.A. Sneath (in [364] and [365]) discussed it again.

By the single-linkage method, two objects are merged into the same group if their distance is smaller than or equal to a certain threshold d which must be defined by the researcher. A single object O_i is attached to a group C_k already existing if its distance to that element of C_k nearest to O_i is not larger than d. Two clusters are joined if their mutually closest elements have a distance at most d. The method is known as "single linkage" since clusters are joined by the single shortest link (or by the mutually closest elements). For any cluster of two or more objects produced by the single-linkage method, every member is more similar to some other member of the same cluster than to any other entity not in that cluster. This holds, since for every object O_i of a cluster, another object O_j of that cluster must exist with $d_{ij} \leq d$; however, every object O_h, being not in the same cluster as O_i, must have a distance $d_{hi} > d$ from O_i. Such **clusters of level** d often are elongated and sickle-shaped or ramified. However, for every two objects of the same cluster which are more than a distance d apart from each other, exists a "chain" of objects of that cluster such that any two successive entities are linked together by a distance smaller than the threshold d. Therefore, the tendency to give long serpentine clusters is called **chaining effect**.

This property is often criticized because entities at opposite ends of a cluster may be markedly dissimilar. However, because of this property, the single-linkage method is one of the very few clustering techniques which can detect nonellipsoidal clusters. The single-linkage technique relies more on the aspect of separation between groups than on the aspect of homogeneity within one group (see (2-6b) and (2-7b)). The properties of single-linkage procedures are discussed in [6], [167], [186], [201], [202], [212], [213], [218], [244], [332], [333]. In [439], C.T. Zahn gives a discussion of the intuitive aspects of single-linkage clustering. He also provides an algorithm to detect chains (or "bridges" as he calls these links between different clusters).

The single-linkage method has a close connection with certain aspects of graph theory. As shortly explained in Chapter 1, a graph is a set of vertices (nodes or points) together with a set of edges (lines or arcs) joining together some or all of the vertices. If the data units are interpreted as vertices of a complete graph and the edges are taken to have the distances of their endpoints as weights, then the **minimal spanning tree** is the "shortest" of all spanning trees of this complete graph with the elements of S as vertices. Omitting all those edges with weights greater than d gives all single-linkage clusters of this level d. Thus, the single-linkage method has interesting connections with the problem of finding the minimal spanning tree of a graph (see [1], [6], [38], [148], [167], [214], [224], [259], [260], [261], [272], and [312]).

The **complete-linkage method** is the logical counterpart of the single-linkage method (see Figures 2-8a and 2-8b): The dissimilarity between two subsets C_k and C_l is defined as the dissimilarity between the most distant objects, one from C_k and one from C_l,

$$(2\text{-}13\text{b}) \qquad\qquad d(C_k, C_l) = \max_{\substack{O_i \in C_k \\ O_j \in C_l}} d_{ij}.$$

By the complete-linkage method, again two objects are merged into a group if their distance is smaller than or equal to a previously defined threshold d. However, a single object O_i is attached to an already existing group C_k if its distance to that element of C_k most distant from O_i (and by this its distances to all other elements of that cluster) does not exceed d. Two clusters are joined if their mutually most distant elements have a distance at most d (see (2-7c)). Thus, the method is called complete linkage because all entities in a cluster are linked to each other at a maximum distance d.

This method was proposed first by T. Sørensen in 1948 (see [383]). It outlines only compact, ellipsoidal clusters as optimization methods do. Contrary to clusters defined by optimization methods, however, every group determined by a complete-linkage procedure has a minimal degree of homogeneity since no pair of objects in a complete-linkage cluster has a proximity larger than the threshold d. Moreover, complete-linkage clusters are not disjoint. In contrast to the single-linkage method, the interpretation of the clusters can be only in terms of homogeneity; no useful interpretation involving the separations between clusters can be given ([201], [202]). Like the single-linkage method, complete-linkage clusters are invariant to monotonic transformations of proximity measures. S.C. Johnson discusses this property in [218] and relates it to the ultrametric property.

Like single-linkage clusters, complete-linkage clusters can also be interpreted on a graph-theoretic basis. Let all elements of S with mutual distances at most d be linked together by edges. This defines a graph. Then every complete subgraph (or clique) of this graph is a complete-linkage cluster. (Every component of that graph, at the same time, is a single-linkage cluster.) By this method, all complete-linkage clusters can be found. Like the single-linkage method, the complete-linkage cluster determination method also can rely on graph-theoretic algorithms of searching spanning trees. The complete-linkage method, however, starts with the pair of elements with the shortest distance, and terminates with the pair with the longest distance. A spanning tree constructed like this has no special interpretation in terms of graph theory. This method also determines only disjoint complete-linkage clusters (see Procedure 2 described below). In [178], P. Hansen and M. Delattre discuss some connections of single-linkage cluster analysis to graph colouring. We will discuss both methods as well as other graph-theoretic cluster determination methods to a greater extend in the subsequent chapter.

Both methods, single linkage and complete linkage, rely on extreme values of proximities between data units. In single-linkage clusters of level d, only the shortest link needed to connect any member of the cluster to some other member should not exceed a threshold d; whereas each complete-linkage cluster is characterized by the longest link needed to connect every element of a cluster to every other element. In [377], R.R. Sokal and C.D. Michener proposed a procedure called **average linkage** (see Figures 2-9a and 2-9b). In this method, a cluster is characterized by the average of all links within it. Thus, this method is a compromise between both the methods discussed above. A distance between two clusters C_k and C_l is defined as the average of all mutual pairs of objects,

$$(2\text{-}13c) \qquad d(C_k, C_l) = \frac{1}{n_k \, n_l} \sum_{O_i \in C_k} \sum_{O_j \in C_l} d_{ij}.$$

By this method, an object is attached to a group if its average distance to the elements already in that group is smaller than a threshold d. Two groups are merged to one single average-linkage cluster if their average distance (2-13c) is small enough.

In the **centroid method**, the distance between two clusters is defined as the distance between their mean vectors or centres of gravidity,

$$(2\text{-}13d) \qquad d(C_k, C_l) = d(\bar{\bar{x}}_{C_k}, \bar{\bar{x}}_{C_l}).$$

J.C. Gower discusses this method and the **median method** in [164]. For the median method, in (2-13d), the cluster centres are replaced by their medians, that is, the distance between two clusters is the mutual distance of their medians. A data unit is linked to a cluster if its distance from the cluster mean (or the cluster median) is small enough, that is, if it is similar to this representative of the cluster, see Figures 2-10a and 2-10b. Both methods are not invariant to monotonic transformations of the data or proximities. Also, both methods emphasize a geometric interpretation of clusters. The idea is that the distance between every object of a group and its representative should not be too large. Long serpentine or sickle-shaped clusters cannot be detected by centroid or median methods.

Obviously, we can define similarities between clusters as well as dissimilarities or distances. This can be done by using (2-5) as well as by a slight modification of (2-13a)–(2-13d). All these definitions of distances and the corresponding linkage methods can not only be used to determine either disjoint or overlapping clusters of certain shapes and at pre-defined levels d of separation or homogeneity (as a comparison of Formulas (2-13a)–(2-13c) with (2-6a)–(2-6c) and (2-7a)–(2-7c) shows), but also are powerful tools in the construction of hierachic or quasi-hierarchic classifications.

Usually, the dissimilarities of the data units are ranked according to their order of magnitude. Depending on the type of clusters, a scientist is interested in, he chooses one of the proximity measures for groups named above. Then, the agglomerative construction of hierachic or quasi-hierarchic classification follows the same scheme for all linkage methods. Therefore, we will describe it for the graph-theoretical based methods of complete or single linkage only. The process starts with the n data units to be clustered as n one-element clusters at the level $d = 0$. The threshold d is increased to min d_{ij}. At this level, those two objects with minimal distance are linked by an edge and merged into a two-element cluster (being both a single- and a complete-linkage cluster). The threshold then is increased to the next higher ranking distance: The edge with this distance as weight is added to the graph. It may link two new objects together, thus forming a new custer (for the single-linkage distance as well as for the complete-linkage distance). It also may link a third object to one of the data units having been connected in the previous step. In this case the analyst gets one three-element single-linkage cluster besides $n - 3$ one-element groups. But if being interested in complete-linkage clusters he now has to differ between two two-element clusters mutually sharing one object. This process is repeated, and at every step all clusters at that level are determined. At the end, the sample $S = \{O_1, \ldots, O_n\}$ is the only cluster. In the following we will denote this procedure as "Procedure 1".

Any distance, at which two groups in a hierarchic procedure are merged together into a new cluster, is called **creation time** or **age** of the new cluster. The difference of the ages of a cluster C_k and the next bigger one, which contains C_k, is the **survival time** or **isolation index** of the smaller cluster C_k. For every hierarchy or quasi-hierarchy, the ages of the clusters can be tagged on an axis as shown in Figures 2-7b, 2-8b, 2-9b or 2-10b. This defines a dendrogram for the clusters. The procedure of amalgamating the clusters can be stopped at an initially defined level d for the cluster distances. By this, we get all the clusters of that pre-defined level at the last step of the procedure. No information about the number of groups is needed; this number is a function of the threshold d: $m = m(d)$.

Procedure 1 is very expensive and time-consuming. In the worst case, the cluster-determination step must be repeated $\binom{n-1}{2} + 1$ times. Therefore, the following agglomerative Procedure 2 usually is applied instead of Procedure 1. It is described in [6] and [38].

(a) Choose one of the distance measures for groups. Each of the n observations is treated as a group with one member. In this first classification step, the classification C consists of $m = n$ one-element clusters.

(b) Examine all pairs of the m groups $C_1, \ldots, C_m$ and find those two which are mutually closest together. These groups are merged together to form a new group. Calculate the distance of this new group to all other groups.

(c) Repeat step (b) for $m = n, n - 1, \ldots, 2$.

Repeating (b) $n - 1$ times leaves the whole sample S as one single cluster at the top of the hierarchy. If the amalgamation of clusters is stopped after reaching a threshold d, the cluster determination process results in a classification at a level d.

For a given sample and distance measure, both procedures result in the same classification if a hierarchy or disjoint clusters of a level d are searched (with the restriction that all distances are assumed to be different). Both procedures, however, result in classifications, which are completely different if they are applied together with (2-13b) to a sample to produce complete-linkage clusters. At every level d, Procedure 1 determines all possibly overlapping complete-linkage clusters. In contrast, by Procedure 2 only the disjoint complete-linkage cluster can be found. This is shown in the following example with fictive data.

Example 2-2 Let $S = \{\vec{x}_{1*}, \vec{x}_{2*}, \vec{x}_{3*}, \vec{x}_{4*}\}$ with the six L^2-distances as shown in Figure 2-6a. Procedure 1 together with the complete-linkage distance (2-13b) gives the correct, complete quasi-hierarchy of Figure 2-6b. At the level $d = \sqrt{1.3}$, we get three overlapping two-element complete-linkage clusters. However, Procedure 2 results in the hierarchy shown in Figure 2-6c. At the level $d = \sqrt{1.3}$, the classification C does not contain the clique $\{O_2, O_3\}$. Procedure 2 is a method to find a spanning tree. From this rather artificial example, it can be already seen that the procedure terminates with the distance between the most distant objects. •

This important difference between the procedures is not pointed out in the manuals [343] and [428] of the CLUSTAN program package. The subroutine HIERARCHY together with the distance function FURTHEST NEIGHBOUR (the complete-linkage distance) only finds the — disjoint — hierarchy of Procedures 2.

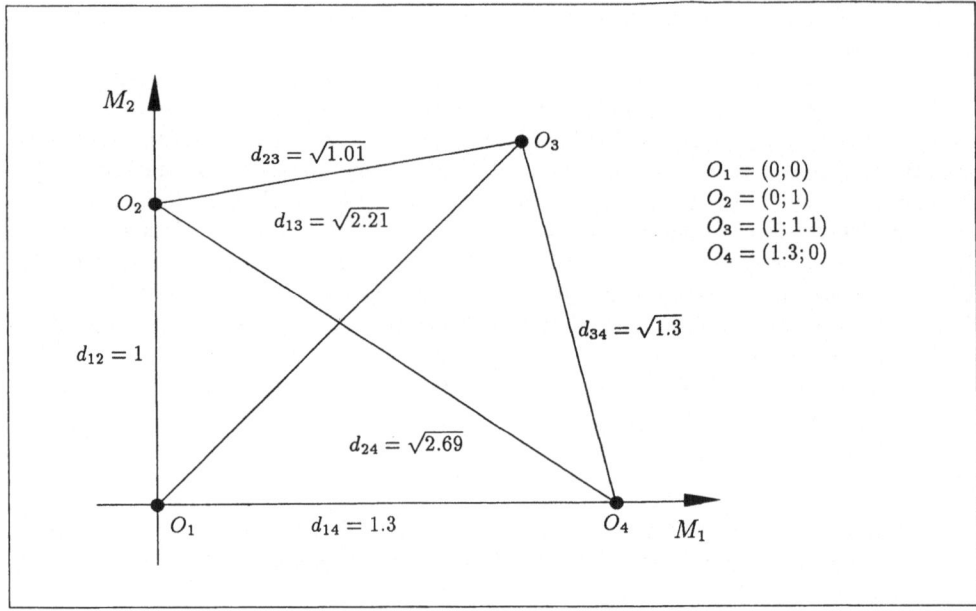

Figure 2-6a *Data for Example 2-2: Classification of four objects by the complete-linkage method, L^2-distance.*

From the distance measures (2-13a) and (2-13b), it is evident that the threshold d should be not too large for the construction of single-linkage clusters of a level d; however, it should be not too small for the construction of complete-linkage clusters of a level d. In the first instance, we possibly get the whole sample as the only single-linkage cluster if d is too large; the second case, for a small value of d, can result in too many complete-linkage clusters which are mutually overlapping to a very high degree, leaving no room for meaningful interpretations. A "natural" distance threshold seldomly is known. However, a favourable threshold often can be determined by practical or statistical means (see Chapter 6 and [439]). If no preferable threshold is known a priori, then the best way is to vary d to get the dendrogram. Especially for single-linkage clusters, S. Krolak-Schwerdt proposes an agglomerative procedure which includes a stopping rule: A favourable threshold is determined by comparing the minimum spanning tree and the maximum spanning tree (that is, the spanning tree with the maximum sum of distances). The following example shows the dendrograms for different linkage methods using the data of Example 2-1.

Example 2-3 Using the attributes $M_1, M_2, M_5 - M_7, M_{14}$ and M_{15} of the data set of Example 2-1 and Table 2-1, cluster analyses have been performed to study the properties of four linkage methods. (We omitted the centroid method; it did not differ from the median method.) For all these methods we used the procedure HIERARCHY of the CLUSTAN package together with the four distance functions needed to produce single-, complete- and average-linkage clusters and median clusters, respectively. The data vectors of the third, 13th, 15th, and 17th proband could not be used because of missing data. The remaining rows in the data matrix have been renumbered from 1 (corresponding to O_1), 2 (corresponding to O_2), 3 (previously O_4) up to 17 (previ-

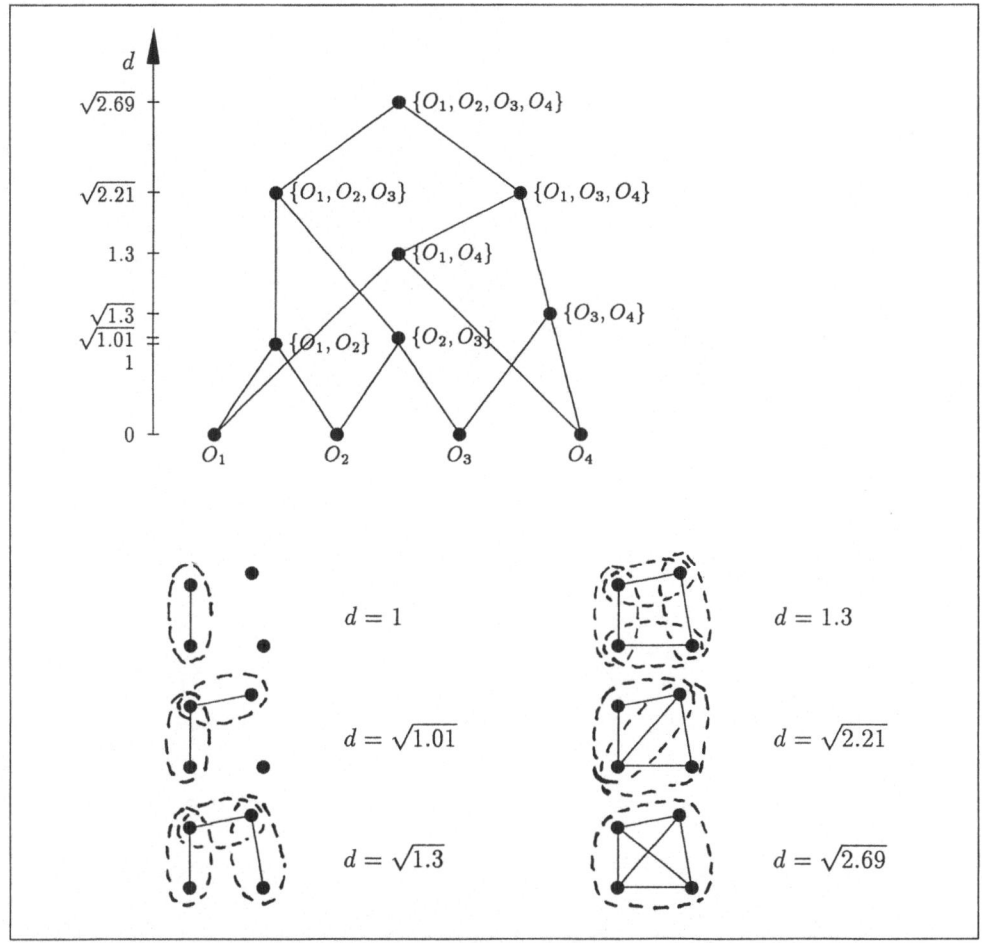

Figure 2-6b *Quasi-hierarchy and dendrogram as given by Procedure 1.*

ously O_{21}). Before performing the cluster analyses, the variables have been normalised to $[0; 1]$. Figures 2-7a–2-10b show the dendrograms given by Procedure 2 and classifications at varying levels d. The latter are depicted as two-dimensional projections of the minimum spanning tree. A typical difference between the dendrograms of the four methods is shown fairly well by the figures: For single-linkage clusters, the process of merging clusters is very fast at the beginning and slows down for larger values of d; in contrast, it is a uniform process for complete-linkage clusters. The dendrograms of the other two methods show that these methods are compromises between these extreme situations as we said earlier. It should be noticed that proband 16 (that is, O_{20} in Table 2-1) remains isolated in all four dendrograms. The reason probably is the body weight of 129 kg, which differs significantly from all other body weights. •

The centroid and the median method both have a property which can be seen in the dendrogram of Figure 2-10b: It may occur at different levels $d_1 < d_2$ that a cluster of level d_1 arises after clusters of level d_2 already came into existence. This phenomenon

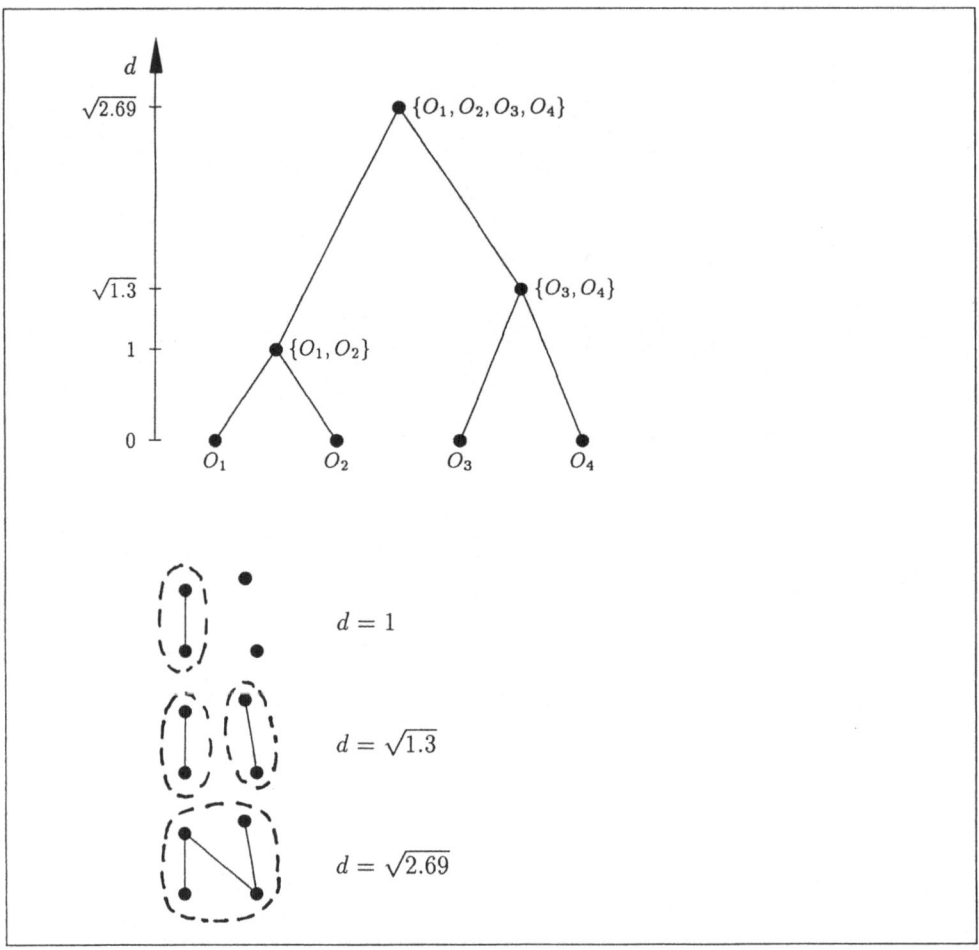

Figure 2-6c *Hierarchy and dendrogram as given by Procedure 2.*

is called **inversion**. This property is often criticized, and the clusters are hard to be interpreted because it is expected intuitively that groups which arise at a level d_1 with $d_1 < d_2$ must be more homogeneous than those clusters of a level d_2 (see [38]). Thus, both methods are hardly used in hierarchical classification. In the following example, we illustrate this property using three artificial data vectors of the $\mathcal{R}^2$.

Example 2-4 Let $S = \{\vec{x}_{1*}, \vec{x}_{2*}, \vec{x}_{3*}\}$ with the L^2-distances as shown in Figure 2-11. For growing d, at first O_1 and O_2 are merged into a median cluster when d reaches a value of 2.5. Thereafter, O_3 is attached to this group. However, its distance from the group centre of $\{O_1, O_2\}$, that is, its centroid or median distance from the group $\{O_1, O_2\}$ according to Formula (2-13d), is only 2.280. That means that O_3 is more similar to the whole group $\{O_1, O_2\}$ than to any of its elements. The cluster $\{O_1, O_2, O_3\}$ has an age of 2.280 but cannot be detected before the cluster $\{O_1, O_2\}$ of level $d = 2.5$ has been detected. An explanation of this phenomenon is that the median method and the centroid method both emphasize a geometric interpretation of

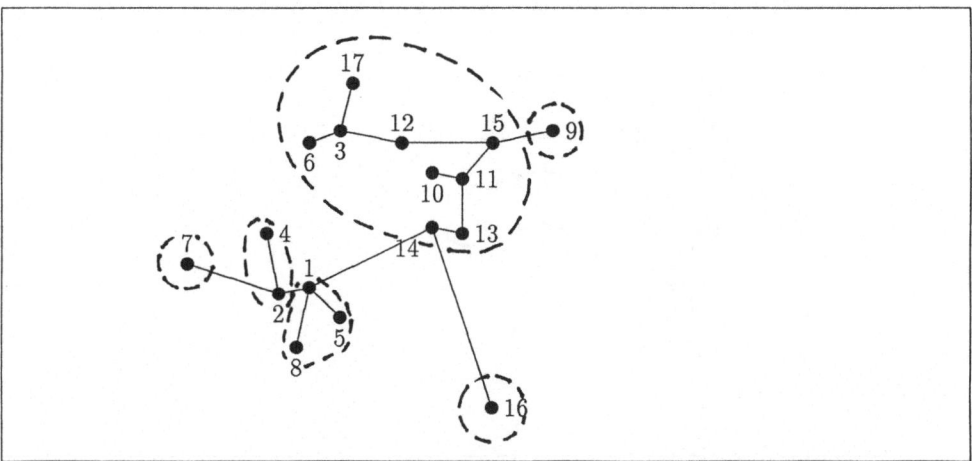

Figure 2-7a *Plot of the data from Example 2-1 on the first two principal components of the continuous measurements. The single-linkage clusters of the level $d = 0.725$ are encircled (see Figure 2-7b for the dendrogram).*

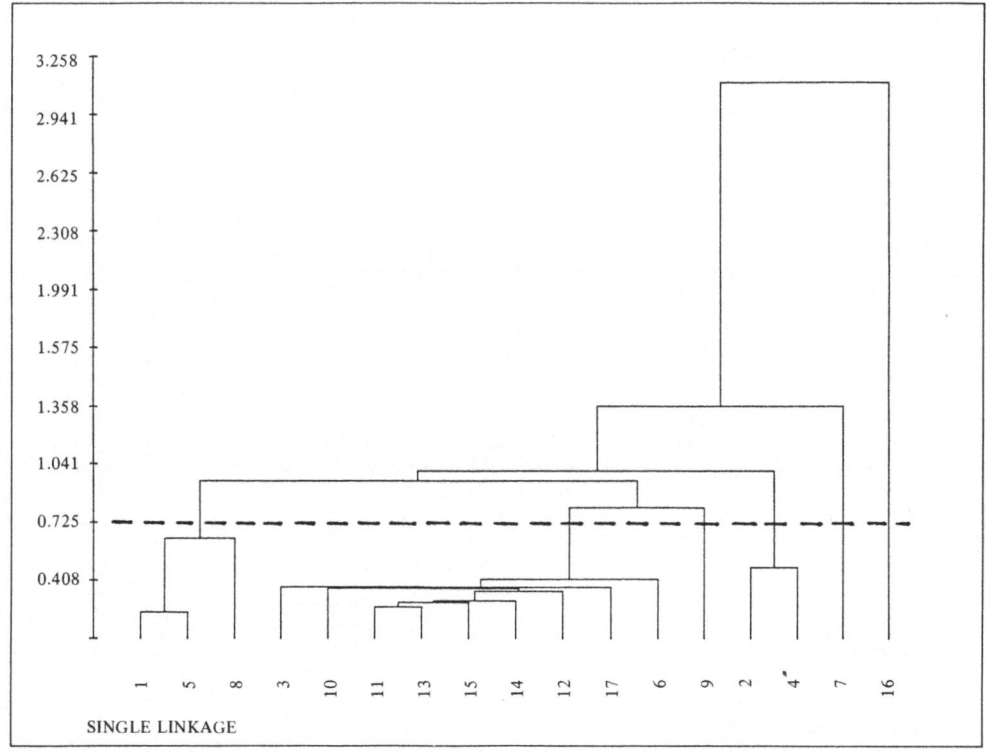

Figure 2-7b *Single-linkage dendrogram of the data of Example 2-1 for the continuous measurements only. The level $d = 0.725$ is indicated by a dashed line.*

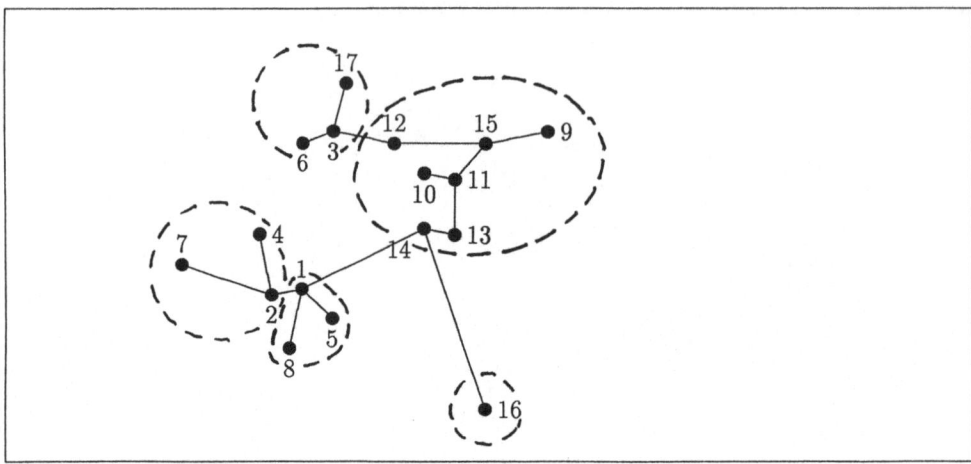

Figure 2-8a *Plot of the data from Example 2-1 on the first two principal components of the continuous measurements. The complete-linkage clusters of the level d = 2.0 are encircled, see Figure 2-8b for the dendrogram.*

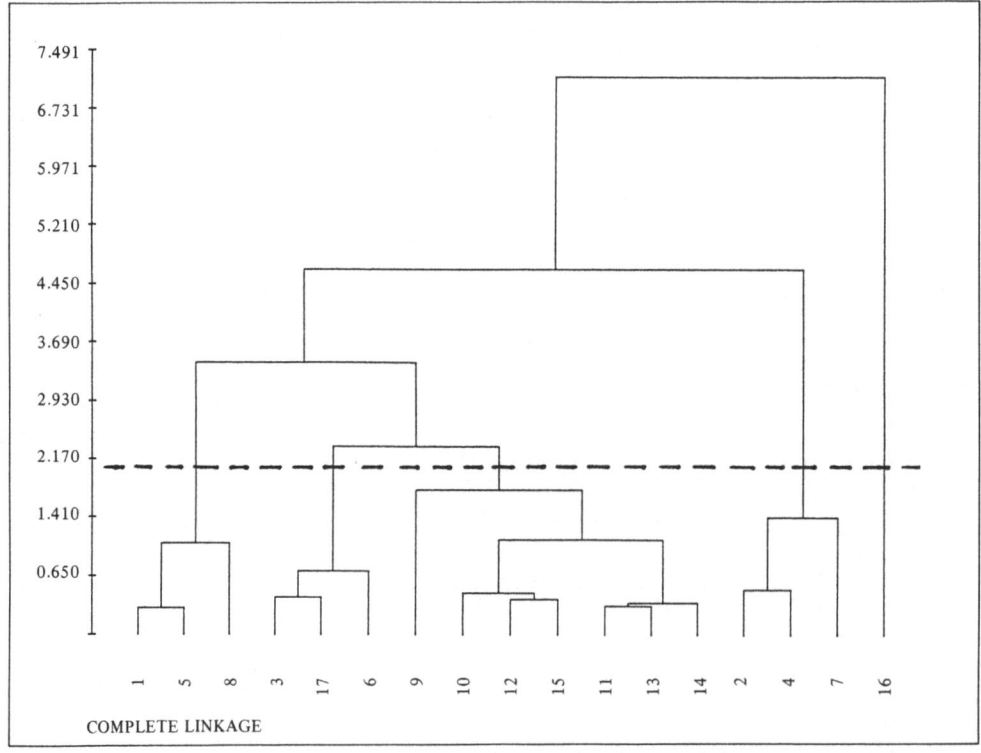

Figure 2-8b *Complete-linkage dendrogram of the data of Example 2-1 for the continuous measurements only. The level d = 2.0 is indicated by a dashed line.*

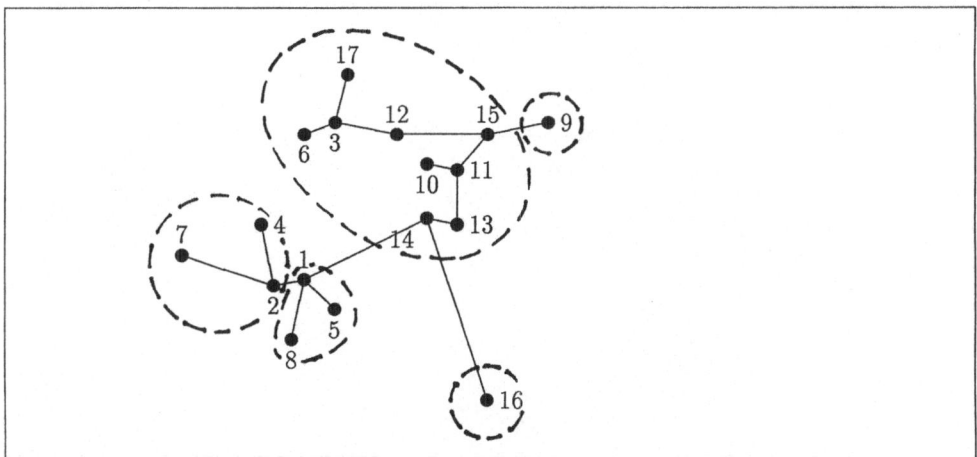

Figure 2-9a *Plot of the data from Example 2-1 on the first two principal components of the continuous measurements. The average-linkage clusters of the level d = 1.5 are encircled, see Figure 2-9b for the dendrogram.*

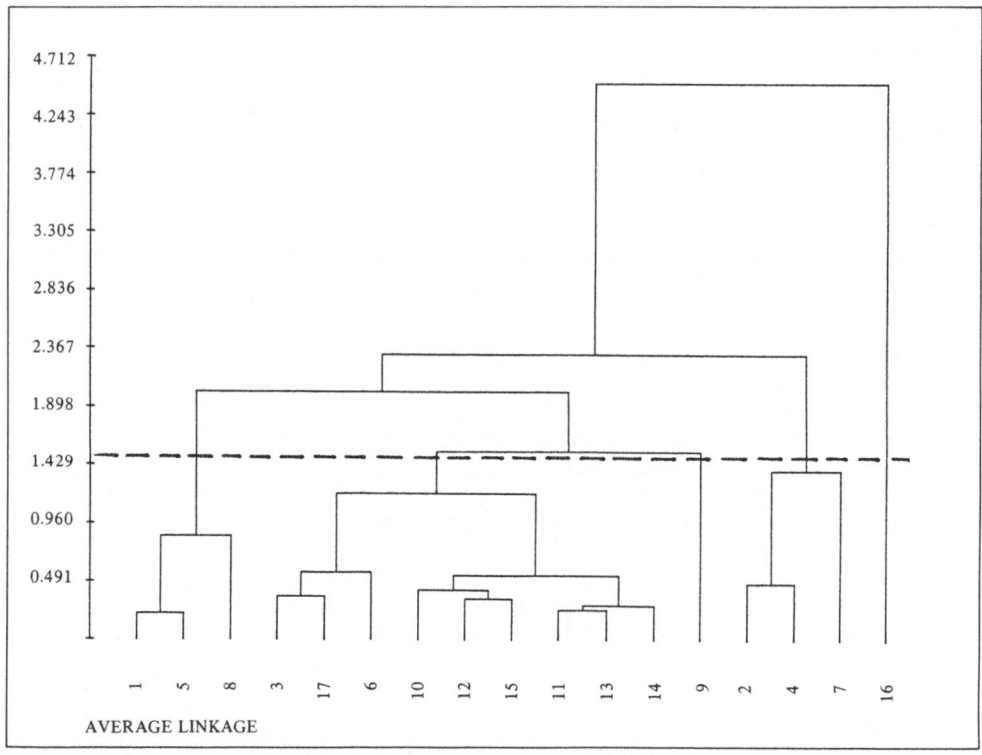

Figure 2-9b *Average-linkage dendrogram of the data of Example 2-1 for the continuous measurements only. The level d = 1.5 is indicated by a dashed line.*

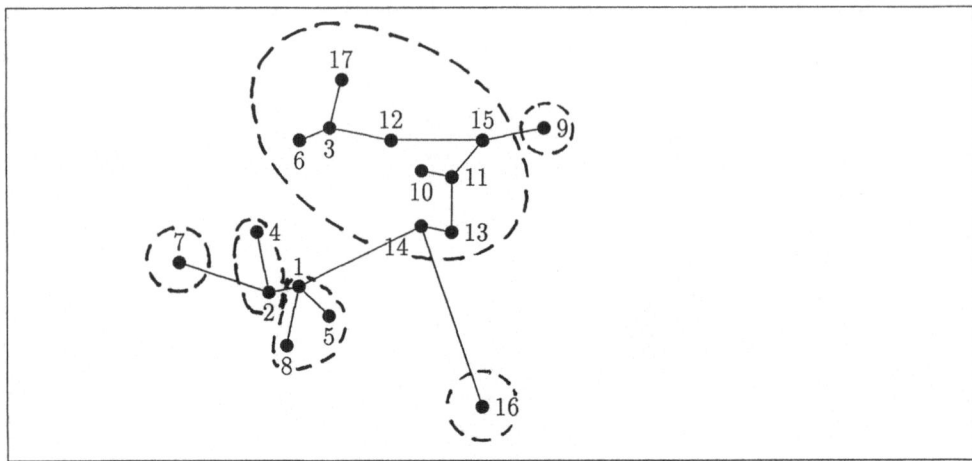

Figure 2-10a *Plot of the data from Example 2-1 on the first two principal components of the continuous measurements. The median clusters of the level $d = 1.0$ are encircled, see Figure 2-10b for the dendrogram.*

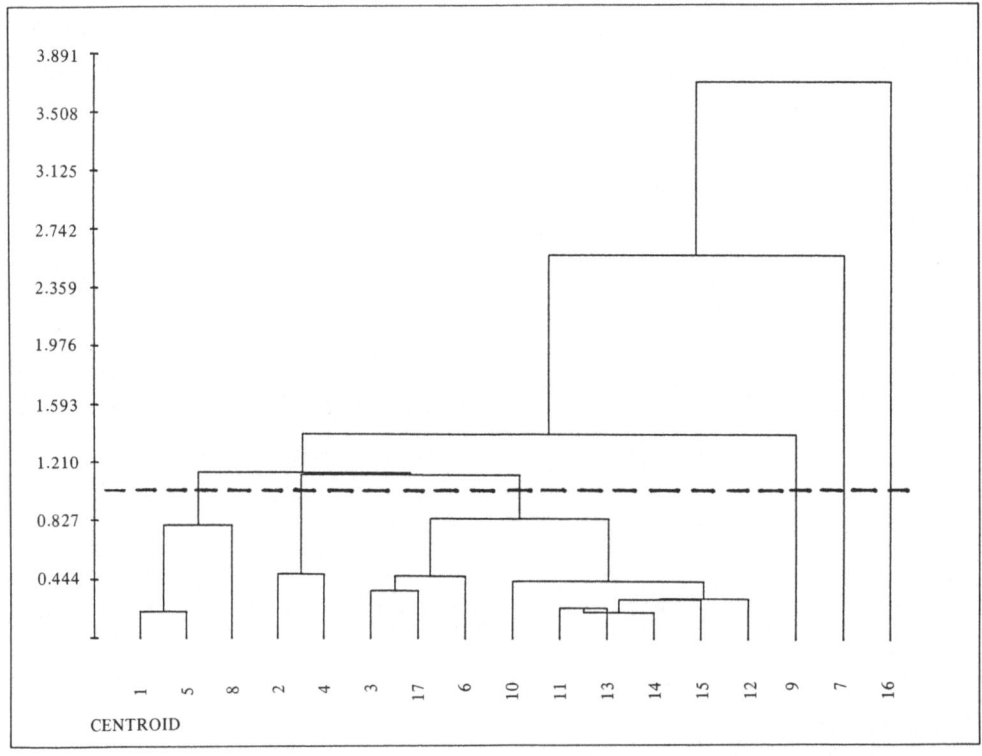

Figure 2-10b *Median dendrogram of the data of Example 2-1 for the continuous measurements only. The level $d = 1.0$ is indicated by a dashed line.*

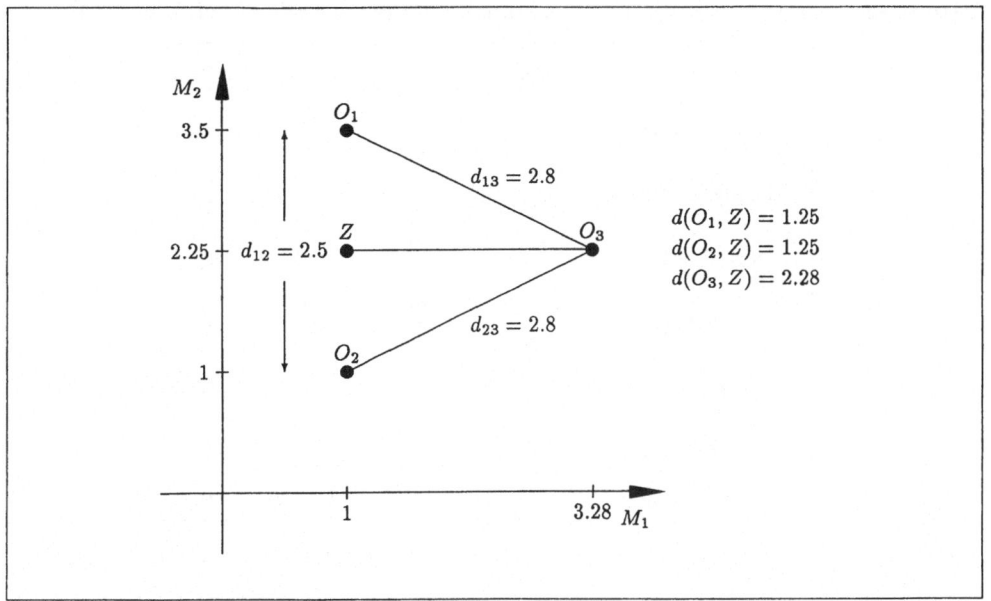

Figure 2-11 *Data for Example 2-4: Classification of three objects by the centroid or median method, L^2-distance.*

clusters. Here, clusters are not understood as subsets of a sample $\mathcal{S}$ but as solid forms in the t-dimensional space $\mathcal{R}^t$, which can be characterized by their centres of gravity and extension. •

In [164] and [251], the authors show that all linkage methods can be represented by one single algorithm. Thus, they can be considered as belonging to one type of classification procedures. If two clusters $\mathcal{C}_k$ and $\mathcal{C}_l$ are merged into a new cluster in Step (b) of Procedure 2, then the distance of the new cluster $\mathcal{C}_k \cup \mathcal{C}_l$ from every other cluster $\mathcal{C}_m$ must be calculated. For all methods named in this Subsubsection 2.2.4.4, this distance simply can be computed recursively from the distances between $\mathcal{C}_k$, $\mathcal{C}_l$, and $\mathcal{C}_m$, respectively. The relation between the old and new distances is given by

$$
(2\text{-}14) \qquad
\begin{aligned}
d(\mathcal{C}_k \cup \mathcal{C}_l, \mathcal{C}_m) = {}& \alpha_k\, d(\mathcal{C}_k, \mathcal{C}_m) + \alpha_l\, d(\mathcal{C}_l, \mathcal{C}_m) + \beta\, d(\mathcal{C}_k, \mathcal{C}_l) \\
& + \gamma\, |d(\mathcal{C}_k, \mathcal{C}_m) - d(\mathcal{C}_l, \mathcal{C}_m)|
\end{aligned}
$$

with constants α_k, α_l, β and γ. Choosing $\alpha_k = \alpha_l = \frac{1}{2}$, $\beta = 0$ and $\gamma = -\frac{1}{2}$, we get the single-linkage method; changing only γ to $\gamma = \frac{1}{2}$, we end in the complete-linkage method (see Table 2-2).

In [426], D. Wishart shows that the WGSS criterion of J.H. Ward also fits in this pattern. It therefore can be considered as a linkage method and can be used to derive hierarchic classifications. For the WGSS criterion, the distance between two groups is defined by

$$
d(\mathcal{C}_k, \mathcal{C}_l) = \frac{n_k\, n_l}{n_k + n_l}\, d(\bar{\bar{x}}_{\mathcal{C}_k}, \bar{\bar{x}}_{\mathcal{C}_l})^2 .
$$

This is a weighted distance between the group centres derived from (2-10). In [272] and [274], D.W. Matula discusses the pros and contras of agglomerative and divisive

Table 2-2 *Constants of several agglomerative linkage methods (cited from [38]).*

Method	α_k	α_l	β	γ
Single linkage	$\dfrac{1}{2}$	$\dfrac{1}{2}$	0	$-\dfrac{1}{2}$
Complete linkage	$\dfrac{1}{2}$	$\dfrac{1}{2}$	0	$\dfrac{1}{2}$
Average linkage	$\dfrac{n_k}{n}$	$\dfrac{n_l}{n}$	0	0
Centroid	$\dfrac{n_k}{n}$	$\dfrac{n_l}{n}$	$-\dfrac{n_k n_l}{n}$	0
Median	$\dfrac{1}{2}$	$\dfrac{1}{2}$	$-\dfrac{1}{4}$	0
Ward's method	$\dfrac{n_k + n_i}{n + n_i}$	$\dfrac{n_l + n_i}{n + n_i}$	$-\dfrac{n_i}{n + n_i}$	0
Flexible strategy	α	α	$1 - 2\alpha$	0

methods for the determination of average-linkage clusters. In [38], H.H. Bock lists another linkage method, the so-called "flexible strategy".

2.3 A SHORT REVIEW OF CLASSIFICATION METHODS

When clustering the objects of a sample by visual inspection, most persons intuitively prefer methods which can be considered as "hierarchical optimization". However, unlike Ward's method, divisive algorithms usually are applied. Seldomly a person starts a classification process by searching the most similar pair of objects, and goes on by merging groups. Most people at first divide the sample roughly into two or three classes, which then are split more and more according to a personal, intuitive optimization criterion. This criterion seems to become the sharper and more distinctive, the finer and smaller the subgroups are. As a logical consequence, the first procedures of numerical taxonomy were based on optimization criteria like the WGSS criterion (see [107]). For these procedures, the term "cluster" usually is left undefined. It is not possible to say in abstract terms what a cluster is. Clusters are defined constructively through statement of the criterion and an implementing algorithm. Thus, the choice of a clustering criterion is tantamount to defining a cluster as M.R. Anderberg says in [6]. Rarely special prepositions of homogeneity or separation must be satisfied by those "clusters".

The majority of cluster determination procedures has been designed to be used with continuous data only. They can uncover ellipsoidal, compact groups in a data sample. Only few procedures like the valley-seeking technique of W.L.G. Koontz and K. Fukunaga, or the single-linkage approach can detect groups of nonellipsoidal shapes or groups of different shapes in the same sample. Most of the classification methods are based purely on heuristics and don't rely on theoretical approaches. Therefore no correct model validation is possible. The behaviour of such methods — if applied to samples with a group structure not "fitting" to the heuristical base — can be studied only by applying them to real or simulated data of structures which are known in advance. Such validation or robustness studies must replace statistical considerations (see [31], [32], [78], [85], [101], [134], [197], [240], [247], [279], [280], [366], [368], [431]). Moreover, all clustering algorithms must produce clusters even if a sample lacks any group structure and is homogeneous. Such empirical methods, however, don't allow the performance of statistical tests to prove hypotheses of "randomness" versus "reality" of such **cluster candidates**. The quality of the resulting groups only can be judged by some plausible conditions which they must satisfy ([122], [200], [319], [376]).

This is no disadvantage if the analyst is interested in reducing a very large body of data to a relatively compact description through cluster analysis instead of looking for natural classes. The grouping suggested by an algorithm then is adopted for operational use only. However, when it comes out to detect natural clusters in real data, then the term "cluster" must bear a definite meaning. Here, from a methodological point of view, it is better to establish certain mathematically convenient and evident conditions. These conditions then control the homogeneity or separation of subsets of $S = \{O_1, \ldots, O_n\}$. Only those maximal subsets of S are called clusters, which satisfy several conditions. Thus, the classification problem is reduced to that of uncovering well-defined clusters in a data set. A first practical realization of this idea is to define the homogeneity of an arbitrary subset A of S by one of the formulas (2-6a)–(2-6c), together with the separation of this subset from $S - A$, defined by (2-8) and one of the formulas (2-7a)–(2-7c). Only those maximal subsets of S are detected as clusters by an algorithm, which satisfy the condition $\hom(A) \leq d$. Also, the values of $\text{sep}(A) > d^*$ can be used for deriving classifications, or a B-coefficient $b(A) = \hom(A)/\text{sep}(A)$, which is small enough (see [38], [42], [385]). Here, **every** cluster must satisfy certain conditions, whereas for an optimization criterion, only the sum of the B-coefficients of all subsets in a partition must take a minimum for an optimal classification. Nothing is said about the coefficient of a particular subset.

Classification methods based on the optimization of a global classification criterion are fairly good when used to outline ellipsoidal clusters. Thus, they can be used if either a sample really consists of a mixture of normal distributions as in Figure 2-12, or the analyst is interested in a classification for practical or operational use only. However, they are very sensitive to violations of certain prepositions (which are implied by the choice of a wrong number of group or the existence of nonellipsoidal groups in a sample). In case of sickle-shaped clusters, all conventional optimization criteria reach their minimum for the classification shown in Figure 2-13b. This is inconsistent with any intuitive understanding of the term "cluster". Only one criterion reaches its minimum at the partition shown in Figure 2-13a, namely the sum of the inverse separation indices (2-7b): A criterion derived from single linkage. For these reasons, optimization methods cannot be recommended as methods of choice in exploratory data analysis. Cluster analysis as a method for exploring structures and generating hypotheses is used primarily in those cases, where too less is known about the sample to rely on prepositions which are mandatory for optimization methods ([38], [42], [117], [148], [243], [259]).

Many optimization methods are not invariant to scale transformations or monotonic transformations of the distances. Moreover, all these methods suffer from three fundamental drawbacks. First, the global optimization criterion for a partition is an average over homogeneity or separation indices of all clusters of that partition. Thus, single inhomogeneous groups besides more homogeneous ones must be tolerated. The second disadvantage is that optimization algorithms find a best fitting structure for a

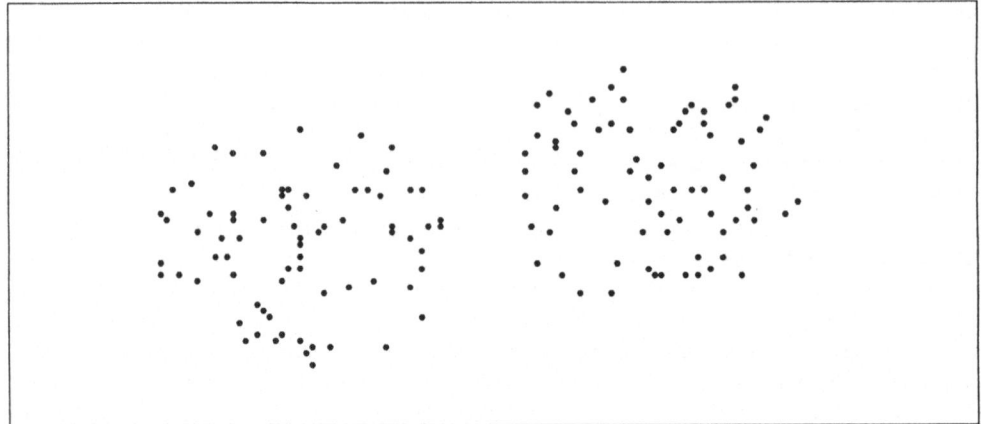

Figure 2-12 *Mixture of two normal distributions: Disjoint compact groups in the $\mathcal{R}^2$.*

given number of clusters. However, deciding on the number m of clusters in the data is a substantial problem in performing a cluster analysis. The literature contains a lot of wishings for mechanical methods of determining the number of groups. It can be a difficult choice. The third and major drawback for optimization methods is the number of ways of sorting n observations into m nonempty groups. It is a Stirling number of the second kind,

$$S_2(n,m) = \frac{1}{m!} \sum_{\mu=0}^{m} (-1)^{m-\mu} \binom{m}{\mu} \mu^n.$$

For every such partition, the value of the validity function is to be computed to find the optimum. To give an impression of the order of magnitude of those numbers: For sorting $n = 25$ objects into $m = 5$ nonempty groups, $S_2(25,5) = 2\,436\,684\,974\,110\,751$ cases must be taken into consideration. For even the relatively tiny problem of sorting 25 objects into 3 groups, the number of possibilities is the astounding quantity $S_2(25,3) = 141\,197\,991\,025$. It would take an inordinately long period of time even for medium sized samples and on fast computers to examine so many alternatives to find the optimal one. Thus, only quasi-optimal or hierarchical optimization methods are of practical interest.

Recursive methods also favour ellipsoidal groups. Their advantage is that no knowledge about the true number of clusters is needed. Some authors especially recommend recursive methods for classifications of objects with mixed data. However, as with optimization methods, at the moment no appropriate statistical tests exist for testing the hypothesis of artificial clusters in a homogeneous sample against the alternative that the groups outlined by the procedure are real clusters — or at least centres of real clusters — in an inhomogeneous sample. (Some work is done showing that direction as we shall discuss in Chapter 4.) We can recommend the use of recursive or iterative (quasi-optimal) classification methods to analysts who want to perform a cluster analysis for practical or operational reasons only. In most cases, they are interested in compact, round groups. The classification criterion and the algorithm completely determine both type and shape of the clusters. Here, a researcher who only is looking for a classification for practical or commercial reasons is in advantage as against a bi-

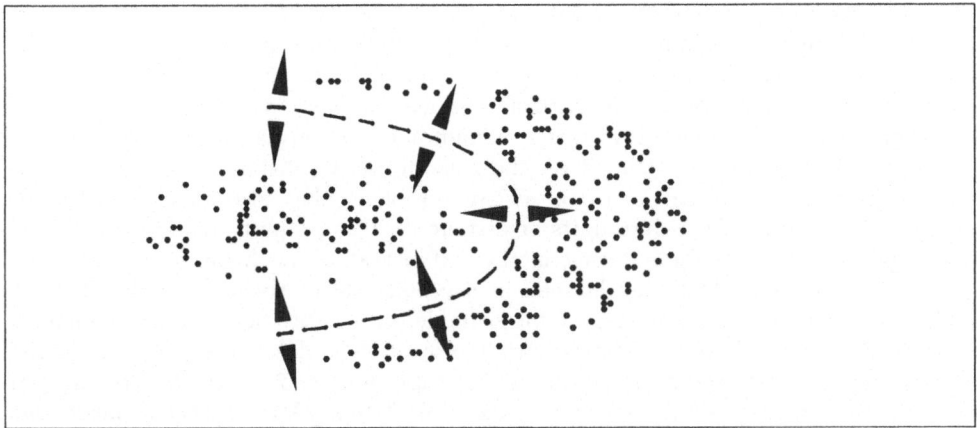

Figure 2-13a *Sickle-shaped groups in the $\mathcal{R}^2$, correct partition.*

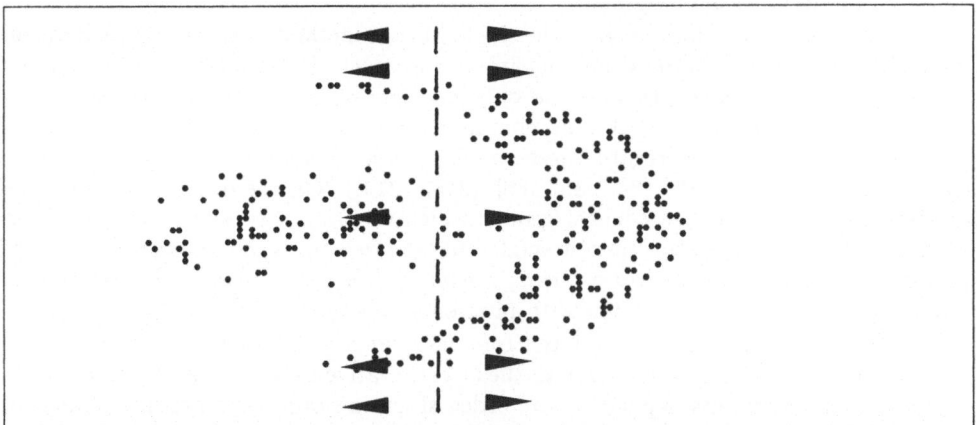

Figure 2-13b *Sickle-shaped groups in the $\mathcal{R}^2$, wrong partition.*

ologist who must know the type of the natural classes in his data to be able to choose the best classification procedure. Nevertheless, this type is not known but sought for in exploratory data analysis.

Classifications by mode seeking techniques require assumptions on the joint distribution of the random variables (the attributes in the raw data vectors). Usually, mixtures of t-dimensional normal distributions are assumed (compare Subsubsection 2.2.4.3). Such methods therefore can be applied to continuous data only. The number m of clusters must be known in advance, or it must be estimated; but only an approximative method to estimate m exists (see [37], [38], [430]). Simulation studies proved that parametric techniques like mode seeking seldomly are robust against violations of the condition of mixed normal distributions. Therefore, nonparametric procedures have been developed, most of them based on nonparametric density estimations using fixed or variable kernels. W.L.G. Koontz's and K. Fukunaga's method is one of the few procedures of this type that can be used without any knowledge of the number of groups ([235], [236], [237]).

The fewest prepositions to the data are required for R.A. Jarvis' and E.A. Patrick's method of k nearest neighbours ([217]) as well as for several linkage methods. Thus, these procedures can be used to uncover ellipsoidal clusters and sickle-shaped clusters likewise (with the exception of the centroid and median method). Both the single- and complete-linkage clusters are invariant to monotonic transformations of similarities or distances; the results only depend on the ranking of the distances. Both methods are non-parametric approaches to the cluster problem, they are independent of the original distribution of the data units or distances. This property renders possible to use single-linkage or complete-linkage methods if instead of the distances d_{ij} only their rankings are known. This may happen in psychological or sociological classification problems, but also in medical ones (if, for example, similarities between probands must be defined with reference to the attribute "nociception"). Here, the pair of mutually closest probands gets the rank number 1, the rank number 2 is attached to the pair with the next highest similarity, and so on ([38], [148], [261]). Either a mean rank or subsequent ranks are attached to different pairs with equal similarities (subsequent ranks should be attached randomly to the pairs). This useful property does not hold for the centroid or median method.

N. Jardine and R. Sibson regard the single-linkage method as the only classification method, which should be used for disjoint or hierarchical classification at all. They argue that single linkage provides the only method satisfying certain mathematical criteria which should be regarded as minimum criteria to all classification methods. N. Jardine and R. Sibson regard these conditions as most necessary for hierarchical classification methods ([83], [84], [85], [86], [213], [214]). On the other hand, **bridges** or **chains** of data units can connect groups which are otherwise well separated. Such groups are always merged together by single-linkage procedures. Some authors therefore prefer other linkage procedures as being better in practical applications than single linkage. They justify their opinion with Monte Carlo studies ([34], [117]).

However, for the median and centroid method as well as for the non-disjoint complete-linkage method, together with the agglomerative Procedure 2, it is well known that classification results depend on the order of adding objects to groups. This is of great importance to know, since pairs of objects of equal similarity occur often enough in real data. Examples for that phenomenon are easy to construct in analogy to Examples 2-2 or 2-4 (see [164]). The complete-linkage distance between clusters together with Procedure 1 usually results in many, heavily overlapping, classes which virtually cannot be interpreted. For the average-linkage method, and for Ward's method, too, big clusters tend to "swallow" small ones if they are too close together ([148], [259]).

The struggle for the "best" classification method has not always been fought using scientific and matter-of-fact arguments as can be seen from the correspondence in form of several articles and discussions between the two groups with N. Jardine and R. Sibson on one side, and with G.N. Lance and W.T. Williams on the other side (see [360], [421]). This discussion also shows clearly that no uniform understanding of the term "cluster" exists. Probably, such a uniformly accepted definition never will come to existence, since the understanding of what a cluster shall be, depends on the type of the data as well as on the interests of the researcher.

Single linkage and complete linkage are first steps in the direction of a researcher-independent cluster definition, which is mathematically based (see [1], [16], [38], [42], [114], [135], [148], [167], [186], [202], [213], [214], [224], [259], [260], [261], [262], [272]). In this graph-theoretic model of cluster analysis, objects are represented by vertices, and those pairs of objects satisfying a particular similarity relation are termed adjacent and linked together; they constitute the edges of a graph. Clusters are then characterized by appropriately defined subgraphs. Single-linkage clusters of level d are the components of a graph where any pair of objects is linked by an edge if these objects have a mutual distance not larger than a threshold d. Complete-linkage clusters of level d are the cliques of such a graph as we already mentioned. In contrast to the numeric-matrix orientated cluster analysis procedures, graph-theoretic cluster analysis provides a simpler combinatorial cluster analysis model. This is most appropriate where either the raw data are only available in form of a similarity matrix or where the number of objects is too large for distance matrix methods to be computationally tractable. Because of the simple model, graph-theoretic methods are a fairly good approach to the classification of mixed data. Let us shortly summarize several instances where this type of cluster analysis is most appropriate.

(a) **Relational association data:** The association data between objects are a single (algebraic) relation on the objects. (In a sociological study of employee work habits, for example, the data are the algebraic relation indicating each pair of employees that work well together.)

(b) **Sparse association data:** For each object there are data given only on those relatively few objects that are most similar to this object. (In a study of economic activities, for example, each firm provides information on those other firms that are felt to be significant competitors.)

(c) **Computationally intractable distance matrix:** Suppose that the number of data units is very large, so that the full distance matrix is too large to compute or store efficiently (for a sample size of $n = 5\,000$, the full distance matrix has $25\,000\,000$ elements). In this case it may be possible to determine only a limited number of pairwise distances corresponding to all distances below some threshold d. If a procedure allows the resulting number N of sufficiently similar object pairs to be computed in time $O(N)$, and N satisfies $N = o(n^2)$ or $N \ll n^2$, computational tractability may be achievable.

(d) **Ordinal pairwise association data:** A ranking of all pairs of objects in their order of similarity is available either fully or to some threshold level. The data may then be considered as a hierarchy of graphs or by a **proximity graph** as shown by D.W. Matula in [272]. (A proximity graph is a graph where the edge set satisfies an order relation.) Suppose that the (n, t)-matrix of raw data in a taxonomy application contains considerable nonnumeric data, such as colour, shape, or other nominally scaled data. Computation of any meaningful real-valued distance function between each pair may be considered too subjective and therefore unreliable for the application of distance matrix cluster analysis methods. However, ranking of all object pairs that are sufficiently close by some acceptably objective criteria may be possible, and thus provide the basis for application of graph-theoretic cluster determination methods on the resulting proximity graph.

Furthermore, null hypotheses on the randomness of clusters as results of a complete- or single-linkage classification procedure can be stated and tested because of this analogy between clusters and subgraphs of a graph. This has been seen by C. Abraham already in 1962. (At that time, the theory of **random graphs** had been founded by P. Erdős and A. Rényi, see [1], [148], [224].) The single-linkage method emphasizes a local aspect of classification: Only the distances between an object and those other

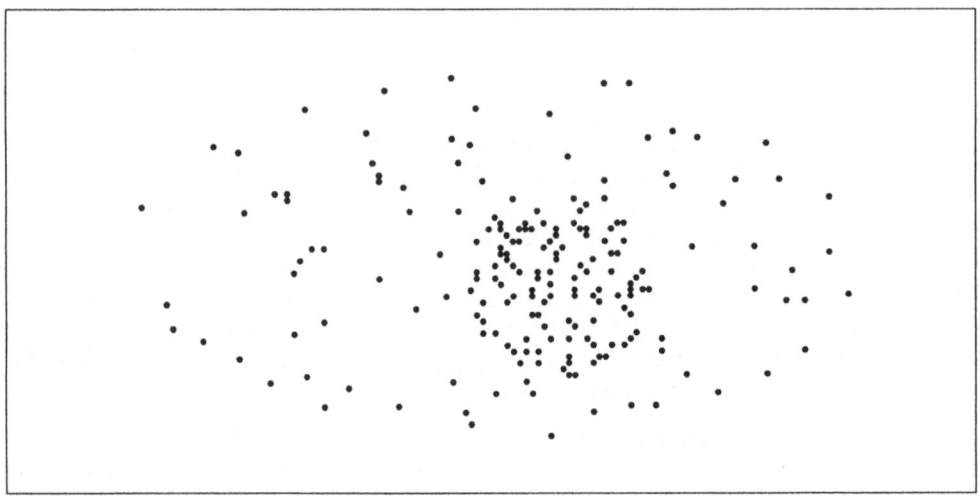

Figure 2-14 *A cluster* $\mathcal{A}$ *of higher density within uniformly distributed points of lower density.*

data entries which are close enough, constitute a cluster. It therefore provides a tool for the detection of groups if the points were distributed as in Figure 2-14. This type of groups neither can be outlined by optimization methods nor by analysing the point densities.

For these reasons, single-linkage procedures in our opinion can be recommended as being the best cluster determination methods to **explore** previously unknown group structures of a sample of multivariate data. For if the relations between the data entries cannot be plotted satisfactory in the one-, two-, or three-dimensional space, then it is of great importance to have methods not only for simply constructing and defining groups by algorithms but also for detecting natural clusters in a data set, that means, for testing the hypotheses of randomness against realness of those cluster candidates outlined by an algorithm. However, if the objects to be classified can be represented satisfactory by points of a low dimensional Euclidean space (possibly after normalisation of the data vectors, see [38], [245], [246], [385], [403]), then often no numerical classification algorithm is as effectiv as the method to plot the scattergram of the data and to choose the groups by visual inspection.

The choice of the appropriate classification method and algorithm also depends on the sample size, for storage space and computation time are the critical requirements that eventually render a cluster determination algorithm intractable as the size of the application grows. Until recently, large data sets allowed only iterative or recursive methods to be applied. Hierarchical or linkage methods could be used for samples with at most 500 entries ([443]). However, today even sample sizes of more than 1000 objects do not constitute barriers for the application of hierarchical procedures (see [391]). In [284], A. Morineau and L. Lebart propose a very fast **reciprocal neighbours chain search algorithm**, a graph-theoretic method related to the construction of minimum spanning trees as applicable to medium-sized samples (see [38]). In the same paper, they recommend a mixed clustering algorithm for large data sets, with the detection of so-called **stable groups** by iterative methods, followed by an agglomerative method.

Usually, this procedure merges ca. 2000 objects into about 200 stable groups, which constitute the basis for a linkage procedure to be applied.

Graph-theoretic classification procedures also can improve efficiencies in storage space and execution time, thus allowing investigations in problems with a relatively large number of objects. Efficient algorithms exist to compute minimum spanning trees ([6], [286], [287], [289]). Because of the need of only the distances between objects lying close together, no information is lost by omitting distances greater than a threshold d. Thus, we can use the sparse association matrix, and no storage space is wasted.

In the following chapters, we want to discuss the developments in graph-theoretic cluster analysis. We propose a multigraph model and discuss its advantages in deriving test statistics for testing the null hypothesis of homogeneous data. Our model allows the definition of multiple distances between every pair of objects. These multiple distances are regarded simultaneously during the classification process. This is of special interest in case mixed data. Furthermore, problems in connection with the normalisation of variables can be avoided (see Subsection 2.2.2).

2.4 Preparation and Presentation of Results

To provide a meaningful interpretation of the result of a cluster analysis, a computer program should not only give the raw data and the number of the group of every data unit, but should supply the researcher with more information. If the routine is part of a whole package of classification programs like CLUSTAN, then it is essential to know the type of classification which has been performed, since the stucture of the detected classes does not only depend on the data but also on the algorithm chosen by the researcher. If a proximity or similarity measure is used instead of the raw data for the analysis, then this measure must be indicated. And it should be evident from the output whether the data have been standardised or not. Moreover, a program should be able to print the similarity or dissimilarity matrix if desired.

This information is the absolute minimum the analyst must be provided with. This holds for the more practical orientated researcher interested only in a good classification of his data as well as for the consultative biometrician being more interested in the appliability of mathematically based methods. Moreover, in case of disjoint classifications informations about the characteristic representatives like mean vectors and median vectors are printed together with standard deviations, covariance and correlation matrices of each group. For hierarchical or quasi-hierarchical classifications a plot of the dendrogram is essential. The isolation indices hint to the fact whether cluster candidates are stable and can be interpreted as real clusters or not: The longer a group remains unchanged the higher is the probability that it forms a natural cluster. If the creation time of a cluster with not too few elements is small, then this may hint to a "cluster kernel".

For quantitative data units $O_1, \ldots, O_n$, the n vectors $\vec{x}_{1*}, \ldots, \vec{x}_{n*}$ of the t measurements can be plotted in the Euclidean space $\mathcal{R}^2$ by projecting them onto the plane spanned by their first two principal components. However, groups can be merged by this projection as Figure 2-2 of Subsection 2.2.1 shows (see also [6], [10], and [243]). For other data (qualitative or mixed scales), similar plots can be produced by multidimensional scaling ([190], [385]). If a matrix $\mathbf{S}$ of similarities or a matrix $\mathbf{D}$ of proximities or distances is available for computation instead of the raw data, then a matrix $\mathbf{X}$ of quantitative data can be constructed by multidimensional scaling methods. This can be used for a scattergram of relational association data, sparse association data, or ordinal pairwise association data.

A method of graphical representation of similarities instead of the raw data is the **shaded diagram**. The graphic principle used is the presentation of the elements of a similarity or dissimilarity matrix by printed symbols of various shades of darkness, where a dark symbol in cell (i, j) corresponds to a high similarity or small distance between the two objects O_i and O_j. These plots, applied to a data matrix before clustering and to the rearranged matrix after clustering, show at a glance whether clustering brought forth any distinctive clusters (see [259], [262]).

Another approach to the search for classifications or typologies of objects or persons is through the use of **graphical methods**. Graphical methods provide an extremely useful and flexible medium for explaining, interpreting, and analysing data by means of points, lines, areas, faces, or other geometric forms or figures. Well-known graphical methods for representing multivariate data are **glyphs** and **metroglyphs**, **Fourier series**, and **Chernoff faces**. These graphical methods differ from the hierarchical and partitioning clustering procedures in that no formal computational algorithm is employed for uncovering the clusters.

In [8], E. Anderson developed a technique for displaying the data on a t-dimensional response vector, using the names "glyphs" and "metroglyphes" for his displays. Since their introduction, many variations have been proposed. This technique is discussed in [147] (see also [93]).

In [9], D.F. Andrews suggested transforming a t-dimensional data vector $\vec{x}^T = (x_1, \ldots, x_t)$ by the Fourier series

$$ y(t) = f_{\vec{x}}(t) = \frac{x_1}{\sqrt{2}} + x_2 \sin(t) + x_3 \cos(t) + x_4 \sin(2t) + x_5 \cos(2t) + \cdots $$

over the range $[-\pi; \pi]$. This method has a number of useful properties that make it particularly well suited for exploratory data analysis: The function $f_{\vec{x}}(\cdot)$ preserves means, distances, and variances as well (see [93]). Close points in a Euclidean space will appear as close functions, and distant points as distant functions. This is particular useful when using Andrews plots for uncovering multivariate clusters and outliers.

H. Chernoff suggested in [73] a novel way to represent multivariate data: He associated the values of the data vectors with facial features. In its original form, Chernoff allowed for up to 18 dimensions. The first dimension of the vector gives the distance between the eyes, the second dimension is associated with the eyes position and so on. In [60], L.A. Bruckner gives a program for computing and displaying such stylized faces. Considering the plotted faces, the analyst can distinguish between mutually similar and dissimilar faces. Thus, by pure visual inspection, the objects can be divided into different — even overlapping — classes. Since its introduction, creative application has accelerated, and today it has become a useful tool in exploratory analysis with application in cluster analysis as well as in outlier detection and multidimensional time series analysis, as W.R. Dillon and M. Goldstein report in [93].

Methods of graphical representation of multivariate data like these are discussed in [17], [72], [116], [138]. In [90], [93] and [130], the authors illustrate the method of Chernoff faces using real data. All these graphical methods differ from those procedures discussed in the previous sections in that they by no means provide algorithms to perform a classification or to uncover data structures numerically. They only provide a medium for the graphical representation of usually high dimensional data in the 2-dimensional plane by which the merging effect known from the principial components analysis should be overcome. The final partition of the data units into clusters is left to the analyst's experience and preference. The judgement of facial expressions, for example, depends to a high degree on the personal preferences of the spectator. Thus, the classes can vary markedly depending only on the reviewer or just on the ordering of the attributes because the eyes are more important for one researcher, another one perhaps looks closer to the mouths of the Chernoff faces. Therefore, we cannot recommend these methods as stand-alone procedures without a word of cautiousness. However, methods like these and the shaded diagrams can be recommended as additional aids to the methods of numerical classification. They can provide useful hints to the interpretation of the outlined clusters, thus serving as a kind of validation. And they are superior to the principal components analysis.

Rarely it happens that the true group memberships of the objects become evident by some external criterion as in the example from Section 1.4 or in Example 2-1. Thus, the clusters outlined by a cluster determination algorithm cannot be validated directly. Some authors recommend to apply three or more algorithms to the data to be classified. Thus, the analyst at least can uncover stable groups or cluster kernels. Here, dendrograms, and diagrams of group homogeneities versus separation indices can provide helpful hints for the interpretation. It would be better to test the hypothesis of inhomogeneous data, that is, of real clusters in a sample to be able to exclude the fact that the clusters outlined are only the result of a numerical procedure, randomly drawn from a homogeneous sample. However, it depends on the classification procedure whether such a test can be performed or not. Moreover, the development of tests for cluster validation postulates conditions to the distribution of the data or the proximities which seldomly can be proved.

In [32], [33] and [34], R.K. Blashfield, M.S. Aldenderfer and L.C. Morey give an overview of the literature and software on cluster analysis. H.H. Bock reviews cluster analysis software in [39], and, together with H.P. Ohly and D. Bender, in [50]. Linkage methods are contained in the CLUSTAN software package, SPSS-X, and in SAS. At the moment, BMDP allows single-linkage methods only for the classification of variables, not of objects. We also can recommend R.F. Ling's method, discussed in [259], [260], [261], which is available as source code (the program is called k-CLUSTERING). Another single-linkage procedure is SLINK from R. Sibson (see [361]). Other programs are described in [82], [286], [291], and [339]. Today, programs are available which can be run on personal computers.

Many approaches to graphical representations of data exist. Subroutines for the generation of dendrograms are included in most programs for hierachical classification, for example, in CLUSTAN, in R.F. Ling's procedure k-CLUSTERING ([260]), and in the programs published by M.R. Anderberg ([6]), J.A. Hartigan ([183]), and H. Späth ([385]). In [438], F.W. Young, T. Edds, D. Kent and W.F. Kuhfeld propose interactive hypergraphics which can be used as "macro" or "user written procedure" for the SAS package. Programs for the graphical representation of data can be found in the "JOURNAL OF CLASSIFICATION" and its book reviews, too. Another good sources for programs are the "JOURNAL OF THE ROYAL STATISTICAL SOCIETY, SERIES C: APPLIED STATISTICS", and the monographs of M.R. Anderberg ([6]) and J.A. Hartigan ([183]). Some listings are printed in the proceedings of several meetings recently organised by the AMERICAN STATISTICAL ASSOCIATION.

CHAPTER 3

GRAPH-THEORETIC METHODS OF CLUSTER ANALYSIS

Two ways of defining clusters exist. Clusters can be defined constructively, by statement of a criterion and choice of an algorithm. The term "cluster" usually is left undefined. On the other hand, clusters can be defined as subsets of a sample $\mathcal{S}$, satisfying certain mathematically convenient and evident conditions. In this **axiomatic** approach to classification theory, the term "cluster" is well-defined. Such pre-defined clusters are usually uncovered using measures of similarity, disparity, dissimilarity or distance, respectively. Rarely the raw data are used directly to find classes by outlining their shapes. In this chapter, we introduce cluster definitions, which are based on graph theory.

Graph theoretical concepts are often helpful in problems of taxonomy. A graph is a set of points (called **vertices**) and some lines (called **edges**) connecting pairs of vertices. The objects $O_1, \ldots, O_n$ of a data set to be clustered can be interpreted as vertices $\xi_1, \ldots, \xi_n$ of a graph. Two vertices are connected by an edge if and only if the related objects are similar enough. The components of such a graph are known as **single-linkage clusters**, and the cliques become the **complete-linkage clusters** (see Section 1.2 and Subsection 2.2.4.4). In [259] and [261], the single-linkage clusters are generalized to (weak) k-**linkage clusters** (or simply k-clusters), which means that an object must be similar to at least k objects of a cluster in order to belong to that class. In [213] and [214], a weak point of the complete-linkage cluster (too many heavily overlapping groups) is corrected by the introduction of the k-**overlap cluster**. Here, a maximum of $k - 1$ objects may belong to the intersection of two clusters. This type is also called B_k-cluster.

These definitions allow the statement and testing of **hypotheses** on a class structure of a sample $\mathcal{S}$ in an appropriate and mathematically convenient way. Results like those proved in [111] or [112] have been used to derive test statistics for testing the hypothesis of randomness of such clusters, assuming that randomness of clusters means randomness of distances between the objects to be clustered, and can be interpreted as random choice of the edges in the graph model (see [1], [148], [152], [154], [155], [158], [224], [259], [261]). This interpretation is obvious because the distance d_{ij} between two objects O_i and O_j can be thought of as the weight of the edge κ_{ij} connecting the two vertices ξ_i and ξ_j in the related graph.

In the first section of this chapter, we shortly describe the construction of single-linkage clusters of level d and give the exact definitions of the commonly used graph-theoretical notion of clusters. These are the definitions of (weak) k-linkage clusters, strong k-linkage clusters, k-overlap clusters (or B_k-clusters), and of complete-linkage clusters, respectively. (The single-linkage and the complete-linkage clusters have already been discussed in Subsubsection 2.2.4.4 and in Subsection 2.3.) We also discuss some properties of these clusters.

In the second section, we propose a new approach to the cluster analysis of multidimensional data. For this approach, we use the concept of undirected, completely labelled multigraphs which we defined in [149]. Here, as in the other graph-theoretic cluster definitions, a cluster is a well-defined subset of a sample $\mathcal{S}$. It is completely determined by a similarity or dissimilarity matrix, a vector threshold $\vec{d}^T$ and two other parameters, k and s. These parameters define the internal structure of a cluster, they describe the way how the different objects of a group are interconnected. All graph-theoretically based cluster definitions named above can be generalized by our multigraph approach. We especially define the (weak) k-linkage cluster of level $\vec{d}^T = (d_1, \ldots, d_t)$ and degree s of connectivity, the $(k, \vec{d}^T; s)$-cluster. In terms of graph theory, it is an s-**component** of a multigraph Γ_{tnN}. The mathematical definitions of a multigraph Γ_{tnN} and an s-component follow in Section 3.2 and are discussed in Chapter 5.

Some properties of this multigraph-based definition of single-linkage clusters are discussed in Subsection 3.2.3. In this section, the other graph-theoretic cluster definitions listed at the beginning of this chapter are generalized to our multigraph concept, too. In the third section, we develop an algorithm for the construction of $(k, \vec{d}^T; s)$-clusters. The last section contains hints for the construction of hierarchies and dendrograms of $(k, \vec{d}^T; s)$-clusters.

The introduction of this cluster definition was motivated by problems from medical and psychological inquiries. In studies of medical research, the scale levels often vary considerably between the different items, that is, between the dimensions of the data vectors. It is therefore questionable if not impossible to compute overall, or global similarities s_{ij} or dissimilarities d_{ij} between the elements of a sample $\mathcal{S}$. Here, a researcher wants more flexibility. He optionally wants to calculate dissimilarities between pairs of objects for every scale level. and sometimes he simply cannot define a single "global" dissimilarity for each pair of data units instead of "local" dissimilarities for each scale level. The structure of a data set consisting of n t-dimensional vectors then can be described better by a multigraph as we show in this chapter.

3.1 CLASSIFICATIONS BY GRAPHS

3.1.1 THE CLASSIFICATION AT LEVEL d

For the n objects $O_1, \ldots, O_n$ of a sample $\mathcal{S}$, let $\mathbf{D} = (d_{ij})$ be a dissimilarity matrix describing the similarity or dissimilarity structure of $\mathcal{S} = \{O_1, \ldots, O_n\}$ properly. For an axiomatic establishment of the term "cluster", a subset of $\mathcal{S}$ is called **cluster** (or class, group) if, with every element O_i, it contains all other elements similar to O_i. Two objects O_i, O_j are **similar** if their mutual dissimilarity d_{ij} does not exceed a given **threshold** $d > 0$. Thus, the following definition from P.H.A. Sneath makes sense ([363], [364], compare also [38]).

Definition 3-1 *A subset $\mathcal{C}$ of $\mathcal{S}$ is called a **single-linkage cluster of the level** d if*

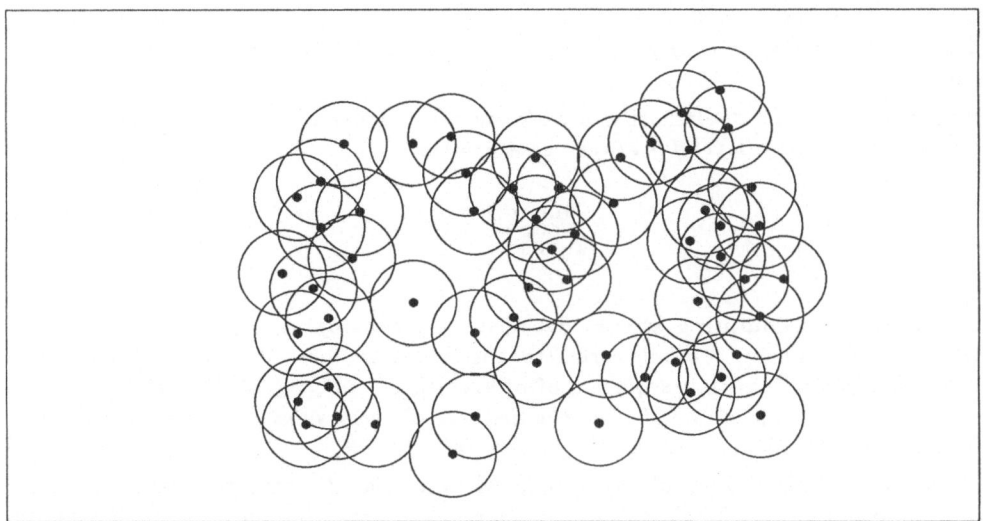

Figure 3-1 *The construction of single-linkage clusters of level d in the* $\mathcal{R}^2$.

(a) $\mathcal{C}$ is not empty;

(b) for every $O_i \in \mathcal{C}$, all O_j with $d_{ij} \leq d$ also belong to $\mathcal{C}$;

(c) $\mathcal{C}$ is maximal with respect to conditions (a) and (b) (that means, if no subset $\mathcal{C}^0 \subset \mathcal{C}$ already satisfies (a) and (b)). •

Figure 3-1 illustrates this definition and the method how to find single-linkage clusters of level d for objects with $t = 2$ attributes. (For this figure, the Euclidean distance provides the dissimilarity measure, and the n data units $O_1, \ldots, O_n$ are represented by their data vectors $\vec{x}_{1*}, \ldots, \vec{x}_{n*}$ in the $\mathcal{R}^2$.) Starting with an object O_i arbitrarily chosen, all objects O_j are added to the cluster $\mathcal{C}_1$ containing O_i if and only if their data vectors $\vec{x}_{j*}$ lie within a disk with radius d and centre $\vec{x}_{i*}$. This procedure is continued for every element O_j within the disk, and for all elements within the disks round every O_j, and so on, until no further element is added to the cluster containing O_i. Then a new element $O_{i'} \in \mathcal{S} - \mathcal{C}_1$ is chosen, and the same procedure is repeated to uncover the cluster $\mathcal{C}_2$, and so on. The algorithm stops if every element is attached to a class. If no further element lies within the disk with an element $O_{i''}$, then $O_{i''}$ forms a one-element cluster. Usually, this algorithm is not applied to the vectors $\vec{x}_{i*}$ of the raw data but to the vectors $\vec{z}_{i*}$ of the normalised or standardised data.

In several papers of L.L. McQuitty, R.M. Needham, F.J. Rohlf and P.A.H. Sneath, this method has been discussed extensively ([275], [295], [331], [332], [333], [334], [363], [364], [365], [373]). Since the 1950s, single-linkage clustering often has been used in exploratory data analysis, especially in problems of biological taxonomy. Today, it is known to nearly all researchers working with methods of exploratory data analysis. In 1962, C. Abraham used the analogy between the definition of single-linkage clusters and the mathematical calculus of graphs to derive and test null hypotheses of **randomness** of single-linkage cluster "candidates" as pure results of applying a clustering algorithm to a homogeneous sample ([1]). Some important properties of single-linkage clusters of Definition 3-1 are given in the following theorems.

Theorem 3-1 *The set* $\mathcal{S} = \{O_1, \ldots, O_n\}$ *of n objects is split into m disjoint clusters of* *level d. The number of clusters is uniquely determined by the threshold d: $m = m(d)$. The resulting partition is independent of the order of choosing the objects to form the clusters.* •

Theorem 3-2 *The dissimilarity or distance between two objects from different clusters C_k and C_l of the level d is always greater than d,*

$$\min_{\substack{O_i \in C_k \\ O_j \in C_l}} d_{ij} > d. \quad •$$

Theorem 3-3 (Chaining effect) *Between every two objects O_{i_1} and O_{i_m} of a cluster C_k, for which $d_{i_1 i_m} > d$ holds true (that is, which are more than a distance d apart from each other), exists a sequence of objects $O_{i_2}, \ldots, O_{i_{m-1}}$ of C_k with $d_{i_1 i_2} \leq d$, $d_{i_2 i_3} \leq d, \ldots, d_{i_{m-1} i_m} \leq d$. That is, the distance of every subsequent pair of elements in this* **chain** *$O_{i_1}, O_{i_2}, \ldots, O_{i_{m-1}}, O_{i_m}$ never exceeds d.* •

Theorem 3-4 *If a threshold d_2 is reduced to a value d_1, then the classification at level d_1 results from the classification at level d_2 by splitting all or some clusters of level d_2 into subgroups. On the other side, every cluster of level d_2 arises from the classification at level d_1 by merging different classes of level d_1.* •

The name "single-linkage cluster" in Definition 3-1 has been derived from the property that subsets of S are combined to one cluster by the single shortest link (the mutually closest elements), and that an object is attached to a group if its distance to a single element of the group is small enough. For $\min_{\substack{O_i \in C_k \\ O_j \in C_l}} d_{ij}$ (the single-linkage distance (2-13a) between clusters), two clusters are amalgamated into a single cluster, for increasing thresholds d, if d_{ij} is lower than the threshold even for a single pair (O_i, O_j) of objects, one from each cluster. This property often yields long serpentine clusters with markedly dissimilar entities at opposite ends of a cluster. The usefulness of the single-linkage cluster definition thus is limited by the fact that a single-linkage cluster does not always match the intuitive feeling for a "homogeneous set of mutually similar objects". On the other side, this definition enables us to detect nonellipsiodal classes, and to control the maximum size of "gaps" within a cluster by Theorem 3-2. Moreover, this theorem guarantees a certain degree of separation between different clusters of level d.

For several types of clusters, examined in Subsection 2.2.4, "separation" can be defined adequately; but for none of them exists an analogon of Theorem 3-2. No lower bound for the dissimilarities of objects from different groups is known. The results of Theorems 3-1 and 3-4, however, hold for all linkage clusters as has been mentioned in Subsubsection 2.2.4.4. From Theorem 3-4, it follows that linkage procedures not only can produce (disjoint or nondisjoint) classifications for a given level d, but are appropriate to generate hierarchies or quasi-hierarchies, too (see also Subsubsections 2.2.4.4 and 3.1.3).

3.1.2 Single-Linkage Clusters as Components of a Graph

For given dissimilarity matrix **D**, the graph-theoretic terms of Section 1.2 offer an easy approach to construct single-linkage clusters of level d. The method is as follows. The n objects $O_1, \ldots, O_n$ are defined as vertices $\xi_1, \ldots, \xi_n$ of a graph. The dissimilarities d_{ij}

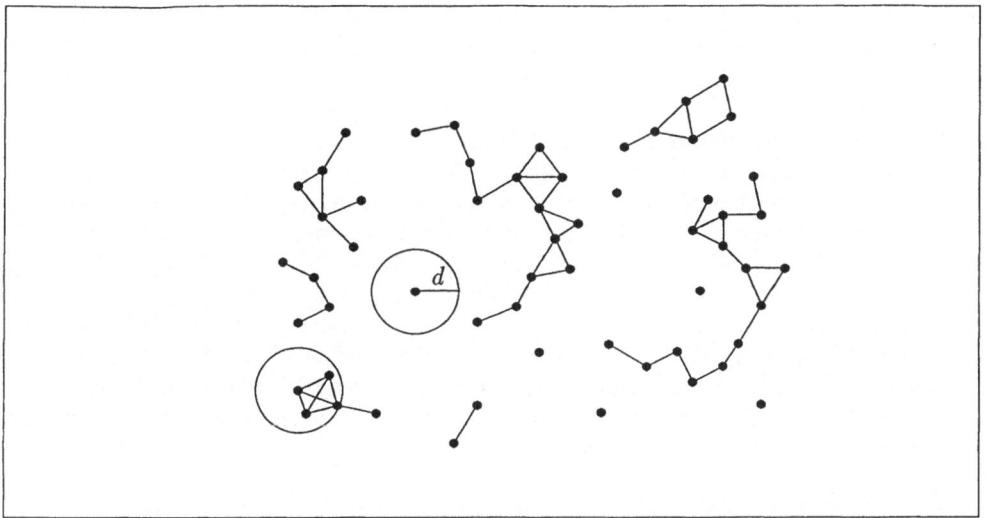

Figure 3-2 *Single-linkage-clusters of level d as components of $\Gamma(d)$ (the clusters are those of Figure 3-1).*

or their ranks are the weights of the edges κ_{ij} as we mentioned in Section 1.2 and Subsubsection 2.2.4.4. Two vertices ξ_i and ξ_j are linked by an edge κ_{ij} if $d_{ij} \leq d$ holds for a given threshold d. This gives a graph $\Gamma = \Gamma(d)$ with n vertices and $N = N(d)$ edges. The analogy between the components of this graph and single-linkage clusters of level d immediately implies the following definition.

Definition 3-2 *Every **single-linkage cluster** C **of level** d consists of the vertices of a component of the graph $\Gamma(d)$.* •

The geometric arrangement of the vertices does not affect the resulting clusters because only the presence or absence of an edge indicates whether two vertices are mutually "close together" or not. When plotting the objects in a scattergram, however, the visual impression is improved if objects with small distances are close together. Thus, projections of the data vectors $\vec{x}_{1*}, \ldots, \vec{x}_{n*}$ onto the $\mathcal{R}^2$ are generally used to support imagination (see Figure 3-2 which is the graph-theoretic equivalent to Figure 3-1). In this form, Definition 3-2 was proposed by N. Jardine and R. Sibson in [213] and [214].

This graph-theoretic cluster definition, however, can be traced back to a definition already given by J. Perkal in 1951 and a practical example from K. Florek, J. Łukaszewicz, J. Perkal, H. Steinhaus and S. Zubrzycki (see [128], [129] and [312]). K. Florek, J. Perkal and co-authors used the more constructive — but equivalent — definition of the single-linkage cluster by the minimum spanning tree: According to A. Cayley ([69], [70]), n vertices can be connected in n^{n-2} different ways by $n-1$ edges to a tree. Attaching the dissimilarities d_{ij} as weights to the edges κ_{ij} of such a tree, K. Florek and co-authors determined the tree with the minimum sum of weights of the edges. Omitting all edges with weights $d_{ij} > d$ from this **minimum spanning tree**, they got the single-linkage clusters of level d as components (or more precisely, as the vertex sets of the components). The vertex sets of these components are identical to the vertex sets of the components of the graph with all edges of weights $d_{ij} \leq d$. Here, the number of clusters depends only on d: $m = m(d)$. Omitting the $m-1$ edges with the greatest weights, yields m groups ([38], [167], [244], [334], [439]).

By means of this graph-theoretic definition of a cluster, it can immediately be proved that single-linkage clusters are invariant to monotonic transformations of the dissimilarity measure $d(\cdot, \cdot)$. Thus, the single-linkage method can be also applied in such cases where no matrix $\mathbf{D} = d_{ij}$ of dissimilarities is given but a matrix $\Delta = \delta_{ij}$ of ranks of dissimilarities (see Section 2.3). Since its introduction, the heuristic as well as the statistical properties of this method of uncovering clusters by computing the minimum spanning tree and related procedures have been studied in a number of papers (see [15], [38], [42], [108], [148], [149], [152], [154], [155], [169], [186], [201], [202], [211], [212], [213], [214], [224], [244], [259], [261], [265], [270], [272], [273], [350]). Well-documented programs for uncovering the single-linkage clusters of a sample S exist (see [34], [259], [260], [262], [286], [291], [334], [361], [428], and Section 2.4).

In sociometric inquiries, disparity measures often must be used for a classification instead of dissimilarities or distances. Here, digraphs, as defined in Section 1.2, can be used as classification models (see [68]).

3.1.3 MODIFICATIONS OF THE CLUSTER DEFINITION

By single-linkage clusters, separation between groups is emphasized more than homogeneity within a cluster. Thus, not only ellipsoidal clusters can be uncovered by single-linkage procedures, but also ramified, sickle-shaped or elongated groups. Groups are not defined by great similarities between **every** pair of their objects, but are determined by the existence of chains between the different pairs of data units. A weak point of this cluster definition is that a chain or **bridge** of few elements, lying between two otherwise well separated classes, can prevent a cluster determinating procedure from separating these two clusters (see Figure 3-3). This critical point can be by-passed by using procedures for the detection and elimination of the elements of the bridge, as proposed by D. Wishart and C.T. Zahn ([424], [439]).

Another way to avoid this problem is offered by modified cluster definitions. The objects forming a bridge between two otherwise well-separated clusters usually lie in regions of low point density. Only few other data entries are adjacent and thus linked to them by edges in the graph $\Gamma(d)$. This consideration implies the following variations to the term "cluster" (see [38], [42]).

Definition 3-3 *A **weak k-linkage cluster** C **of level** d consists of the vertices of a (weak) component of degree k of the graph $\Gamma(d)$, that is, of a connected subgraph of $\Gamma(d)$ which is maximal with respect to the fact that the minimum degree of the vertices is at least k. We speak of (k, d)-clusters.* •

This definition has been proposed by R.F. Ling in [259]. The case $k = 1$ gives single-linkage clusters, but without one-element classes (which are called **isolated vertices** in graph theory). For $k = 2$, "branches" are separated from a cluster. For $k = 3$, the bridge in Figure 3-3 is deleted because some of its elements have only two neighbours. We thus get two disjoint $(3, d)$-clusters (at the left side and at the right side of Figure 3-3) instead of one large $(2, d)$- or $(1, d)$-cluster. Obviously, for given threshold d, the (k, d)-groups are the more compact (and smaller), the greater k is. Furthermore, every (k, d)-cluster must consist of at least $k + 1$ objects.

Like the weak k-linkage method, for $k = 1$, the strong k-linkage method of Definition 3-4 gives single-linkage clusters of level d, too.

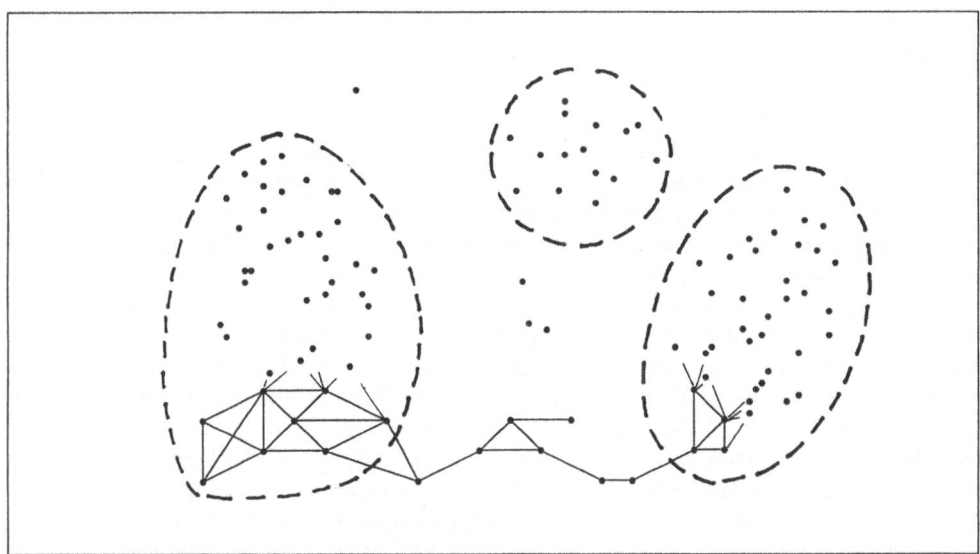

Figure 3-3 *Chaining effect: Two different clusters are connected by a bridge and thus are merged into one single cluster by single-linkage procedures.*

Definition 3-4 *Every **strong k-linkage cluster** C **of level** d consists of the vertices of a strong component of the degree k of the graph $\Gamma(d)$; it is a k-component of $\Gamma(d)$.* •

For the remainder, strong k-linkage clusters of level d, that is, k-components of a graph $\Gamma(d)$ whose edges have been chosen according to a dissimilarity matrix $\mathbf{D}$ and a threshold d, are referred to as (k, d)-components. Like (k, d)-clusters, (k, d)-components consist of at least $k+1$ vertices (data units), and for $k = 1$, no difference exists between $(1, d)$-clusters and $(1, d)$-components. For $k \geq 2$, however, the conditions to (k, d)-components are more restrictive than those to (k, d)-clusters. The bridge in Figure 3-3 together with the two sets encircled is a $(2, d)$-cluster. It is no $(2, d)$-component, since only one path exists between several vertices of the bridge. Classifications by means of (k, d)-components are not commonly used.

The graph-theoretic cluster definitions given above can be used to generate disjoint classifications only. With the following two definitions of the k-overlap, and of the complete-linkage cluster, respectively, we are able to uncover overlapping classes as well (for the complete-linkage cluster, see also Subsubsection 2.2.4.4).

Definition 3-5 *A **k-overlap cluster** C **of level** d consists of the vertices of a k-block of $\Gamma(d)$.* •

Definition 3-6 *A **complete-linkage cluster** C **of level** d is the vertex set of a clique of $\Gamma(d)$ (it is a maximal set of data units so that every pair of objects has a mutual dissimilarity $d_{ij} \leq d$).* •

Definition 3-5 was given by N. Jardine and R. Sibson in 1968; they called their method "B_k" and used it for hierarchical classifications ([213], [214]). Again, for $k = 1$, no difference exists between 1-overlap clusters and the clusters of Definitions 3-2–3-4: With the exception of isolated vertices, 1-blocks are the same as components of a graph. However, for $k \geq 2$, two k-blocks of a graph can share up to $k - 1$ vertices. Moreover, a graph cannot contain more than $(n - k + 1)/2$ k-blocks ([212], [213], [214], [270], [272], [273], [333]). Thus, the degree of overlapping and the maximum number of k-overlap clusters of a level d can both be controlled by the parameter k.

In Definition 3-6 of T. Sørensen ([383]), the number of objects shared by two complete-linkage clusters (cliques) is not controlled. Furthermore, examples have shown that, in practical applications, too many complete-linkage clusters are outlined which are nearly identical. (The maximum number of cliques in a graph with n vertices is $3^{n/2}$, see [178], [271].) Thus, for practical applications, the use of complete-linkage procedures cannot be recommended.

We speak of (k, d)-clusters, (k, d)-components, (k, d)-blocks or d-cliques, respectively, if we want to emphasize their dependence on a threshold d. However, d is not uniquely determined. For example, a (k, d_1)-cluster can be a (k, d_2)-cluster at the same time if, by an increase of the threshold d_1 to d_2, no further objects are added to that cluster. In contrast, in [259], R.F. Ling defines the parameter d as that level, at which exactly the (k, d)-cluster arises. We refer to this parameter as the "creation time" or "age" of a cluster $\mathcal{C}$, and denote it by $e(\mathcal{C})$ (see Subsubsection 2.2.4.4). On the other side, we simply speak of single-linkage clusters without specifying a level d if there is no reason to emphasize a level or the creation time of the groups. The same holds if we speak of (weak) k-linkage clusters or k-clusters, strong k-linkage clusters, k-overlap clusters, or complete-linkage clusters, respectively.

It can be shown ([42]) that Definitions 3-3–3-6 are stepwise subtilizations or improvements. A cluster $\mathcal{C}$ of any of these definitions must be contained in a cluster of the preceding definition. Moreover, for every graph Γ, it is easy to see that every weak component of degree $k + 1$ must be a subgraph of a weak component of degree k. A $(k + 1)$-component must be a subgraph of a k-component, and a $(k + 1)$-block must be a subgraph of a k-block (and the same relations must hold true for the groups defined by these subgraphs). Theorems 3-1 and 3-4 also hold for clusters given by Definitions 3-3–3-5. In [259], [260], [261], R.F. Ling proves the following relation for k-clusters.

Theorem 3-5 *Let $\mathcal{C}_1$ be a (k, d_1)-cluster with creation time d_1, and $\mathcal{C}_2$ a (k, d_2)-cluster with creation time d_2. Furthermore, let $d_1 \leq d_2$. Then either $\mathcal{C}_1 \cap \mathcal{C}_2 = \emptyset$ or $\mathcal{C}_1 \subseteq \mathcal{C}_2$ holds true. For arbitrary $k < n$, $\mathcal{S} = \{O_1, \ldots, O_n\}$ is the largest k-cluster.* •

This theorem implies that, for fixed k and variable values of d, the class of k-clusters is partially ordered with the set inclusion as order relation. Its hierarchical or nested structure can be represented by a "rooted tree" ([38], [42], [182]). Obviously, this holds also for the other linkage methods. The one-element sets $\{O_1\}, \ldots, \{O_n\}$ are the end points of this tree, and $\mathcal{S}$ is the so-called "root". The (k, d)-clusters of the different levels d are the other vertices (branching points) of that tree. By attaching the creation time d as weight to every (k, d)-cluster, we get a dendrogram or labelled hierarchy. Dendrograms are plotted with the creation times listed on an axis besides them as shown in Figure 2-3c. (For $k = 1$ — the case of single-linkage clusters — the dendrogram easily can be derived from the minimum spanning tree: At the first step, the edge with the greatest dissimilarity as weight is removed; this results in two groups. Then the edge with the second-greatest dissimilarity as weight is removed, and so on. After removing the $(n - 1)$-st — and last — edge of the tree, n one-element clusters are left. In [259] and [261], R.F. Ling derives an **isolation index** $i(\mathcal{C})$ for k-clusters $\mathcal{C}$, which is based on such dendrograms. It is defined as follows.

Definition 3-7 *Let C_1 be a k-cluster, arising at level d_1, and let C_2 be the smallest k-cluster which contains C_1. Let d_2 be the creation time of C_2. That is, $e(C_1) = d_1$ and $e(C_2) = d_2 > d_1$. Then,*

$$i(C_1) := e(C_2) - e(C_1) = d_2 - d_1$$

*is the **isolation index** of the cluster C_1.* •

Remark 3-1 The number of edges in $\Gamma(d_1)$ can be used as creation time of a cluster C_1 instead of the level d_1, at which the cluster arises. The isolation index of a cluster C_1 then is the number of edges, which must be added to $\Gamma(d_1)$ to get the graph $\Gamma(d_2)$ with the cluster C_2 as superset of C_1. These nonparametric indices are independent of monotonic transformations of dissimilarity measures. They are more suitable for applications. These modifications of the creation time and isolation index of a cluster have also been proposed by R.F. Ling. Theorem 3-5 naturally holds if the dissimilarity levels d_1 and d_2 are substituted by the corresponding numbers of edges. •

Remark 3-2 Creation times and isolation indices can be adequately defined for most classification methods to generate hierarchical or quasi-hierarchical classifications. The creation time of a cluster can be interpreted as the weight of this cluster as a vertex in the dendrogram, while the isolation index of a cluster C is the weight of the edge connecting C with the next group which contains C. From the order of magnitude of these indices, the analyst can infer to the "realness" of a cluster. The greater the isolation index of a group is, the greater is probability that this "cluster candidate" is a natural cluster in an inhomogeneous sample. In contrast, clusters with low isolation indices are more probably interpreted as arising at random and without representing a real data structure. Therefore it is of great interest to derive probability distributions for such indices. This would enable biometricians to state null hypotheses for testing the randomness of clusters in dendrograms. For the centroid and median method, this seems to be more difficult than for the other linkage procedures. As shown in Example 2-4 of Subsubsection 2.2.4.4, a median or centroid cluster with a low creation time may arise though clusters of greater creation times already exist (inversion effects). Thus, the isolation index of such clusters may be negative. •

Remark 3-3 In [288], F. Murtagh studies a probability model of random dendrograms. He derives results on the hypotheses of reality versus randomness of groups as vertices of a dendrogram. •

3.2 CLASSIFICATIONS BY MULTIGRAPHS

Having taken t measurements ($t \geq 2$) upon the n objects $O_1, \ldots, O_n$, on which a cluster analysis is to be performed, the data matrix $\mathbf{X} = (x_{il})$ of the raw data or $\mathbf{Z} = (z_{il})$ of the standardised data is computed. From $\mathbf{X}$ or $\mathbf{Z}$, a matrix $\mathbf{D}$ of dissimilarities or $\mathbf{S}$ of similarities is calculated. These matrices $\mathbf{D}$ or $\mathbf{S}$ provide the basis for the graph-theoretic procedures discussed here as they do for most classification procedures (see Subsubsection 2.2.4.4). Computation of $\mathbf{D}$ or $\mathbf{S}$ means that a single dissimilarity d_{ij} between any pair (O_i, O_j) of objects is calculated from their data vectors. (The t variables contributing to the data vectors may be standardised as we mentioned, sometimes weights are attached to them before the dissimilarities are computed, see Subsection 2.2.2.) Here, two objects O_i and O_j can have a dissimilarity $d_{ij} > d$, and thus can be grouped into different clusters as either they differ in all dimensions or they differ significantly in one of their variables and are mutually similar in the remaining $t-1$ dimensions. Thus, it may be problematic to interpret clusters, which have been defined in terms of similarities or dissimilarities.

Moreover, the scale levels often vary considerably between the different items or dimensions of the data vectors. It is therefore questionable if not impossible to compute overall (or global) similarities s_{ij} or dissimilarities d_{ij}, respectively, between the elements of a data set $\mathcal{S}$. The data then first must be transformed by multivariate scaling methods before similarities can be calculated. Using scaling methods, each global dissimilarity d_{ij} is calculated from t local dissimilarities $d_{ij1}, \ldots, d_{ijt}$ — one dissimilarity for each variable between any pair (O_i, O_j) of objects. This may obscure the true structure of a data set.

Here, the calculus of graph-theory provides another way to describe the structure of a data set, which is applicable especially for samples of mixed data. Such samples of n t-dimensional data vectors can properly be described by multigraphs. A first attempt is as follows. In each dimension of the data vectors, a similarity or dissimilarity between every pair of objects is calculated. This gives t dissimilarities between any pair and defines a multigraph Γ_t, again with the n objects as vertices. Every dimension is considered as a layer of a related multigraph. Two vertices now are linked together by an edge in one of the layers if and only if the objects are similar enough in the related dimension, allowing a total of t edges connecting each pair of vertices. For given integer s $(1 \leq s \leq t)$, a single-linkage cluster is a subgraph of that multigraph Γ_t or Γ_{tnN}, a so called s-**component**. (Roughly speaking, in an s-component, every two vertices are directly connected by at least s edges or by a sequence of other vertices which are interconnected in that way. In the notation Γ_{tnN}, the indices give the number t of edges, which may connect each pair of objects or vertices, the number n of vertices, and the number N of similarities or edges to be totally drawn. The exact definitions of a multigraph and an s-component follow in the next subsection.)

Often, several dimensions should be joined to "blocks". For example, one can collect all binary components of the data vectors into a block and calculate similarities between objects for this block by using the matching coefficient or Tanimoto's distance. All continuous items are merged into another block, from which the Euclidean distance between objects can be calculated. If data vectors consist of items, which are either binary or continuous, then this procedure yields two local distances between each pair of objects. For mixed data, this is easier to perform than to calculate one global distance between every pair of objects (see [38], [42].) Another reason for combining dimensions to blocks is that variables can be correlated, or data "belong together" like systolic and diastolic blood pressure or like weight and height. For the remainder of this paper, let t be the number of "local" dissimilarity or similarity measures, that is, the number of blocks, and not necessarily the number of dimensions or items of the data vectors. In the next subsection, we give the mathematical definitions, which are necessary for the understanding of our classification model

3.2.1 UNDIRECTED, COMPLETELY LABELLED MULITGRAPHS

Multigraphs can be defined in various ways. Since, in classification problems of multivariate data, the scale levels usually vary considerably between the dimensions of the data vectors, the following definition of a multigraph seems to be most suitable for application to cluster analysis of multidimensional data. Let $\mathcal{G} = \{\xi_1, \xi_2, \ldots, \xi_n\}$ be a nonempty set, and $\mathcal{K} = \{\kappa_{ij} = (\xi_i, \xi_j) : 1 \leq i < j \leq n\}$ be the set of all two-element subsets (ordered pairs) of $\mathcal{G}$. Furthermore, let $\mathcal{K}_t$ be defined by $\mathcal{K}_t := \mathcal{K} \times \{1, 2, \ldots, t\}$ with elements $\kappa_{ijl} = ((\xi_i, \xi_j), l)$ for $1 \leq l \leq t$. For every subset $\mathcal{H} \subseteq \mathcal{K}_t$, $\Gamma_t = (\mathcal{G}, \mathcal{H})$ is an **undirected, completely labelled multigraph**. The elements of $\mathcal{G}$ are the **vertices**, and the elements of $\mathcal{H}$ are the **edges** of Γ_t (see [149]).

According to this definitions, a maximum of t different edges $\kappa_{ij1}, \ldots, \kappa_{ijt}$ can link two vertices ξ_i and ξ_j together. Thus, it is easy to see that, for the number N of edges of such a multigraph,

$$0 \leq N = |\mathcal{H}| \leq t \binom{n}{2}$$

must hold true. Every nonempty subset $\mathcal{E}_{ij} = \{\kappa_{ij1}, \ldots, \kappa_{ijt}\} \cap \mathcal{H}$ is called a **connection** between the two vertices ξ_i and ξ_j. A connection $\mathcal{E}_{ij}$ is called **s-saturated** or an **s-fold connection** if $|\mathcal{E}_{ij}| \geq s$ holds true, that is, if at least s edges link the two vertices ξ_i and ξ_j together. In this case, the two vertices are called s-**fold connected**. The s-**degree** of a vertex is the number of vertices of Γ_t s-fold connected with it. A vertex with s-degree 0 is s-**isolated**.

Like in Section 1.2, by removing edges from a multigraph or by removing some vertices together with all those edges having one of these vertices as an endpoint, we obtain a submultigraph of Γ_t. A submultigraph is s-**fold connected** if every pair (ξ_{i_1}, ξ_{i_m}) of its vertices is connected by an alternating sequence $\xi_{i_1}, \mathcal{E}_{i_1, i_2}, \xi_{i_2}, \ldots, \xi_{i_{m-1}}, \mathcal{E}_{i_{m-1}, i_m}, \xi_{i_m}$ of vertices and s-fold connections in Γ_t. This sequence is called s-**path**. A submultigraph with m vertices is an s-**component** (of size m) of Γ_t if it is maximal with respect to the fact that any two of its vertices are interconnected by an s-path.

As for simple graphs, it is possible to describe the degree of compactness of submultigraphs by certain modifications of the term s-component (see also Section 1.2 and Chapter 5). An s-component, in which every two vertices are connected by one single s-path, is called an s-**tree**. (In an s-tree of size m are only $m - 1$ s-fold connections linking the m vertices together; s-isolated vertices are s-trees of size 1.) By a **weak s-component of degree k**, we denote an s-fold connected submultigraph of a multigraph Γ_t, which is maximal with respect to the fact that every vertex is s-fold connected with at least k other vertices. In such an s-component the s-degree of every vertex is at least k. A **strong s-component of degree k** of Γ_t is an s-fold connected submultigraph, which is maximal with respect to the fact that every two vertices are connected by at least k different s-paths which mutually share no s-fold connection. We speak of (k, s)-**components**. An s-fold connected submultigraph is a $(k; s)$-**block** if it has at least $k + 1$ vertices and is maximal with respect to the fact that at least k vertex disjoint s-paths connect every two vertices of this submultigraph. (As in Section 1.2, "vertex disjoint" means that the s-paths have no vertex in common with the exception of the endpoints which they connect.) An s-**clique** (or: maximal complete submultigraph) is a maximal submultigraph, where every two vertices are directly linked together by an s-fold connection. Whereas different s-components, weak or strong s-components are vertex disjoint, different (k, s)-blocks or s-cliques can have vertices in common.

The terms defined above make sense for natural numbers t, s with $1 \leq s \leq t$ only. For $s = 1$, they are equivalent to terms, which are already well known in graph theory (connected, isolated, component, and so on). For $t = 1$, a 1-fold connection is the same as an edge and Γ_1 does not differ from an undirected labelled graph Γ as defined in Sections 1.2 and 3.1. For $t \geq 2$, you can visualize a multigraph Γ_t as t undirected

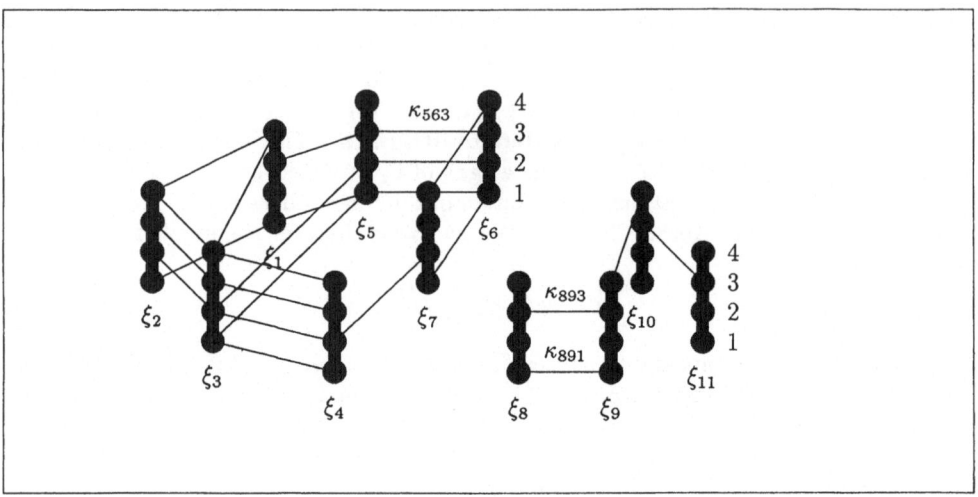

Figure 3-4 *Multigraph Γ_4 with 11 vertices, 24 edges and 8 2-fold connections.*

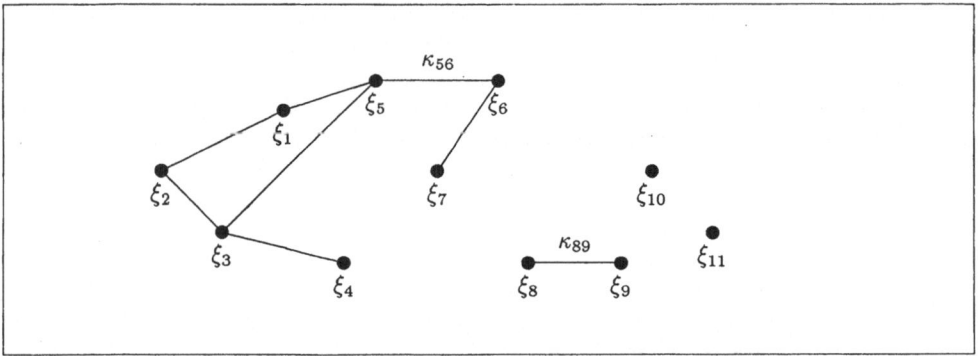

Figure 3-5 *The 2-projection $\tilde{\Gamma}$ of the multigraph Γ_4 from Figure 3-4.*

labelled graphs, each with the same vertex set $\mathcal{G}$, stacked in layers one upon the other, and N_l edges per layer ($N_1 + \cdots + N_t = N$, see Figure 3-4). Thus, the terms, as defined in this subsection, are generalizations of the graph-theoretic terms we mentioned in Section 1.2 (see also Chapter 5). For given natural number s with $1 \leq s \leq t$, every multigraph $\Gamma_t = (\mathcal{G}, \mathcal{H})$ can be mapped onto a simple graph $\tilde{\Gamma}(s) = (\tilde{\mathcal{G}}, \tilde{\mathcal{H}})$ with vertex set $\tilde{\mathcal{G}} = \mathcal{G}$ and an edge set $\tilde{\mathcal{H}}$ defined by $\tilde{\mathcal{H}} = \{\kappa_{ij} : |\mathcal{E}_{ij}| \geq s \text{ in } \Gamma_t\}$. In the image graph $\tilde{\Gamma}(s)$, two vertices are linked by an edge if and only if they are s-fold connected in the original multigraph. Therefore, the number v of edges in the image graph $\tilde{\Gamma}(s)$ of Γ_t is the number of s-fold connections (s-fold connected pairs of vertices) in the original multigraph Γ_t. We call $\tilde{\Gamma}(s)$ the s-**projection** of Γ_t (see Figure 3-5). The components, weak and strong components of degree k, k-blocks, cliques and trees of $\tilde{\Gamma}(s)$ are the images of the s-components, weak and strong s-components of degree k, the s-cliques and s-trees of Γ_t. Thus, it is easy to see that the submultigraphs defined here have virtually the same properties as their s-projections (as simple graphs, see [149] and Chapter 5 for an extensive discussion of these properties).

3.2.2 APPLICATION TO CLASSIFICATION MODELS: THE $(k, \vec{d}^T; s)$-CLUSTER

Let $\mathbf{X}$ be the (n, t)-data matrix of n objects $O_1, \ldots, O_n$, upon which measurements of t attributes $M_1, \ldots, M_t$ have been taken. For every attribute M_l, define a similarity or dissimilarity between every pair of objects O_i and O_j. For quantitative data, this can be done by choosing

$$(3\text{-}1) \qquad d_{ijl} = d_l(O_i, O_j) = d(x_{il}, x_{jl}) := |x_{il} - x_{jl}|$$

as distance measure between the objects for each attribute. This gives t **local** dissimilarity or similarity measures between every pair of data units, thus defining t local dissimilarity matrices $\mathbf{D}_1 = (d_{ij1}), \ldots, \mathbf{D}_t = (d_{ijt})$ or similarity matrices $\mathbf{S}_1 = (s_{ij1}), \ldots, \mathbf{S}_t = (s_{ijt})$. These matrices can be arranged to an (n, n, t)-tensor $\mathbf{D} = (d_{ijl})$, or $\mathbf{S} = (s_{ijl})$, respectively. From such a tensor of local distances, we can compute a matrix of global distances: Let the local distances d_{ijl} be given, for example, by (3-1), then

$$d_{ij} = \left(\sum_{l=1}^{t} d_{ijl}^p \right)^{1/p} \qquad (p > 1)$$

yields an (n, n)-distance matrix with (2-3a) as distance measure. This formula is only another form for (2-3a).

For given distance tensor $\mathbf{D}$ and vector $\vec{d}^T = (d_1, d_2, \ldots, d_t)$ of local distance thresholds, we connect two objects O_i and O_j of the sample $\mathcal{S}$ in their l-th dimension by an edge κ_{ijl} if $d_{ijl} \leq d_l$ holds true $(l = 1, \ldots, t)$. This yields a multigraph $\Gamma_t(\vec{d}^T) = (\mathcal{G}, \mathcal{H})$ with the objects as vertices, $\mathcal{G} = \mathcal{S} = \{O_1, \ldots, O_n\}$, and an edge set defined by $\mathcal{H} = \mathcal{H}(\vec{d}^T) = \{\kappa_{ijl} : d_{ijl} \leq d_l, l = 1, \ldots, t\}$. In practical classification problems, especially in medicine, it is often acceptable for objects to differ in some dimensions of their data vectors; they will be put in the same cluster if they are similar enough in a number of other dimensions. Thus, the following definition is obvious.

Definition 3-8 *Every **single-linkage cluster** $\mathcal{C}$ of level $(\vec{d}^T; s)$ consists of the vertices of an s-component of the multigraph $\Gamma_t(\vec{d}^T) = (\mathcal{G}, \mathcal{H}(\vec{d}^T))$, that is, of a component of the s-projection $\tilde{\Gamma}(\vec{d}^T; s)$ of $\Gamma_t(\vec{d}^T)$.* •

We proposed this classification model 1982 in [152]. According to their data vectors $\vec{x}_{i*}$ and $\vec{x}_{j*}$, two objects O_i and O_j are called "similar" and thus are put into the same cluster if they are similar enough in at least s of the t measurements. For $s < t$, these objects can differ in $t - s$ attributes at most. For studying the structure of a data set, the multigraph model gives a greater flexibility than the simple graph model of Section 3.1. The simple graph model permits different choices of a global distance threshold d only, affecting to all dimensions of the data vectors simultaneously, whereas in our model, it is possible to vary single elements of a vector $\vec{d}^T = (d_1, \ldots, d_t)$ of local thresholds. Furthermore, the homogeneity or separation of the resulting clusters can be controlled not only by the threshold vector $\vec{d}^T$, but also by s. Thus, the structure of the data can be explored more distinctively.

Defining the relation "<" for vectors $\vec{d}^T = (d_1, \ldots, d_t)$ and $\vec{d}^{*T} = (d_1^*, \ldots, d_t^*)$ by

$$(3\text{-}2) \qquad \begin{aligned} & (d_1, \ldots, d_t) < (d_1^*, \ldots, d_t^*) :\Leftrightarrow d_1 \leq d_1^*, \ldots, d_t \leq d_t^*, \\ & d_l < d_l^* \quad \text{for at least one } l, \end{aligned}$$

we immediately get the following theorems as generalizations of Theorems 3-1–3-4 from Subsection 3.1.1.

Theorem 3-6 *The set S of n objects is split into m disjoint clusters of level $(\vec{d}^T; s)$. The number of clusters is uniquely determined by an integer s and a threshold vector $\vec{d}^T = (d_1, \ldots, d_t)$: $m = m(\vec{d}^T; s)$. The resulting partition is independent of the order of choosing the objects to form the clusters.* •

Theorem 3-7 *For more than $t - s$ attributes, the local dissimilarities or distances between the objects from different clusters C_p and C_q of the level $(\vec{d}^T; s)$ always exceed the related elements d_l of the threshold vector $\vec{d}^T$.* •

Theorem 3-8 (Chaining effect) *Between every pair (O_i, O_j) of objects of a cluster of level $\vec{d}^T$, for which $d_{ijl} > d_l$ holds for more than $t - s$ attributes, exists a sequence of other objects of the same cluster which form a chain between O_i and O_j. At least s of the t local distances between every subsequent pair of elements in this chain do not exceed the related thresholds in $\vec{d}^T$.* •

Theorem 3-9 *If a threshold vector $\vec{d}_2^T$ is reduced to $\vec{d}_1^T$, then the classification at level $(\vec{d}_1^T; s)$ results from the classification at level $(\vec{d}_2^T; s)$ by splitting all or some groups of level $(\vec{d}_2^T; s)$ into subgroups. On the other side, every group of level $(\vec{d}_2^T; s)$ arises from the classification at level $(\vec{d}_1^T; s)$ by merging some groups of level $(\vec{d}_1^T; s)$. When reducing s_2 to s_1, clusters of level $(\vec{d}^T; s_1)$ are the result of a fusion of clusters of level $(\vec{d}^T; s_2)$. Every group of level $(\vec{d}^T; s_2)$ arises by splitting groups of level $(\vec{d}^T; s_1)$.* •

The results of these theorems can not only be used to obtain clusters of level $(\vec{d}^T; s)$, but also can produce hierarchical classifications and dendrograms. This can be done by variation of $\vec{d}^T = (d_1, \ldots, d_t)$ as well as of s. The procedure will be described in the following subsection and in Section 3.4. As in Subsection 3.1.3, we again can by-pass the problem of chains or bridges connecting otherwise separated clusters by modifications of the cluster definition. Instead of using Definition 3-8, we define a cluster as follows (compare Definition 3-3).

Definition 3-9 *A **weak k-linkage cluster** C of level $(\vec{d}^T; s)$ consists of the vertices of a (weak) s-component of degree k of the multigraph $\Gamma_t(\vec{d}^T)$. In the following, we speak of $(k, \vec{d}^T; s)$-clusters.* •

Remark 3-4 To this definition, the following is equivalent. A $(k, \vec{d}^T; s)$-cluster consists of the vertices of a maximal submultigraph of $\Gamma_t(\vec{d}^T)$, which is connected in the s-projection $\tilde{\Gamma}(\vec{d}^T; s)$, so that any of its vertices is directly linked to at least k other vertices of the s-projection by an edge. Like (k, d)-clusters, $(k, \vec{d}^T; s)$-clusters consist of at least $k + 1$ vertices (objects). The case $k = 1$ gives the clusters of Definition 3-8, but without the s-isolated vertices (without one-element classes). From $k = 3$ onward, bridges are deleted and groups, which are connected by them, become separated by this definition of a cluster. •

3.2.3 Discussion of the New Cluster Definition

Let s be a fixed integer, $1 \leq s \leq t$, and let the threshold vector $\vec{d}^T$ grow (let the different components of that vector grow). According to Definition 3-8, two clusters C_k and C_l are joined if, for one pair of objects $O_i \in C_k$ and $O_j \in C_l$, at least s of the t distances $d_1(O_i, O_j), \ldots, d_t(O_i, O_j)$ are lower than the respective threshold values of

$\vec{d}^T = (d_1, \ldots, d_t)$. Thus the behaviour of our single-linkage clusters is the same as that of the single-linkage clusters from Section 3.1. As can be seen from Theorem 3-6, separation between two clusters is more emphasized than homogeneity within a cluster. However, for quantitative data and $s = t$, the homogeneity of multigraph-based clusters can be easier controlled by defining a threshold vector $(d_1, \ldots, d_t)$ than the homogeneity of the simple-graph-based clusters using L^p-distances ($p < \infty$) and a threshold d (which, for example, can be calculated by $d = (d_1^2 + \cdots + d_t^2)^{1/2}$ for normalised data and Euclidean distance). In the multigraph model, the dissimilarity between two objects is controlled for each dimension separately, whereas, in the simple graph model, two objects O_i and O_j can have a dissimilarity $d_{ij} > d$, and thus can be grouped into different clusters, if they either differ in all dimensions or they differ significantly in only one of their variables and are similar in the remaining $t - 1$ dimensions.

Another advantage of the multigraph-based cluster definition is that the analyst (the biologist or physician) has an easier task in defining thresholds d_l for every single attribute M_l instead of being forced to look for a global dissimilarity threshold d. In our opinion, this method is more problem orientated, too. In medical investigations, attempting to outline clusters to define syndromes, it is often required that the patients to be clustered must be very similar with respect to certain symptoms to belong to the same group, whereas they can differ quite a lot in several other symptoms. This, in our model, can be controlled by choosing low values of d_l for those attributes M_l where the objects should be very similar when belonging to the same cluster. The other threshold values in $\vec{d}^T$ can be higher. Moreover, it is not necessary to normalise or to scale the data to make them comparable (because a threshold is given for every attribute). The raw data can be used directly to performe a cluster analysis. However, the clusters found by this method are invariant to monotonic transformations of the data.

Furthermore, the analyst can see directly from the multigraph $\Gamma_t(\vec{d}^T)$, constructed from a distance tensor and a threshold $\vec{d}^T$, which properties contribute to the classification and which do not. By variation of $(d_1, \ldots, d_t)$, positive or negative correlations between items can be uncovered: In the corresponding layers of the multigraph $\Gamma_t(\vec{d}^T)$, edges may be present or absent always simultaneously; or the presence of an edge in one layer can coincide with its absence in another layer. The information in the data matrix $\mathbf{X}$ is used in the mathematical model of multigraphs to a higher degree than in the simple graph model or in other models based on global similarities or dissimilarities. Additionally, there are different ways to generate hierarchies with the multigraph model:

(a) Starting with a given threshold vector $\vec{d}^T = (d_1, \ldots, d_t)$ and the related multigraph $\Gamma_t(\vec{d}^T)$ with $N = N(\vec{d}^T)$ edges, the integer s can pass all numbers $t, t-1, \ldots, 2, 1, 0$. In this case, we usually start with some rather small $(k, \vec{d}^T; t)$-clusters. They will be enlarged to $(k, \vec{d}^T; t-1)$-clusters (or amalgamated with other groups, respectively). More clusters are added, when s is reduced. For $s = 0$, the whole sample S forms a single $(k, \vec{d}^T; 0)$-cluster.

(b) Leaving the integer s fixed, we can vary some components of the threshold vector $\vec{d}^T$ either simultaneously or one after the other, some components can even be kept unchanged.

(c) Again choosing an integer s and keeping it fixed, we can start with $\vec{d}^T = (0, \ldots, 0)$ and increase one of the components of $\vec{d}^T$ from 0 to that value, for which all objects are mutually similar in the corresponding attribute (the value, for which all $\binom{n}{2}$ edges in the corresponding layer of the multigraph have been drawn). Then we increase another component of $\vec{d}^T$ and so on, until all $t\binom{n}{2}$ edges have been drawn and the multigraph is complete.

(d) All components in $\vec{d}^T$ are increased simultaneously, again starting from $(0, \ldots, 0)$, so that they are on a straight line through a given vector $(d_1, \ldots, d_t)$ which has to be determined in advance (or is of special interest for the researcher). This value reflects the conditions which the analyst wants to be satisfied for the similarities between the objects in the different attributes.

The hierarchical procedure (b) especially is of interest if clusters are sought for, which reflect mainly the structure of some pre-chosen variables (for which the components in $\vec{d}^T$ remain fixed). By varying the other components in $\vec{d}^T$, the analyst can explore the evolution of subgroups of these clusters and their dependence on the other — maybe secondary — variables.

For procedure (d), we can choose a threshold vector $\vec{d}^T$ with $d_1 = \ldots = d_t = d$ without any restriction: Every threshold $(d_1, \ldots, d_t)$ can be replaced by $(d, \ldots, d)$ if the elements d_{ijl} of every dissimilarity matrix $\mathbf{D}_l$ have been multiplied with the factor g_l for which $g_l \cdot d_l = d$ holds. This property renders possible to determine the s-projection of a multigraph Γ_t directly, without calculating Γ_t itself. The procedure is as follows. Let $(d, \ldots, d)$ be the — weighted — threshold vector. For every pair (O_i, O_j), the weighted dissimilarities d_{ijl} are ordered by magnitude:

$$d_{ij(1)} \le d_{ij(2)} \le \ldots \le d_{ij(s)} \le \ldots \le d_{ij(t)}.$$

Now, $\mathbf{D}_{(1)} = (d_{ij(1)})$, $\mathbf{D}_{(2)} = (d_{ij(2)}), \ldots, \mathbf{D}_{(t)} = (d_{ij(t)})$ are the matrices with the *smallest* local dissimilarities, the *second-smallest* local dissimilarities, ..., the *largest* local dissimilarities between all pairs of objects. Let us consider the matrix $\mathbf{D}_{(s)} = (d_{ij(s)})$ of the s-smallest dissimilarities. If $d_{ij(s)} \le d$ holds true for a pair (O_i, O_j), then these objects are connected by an edge. No edge will be drawn if $d_{ij(s)} > d$ holds true. This gives a graph $\Gamma = \Gamma(d)$. Because of

$$d_{ij(s)} \le d \Rightarrow d_{ij(1)} \le d, \ldots, d_{ij(s-1)} \le d,$$
$$d_{ij(s)} > d \Rightarrow d_{ij(s+1)} > d, \ldots, d_{ij(t)} > d,$$

this graph $\Gamma(d)$ is the s-projection of the multigraph $\Gamma_t(\vec{d}^T)$ with $\vec{d}^T = (d, \ldots, d)$. Thus, the single-linkage procedure for detecting clusters of level d — as described in Subsections 3.1.1 and 3.1.2 — together with the matrix $\mathbf{D}_{(s)}$ yields the single-linkage clusters of level $((d, \ldots, d); s)$ directly, and calculating a multigraph is no requisite for the determination of single-linkage clusters according to Definition 3-8. Of course, the $(k, \vec{d}^T; s)$-clusters of Definition 3-9 can be in the same way uncovered as (k, d)-clusters, using the $d_{ij(s)}$ as dissimilarities.

In fact, the procedure described above does not take full advantage of all the information in the tensor of local dissimilarities for every attribute. However, this algorithm renders possible to use existing programs to uncover single-linkage clusters or weak k-linkage clusters by choosing $\mathbf{D}_{(s)} = (d_{ij(s)})$ instead of the matrix $\mathbf{D} = (d_{ij})$ of global dissimilarities or distances, which these programs originally have been designed for. The following two fact now are obvious:

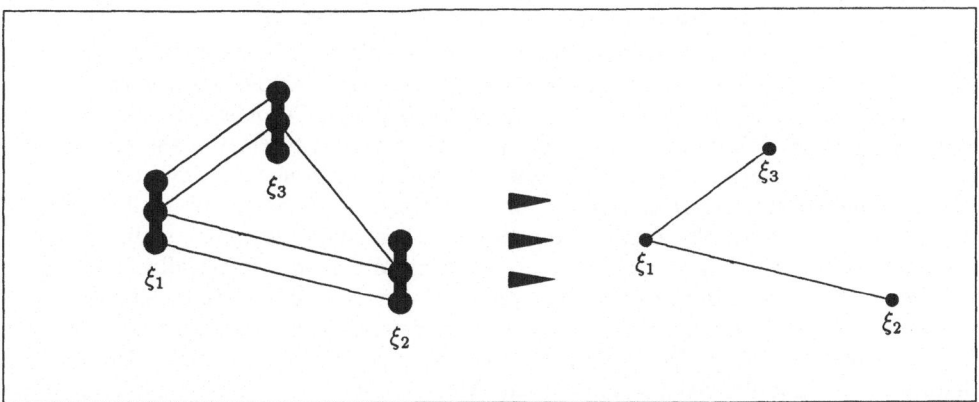

Figure 3-6 *The objects O_1 and O_2, and O_1 and O_3, respectively, are mutually similar in the second attribute. Thus, if local **distances** are used then O_2 and O_3 must be similar in the second attribute (because of the triangle inequality). O_1 and O_2 are 2-fold connected, as are O_2 and O_3. However, O_1 and O_3 are **not** 2-fold connected.*

(a) For $s < t$, the $d_{ij(s)} = d_{(s)}(O_i, O_j)$ usually do **not** satisfy the triangle inequality even if all $d_l(O_i, O_j)$ satisfy this inequality. That is, $d_{(s)}$ usually defines a dissimilarity function only also in those cases where all the d_l are defined as distances ($l = 1, \ldots, t$, see Figure 3-6 for an example).

(b) For $s = t$ and quantitative continuous data, our procedure can be reduced by an appropriate transformation to the single-linkage method with the maximum or Chebyshev distance as distance measure (here, the d_l must be defined as distances, see Subsection 2.2.2).

Sometimes, a clustering procedure cannot be applied to a sample since the number of variables is too large. For such cases some authors suggest that a factor analysis should be performed. They then use the values of the factors (the "scores") of the objects instead of the original — raw or standardised — data as the basis of the cluster analysis. This procedure seems to make sense for reducing dimensionality of high dimensional data. However, its main disadvantage is that in case of inhomogeneous samples, the methodological prepositions are hurt: We cannot assume a linear model. (For this reason, B.S. Everitt asks in [117] that every cluster analysis of scores should be followed by another factor analyses of the items, a separate analysis for the objects of every group of objects. This should be done to prove whether the different factors in every cluster remain unchanged — that is, are composed of the same items as the factors of the whole sample, see also Subsection 2.2.1.) Another problem is that of interpreting the factors in a "correct" way — the problem of defining "intelligence" as composed of several items which then become part of a questionnaire to test the intelligence. This especially holds if the factors are composed of variables of different scale levels. The principal components analysis cannot be applied to reduce dimensionality of mixed data. It supposes continuous data and linearity. For inhomogeneous data, the original group structure can be destroyed by the projection as Figure 2-2 shows. A method for reduction of dimensionality, which does not presuppose a linear model, is that of R. Blomer and co-authors. Nevertheless, from its basic idea it assumes data on at least interval scale level (see [36] and Subsection 2.2.1 for more details).

If the scale levels vary considerably between the different items, then, for methodological reasons, it is questionable to compute global dissimilarities between the objects (see Subsection 2.2.2). For high-dimensional data, it is also a problem to calculate local distances for each attribute, and to use all these distances to classify the data: Storage space is one reason. Another reason — which is of even more importance — is that if some variables (attributes) are correlated then must be the dissimilarity measures, too. This naturally renders impossible to test hypotheses of "realness versus randomness" of the clusters found by a clustering algorithm. Especially for samples with high-dimensional data vectors, composed of variables of different scale levels, the multigraph method provides the researcher with an easy-to-use method for reducing the number t of dimensions. No assumption like linearity must be satisfied. The analyst defines which variables remain isolated and which variables he wants to combine to so-called "blocks". Now, for every block a dissimilarity between the objects is calculated. Thus, the number of dissimilarities for every pair of objects is the same as the number of blocks.

Example 3-1 In Table 2-1 from Example 2-1, the matrix of the raw data of 21 probands is shown. The values of 15 attributes are listed. Here we can combine the variables M_1, M_5 and M_6 to a block. This block, consisting of age, height and weight, may be called "demoscopic variables"). M_9, M_{10}, M_{14}, M_{15}, the blood pressures at rest, are a second block; M_3 and M_4 are the third one, the "family anamnesis". The variables M_{11}, M_{12} and M_{13} together form the fourth block, the "values during exercise"; while M_2, M_7 and M_8, respectively, remain as three blocks of isolated attributes. Thus, the originally 15-dimensional data are reduced to seven blocks. For every pair of objects, seven distances must be calculated. For the first block, this can be the Euclidean distance (after standardisation of the variables), also for the second, fourth, fifth and seventh block. We can choose the matching coefficient together with (2-5) as distance measure of both the third and sixth block. •

Remark 3-5 For this procedure, the number of blocks calculated — that is, the number of dissimilarity matrices $\mathbf{D}_l$ or layers in the corresponding multigraph — is not identical with the number of attributes. It is usually much smaller. For the remainder of this paper, we denote by t the number of blocks instead of the number of dimensions. This is done to avoid unnecessary case distinctions. Obviously, the dimension of the threshold vector $\vec{d}^T$ is also identical to the number of blocks, and not to the original number of attributes. One reason for merging several dimensions of the data vectors into blocks is that variables are correlated or belong together like systolic and diastolic blood pressure or like weight and height. Another intention may be to join all variables of identical scale levels into blocks: All continuous variables form a block, all ordinal scaled data another one and so on. Thus, in samples with mixed data, no complicated procedures like reduction of scale levels must be performed before dissimilarities can be computed. For binary data, we also can combine all those variables to one block, in which the objects must coincide to belong to the same group. For this block, using the matching coefficient (2-4a) and (2-5), a distance can be calculated. Choosing 0 as threshold for this block, we link two objects by an edge in the corresponding layer of the multigraph only if they coincide in all the variables of this block. •

With Definition 3-9, the degree of compactness of the clusters is controlled by the parameter k. If more compact clusters are sought, then the strong k-linkage method can be generalized to multigraphs and then used for classification.

Definition 3-10 *Every **strong k-linkage cluster** C of level $(\vec{d}^T; s)$ consists of the vertices of a strong s-component of degree k of the multigraph $\Gamma_t(\vec{d}^T)$; that is, it is a (k, s)-component of $\Gamma_t(\vec{d}^T)$.* •

Dissimilarity tensors and multigraphs $\Gamma_t(\vec{d}^T)$ can be used to construct nondisjoint classifications, too. For this purpose, Definitions 3-5 and 3-6 can be rewritten as follows.

Definition 3-11 *A **k-overlap cluster** C of level $(\vec{d}^T; s)$ consists of the vertices of a $(k; s)$-block of $\Gamma_t(\vec{d}^T)$.* •

Definition 3-12 *A **complete-linkage cluster** C of level $(\vec{d}^T; s)$ is the vertex set of an s-clique of a multigraph $\Gamma_t(\vec{d}^T)$, wich has been constructed from a dissimilarity tensor $\mathbf{D}$ and a threshold $\vec{d}^T$ (it is a maximal set of objects such that every pair of objects has mutual dissimilarities d_{ijl} with $d_{ijl} \leq d_l$ for at least s blocks of variables).* •

We speak of $(k, \vec{d}^T; s)$-clusters, $(k, \vec{d}^T; s)$-components, $(k, \vec{d}^T; s)$-blocks or $(\vec{d}^T; s)$-cliques, respectively, if wewant to emphasize their dependence on a threshold vector $\vec{d}^T$. As for thresholds d in the simple graph model, the value of $\vec{d}^T$ is not uniquely determined in the definitions of this section. We speak of $(k; s)$-clusters, $(k; s)$-components, $(k; s)$-blocks or s-cliques, respectively, if we do not want to emphasize a classification level $\vec{d}^T$. Like $(k, \vec{d}^T; s)$-clusters, $(k, \vec{d}^T; s)$-components, all $(k, \vec{d}^T; s)$-blocks or $(\vec{d}^T; s)$-cliques can be detected from s-projections, using the matrix $\mathbf{D}_{(s)}$. Thus, programs, already existing for uncovering (k, d)-components, (k, d)-blocks or d-cliques, can be used to find submultigraphs of undirected, completely labelled multigraphs.

Remark 3-6 Creation times $e(C)$ and isolation indices $i(C)$ can be defined in several ways for $(k; s)$-clusters C. A generalization of Definition 3-7 results in a vector-valued index. If we are interested in a scalar as creation time of a group, then we can choose the number of edges to be drawn until this group arises. The number of edges to be added to $\Gamma_t(\vec{d}^T)$ until a cluster C of creation time $\vec{d}^T$ will be "swallowed" by a larger one, can be used as isolation index of C. Instead of the numbers of edges, we also can define the respective numbers of s-fold connections as creation times or isolation indices of clusters. •

Thus, for the multigraph model, Theorem 3-5 can be re-written in several ways. If we decide to use a vector as isolation index, then we immediately get the following theorem as being equivalent to Theorem 3-5.

Theorem 3-10 *Let C_1 be a $(k, \vec{d}_1^T; s)$-cluster with creation time d_1, and C_2 a $(k, \vec{d}_2^T; s)$-cluster with creation time d_2. Furthermore, let $\vec{d}_1^T \leq \vec{d}_2^T$. Then either $C_1 \cap C_2 = \emptyset$ or $C_1 \subseteq C_2$ holds true. For arbitrary $k < n$, $S = \{O_1, \ldots, O_n\}$ is the largest $(k; s)$-cluster.* •

Choosing the number of edges or s-fold connections at which a cluster arises as its creation time, we see that Theorems 3-5 and 3-10 do not differ (see Remark 3-1).

By variation of s, the homogeneity of clusters can be controlled, too. The case $s = 1$ allows objects to belong to the same cluster if they are similar in one block only. The case $s = t$, on the other hand, implies that two objects must be similar in all blocks to belong to the same group. Thus, the multigraph model gives a better insight into structures of the data to be clustered. We see exactly in which layers two objects are similar (connected by an edge). Partly, this information is lost when we switch to the s-projection. Here we only count the number of edges connecting two vertices. However, in our opinion, this is more informative than calculating a single distance between any pair of objects, not knowing whether two objects are in different clusters because they differ either in all dimensions or they differ significantly in only one dimension and are similar in the remaining $t - 1$ dimensions.

3.3 An Algorithm for the Construction
of $(k, \vec{d}^T; s)$-Clusters

By optimization criteria, the groups are not only uncovered but defined by the algorithm. In contrast, every quadruple $(\mathcal{S}, \mathbf{D}, k; s)$ — consisting of a sample $\mathcal{S}$, a dissimilarity tensor $\mathbf{D}$ and parameters k and s — uniquely defines a class of $(k; s)$-clusters with well-defined properties, concerning their degrees of connectedness and compactness. Thhe sole task of a cluster-detecting algorithm then is to uncover these well-defined classes. The following procedure is suitable to detect $(k, \vec{d}^T; s)$-clusters for given parameters k, $\vec{d}^T$ and s. In the form proposed here, it is not appropriate for hierarchical classifications since both s and the threshold vector $\vec{d}^T$ must be given in advance. The procedure consists of three parts:

(a) Input and preparation of the data;
(b) calculation of the dissimilarity tensor;
(c) computation of the multigraph and the clusters.

In detail, the whole program is composed of the following modules:

(a1) Input of the (formatted or unformatted) matrix of the raw data;
(a2) computation of the minimum, maximum, median and standard deviation (or p-th root of the p-th central absolute moment) for all variables;
(a3) transformation of the different variables (optional), either by mapping them onto the interval $[0; 1]$ or by normalisation to variables with mean 0 and p-th central moment 1;
(b1) definition of blocks and determination of those variables which should be merged into blocks, determination of those attributes which should be masked and excluded from the classification procedure (because of missing data, for example), calculation of the number t of blocks (including isolated variables which are not combined to blocks with other ones);
(b2) computation of the dissimilarity tensor either from the original raw data or from the transformed data — for isolated variables, (3-1) is used, for blocks of continuous data, the L^p-distance (2-3a) for arbitrary p, the matching coefficient (2-4a) for binary data, and the generalized matching coefficient for other nominal or ordinal scaled variables can be computed;
(c1) calculation of the multigraph $\Gamma_t(\vec{d}^T)$ for a given threshold vector $\vec{d}^T$ from the dissimilarity tensor;
(c2) determination of the s-projection $\tilde{\Gamma}$ of $\Gamma_t(\vec{d}^T)$ for an integer s (for $t > 1, 1 \leq s \leq t$);
(c3) assignment of objects to clusters, that is, determination of the components of $\tilde{\Gamma}$ (or of the s-components of $\Gamma_t(\vec{d}^T)$), uncovering of the $(k, \vec{d}^T)$-clusters of $\tilde{\Gamma}$ (or of the $(k, \vec{d}^T; s)$-clusters of $\Gamma_t(\vec{d}^T)$);
(c4) output of the number of $(k, \vec{d}^T; s)$-clusters, number of objects in the different $(k, \vec{d}^T; s)$-clusters, number of edges in the $(k, \vec{d}^T; s)$-clusters, membership of the data units to the clusters.

The dissimilarity tensor $\mathbf{D} = (d_{ijl})$ is symmetric in i and j. Thus, for Step (c1), the computation of the multigraph $\Gamma_t(\vec{d}^T)$, the elements d_{ijl} above the diagonal are replaced by elements τ_{ijl}, where

$$(3\text{-}3) \qquad \begin{aligned} \tau_{ijl} &= 1 \quad \text{if} \quad d_{ijl} \leq d_l, \\ \tau_{ijl} &= 0 \quad \text{if} \quad d_{ijl} > d_l. \end{aligned}$$

If $\tau_{ijl} = 1$, then the related edge κ_{ijl} belongs to the edge set $\mathcal{H}(\vec{d}^T)$ of $\Gamma_t(\vec{d}^T)$. Computation of the s-projection $\tilde{\Gamma}$ in (c2) is performed in two steps. At first,

$$(3\text{-}4) \qquad \tau_{ij.} = \tau_{ij1} + \tau_{ij2} + \cdots + \tau_{ijt} = \sum_{l=1}^{t} \tau_{ijl}$$

is calculated for every pair (O_i, O_j) with $i < j$. This is the number of edges in $\Gamma_t(\vec{d}^T)$, which link the vertices ξ_i and ξ_j, or the number of variables or blocks in which two objects are mutually similar. These $\tau_{ij.}$ are arranged in a symmetric matrix $\mathbf{T}$ ($\tau_{ii.} = 0$, $\tau_{ij.} = \tau_{ji.}$). Above the main diagonal of $\mathbf{T}$, the $\tau_{ij.}$ then are replaced by elements u_{ij} with

$$(3\text{-}5) \qquad \begin{aligned} u_{ij} &= 1 \quad \text{if} \quad \tau_{ij.} \leq s, \\ u_{ij} &= 0 \quad \text{if} \quad \tau_{ij.} > s. \end{aligned}$$

The u_{ij} indicate whether a connection between the vertices ξ_i and ξ_j is s-saturated or not, that is, whether ξ_i and ξ_j are connected by an edge in the s-projection $\tilde{\Gamma}$ or not. As default value, $s = t$ is chosen. However, we want to point out that the value of s should be carefully chosen in accordance with the cluster problem.

From the columns of this triangular matrix, in (c3) the components of the s-projection are derived. To uncover the $(k, \vec{d}^T; s)$-clusters, all components with less than $k + 1$ vertices are excluded. Then successively all vertices with a degree less than k (with s-degree less than k in $\Gamma_t(\vec{d}^T)$) are eliminated. The subgraph left then consists of $(k, \vec{d}^T; s)$-clusters only. The degree of compactness of the groups is defined by k. This parameter must be chosen according to the problem. As default, we put $k = 1$, that is, single-linkage clusters are detected by the procedure.

3.4 THE CONSTRUCTION OF DENDROGRAMS OF $(k; s)$-CLUSTERS

Having calculated the dissimilarity tensor $\mathbf{D}$, we can replace $\mathbf{D}$ by the tensor Δ of its ranks. Every layer (or block) of $\mathbf{D}$ is a matrix $\mathbf{D}_l$ of local dissimilarities which is symmetric to the main diagonal. The diagonal itself consists of zeroes. Thus, every matrix is defined by $\binom{n}{2}$ elements d_{ijl}, $1 \leq i < j \leq n$, above the main diagonal; and $\mathbf{D}$ is completely determined by $t\binom{n}{2}$ elements. The rank δ_{ijl} of a dissimilarity d_{ijl} is the position of d_{ijl} in a vector of dimension $t\binom{n}{2}$ with all dissimilarities arranged in increasing — or at least not decreasing — order. Thus, by replacing first the smallest element d_{ijl} ($i < j$) by $\delta_{ijl} = 1$, and d_{jil} by $\delta_{jil} = 1$, then the second smallest dissimilarities d_{ijl} and d_{jil} by the rank 2, and so on, until at last the largest element above and below the main diagonal is replaced by $t\binom{n}{2}$, we get the rank tensor Δ. Taking advantage of the symmetry of $\mathbf{D}$, we can save storage space if we store only those elements d_{ijl} or δ_{ijl} with $i < j$.

We recommend to use not the original distance tensor $\mathbf{D}$, but a suitably normalised one for ranking dissimilarities. We think that the normalisation procedure described in Subsection 3.2.3 is best: Multiplication of every matrix $\mathbf{D}_l$ with a constant g_l such that $g_l \cdot d_l = d$. The entries from different matrices of local dissimilarities then are directly comparable. Tie breaking schemes may be used if not all dissimilarities are different. For the related multigraph, that means that the edges κ_{ijl}, corresponding to dissimilarities d_{ijl} with the same numeric value, are drawn in a random order.)

Using the tensor Δ of ranks of dissimilarities, it is easy to derive a procedure for a hierarchic classification: We can use Parts (a) and (b) from the base algorithm of the last section without any modification. Part (c), however, must be replaced by the following steps.

(c1) Calculation of Δ from $\mathbf{D}$;

(c2) choice of s and k, calulation of the multigraph Γ_t with n objects $O_1, \ldots, O_n$ as vertices $\xi_1, \ldots, \xi_n$, together with those $s\binom{k+1}{2}$ edges which are related to the ranks of the $s\binom{k+1}{2}$ smallest dissimilarities;

(c3) calculation of the s-projection $\tilde{\Gamma}$ of this multigraph (for $t > 1$);

(c4) determination of the components (single-linkage clusters) and k-clusters of $\tilde{\Gamma}$;

(c5) assignment of the objects to clusters, storage and output of relevant quantities (number of $(k, \vec{d}^T; s)$-clusters, number of objects in the different $(k, \vec{d}^T; s)$-clusters, number of the edges in the $(k, \vec{d}^T; s)$-clusters, membership of the data units to the different clasters, creation time) if a new group has been detected;

(c6) examination whether a new detected k-cluster contains another one which has been uncovered at an earlier stage, and for which no isolation index has been calculated before; calculation of this index;

(c7) repetition of Steps (c1)–(c6) for the dissimilarity ranking next, until S is detected as the largest k-cluster.

All k-clusters in S can be uncovered by this algorithm. Since a k-cluster has at least $k + 1$ elements — then with all $\binom{k+1}{2}$ s-fold connections linking the $k + 1$ points together — at least $s\binom{k+1}{2}$ edges are necessary so that such a cluster can arise. This defines the initial value in (c2). Since we use the tensor Δ of ranks of dissimilarities instead of the tensor $\mathbf{D}$ of — possibly normalised — dissimilarities, we are not requested to define a threshold vector $\vec{d}^T$. Performing Step (c7) until N edges have been drawn, we get at the last step a disjoint classification at a level $(\vec{d}^T; s)$ with $\vec{d}^T = \vec{d}^T(N)$. However, this threshold $(\vec{d}^T; s)$ is not uniquely determined by N. (By $\vec{d}^T$, $N = N(\vec{d}^T)$ is defined uniquely but the reverse does not hold.) We write Γ_{tnN} to denote that we have constructed a multigraph with t layers, n vertices $\xi_1, \ldots, \xi_n$, corresponding to the objects to be classified, and those N edges belonging to the N smallest dissimilarities. In the case of simple graphs of Subsection 3.1.2, we write Γ_{nN}.

CHAPTER 4
PROBABILITY MODELS OF CLASSIFICATION

Applying a clustering algorithm to a set of data results in a classification of objects whether the data exhibit a true or "natural" grouping structure or not. This is no problem if clustering is done for obtaining a practical stratification of a given set of objects for organisational purposes. Such purposes justify even purely artificial groupings (**random clusters**). In exploratory data analysis however, interest lies in uncovering an unknown clustering structure of the data. Here, the result of a clustering procedure should reflect the real structure (**real** or **natural clusters**). From the group structure of the objects of a sample S, we usually derive probability models on a population. Here, an artificial clustering is not acceptable. The classes resulting from the algorithm must, in addition, be investigated for their **relevance** and their **validity**.

Both homogeneity and separation of clusters can be used as internal criteria for investigations into the realness of clusters, the composition of objects provides external criteria. For the most part, descriptive, graphical or other other exploratory methods are used for such investigations (compare Section 2.4).

Important criteria for choosing an appropriate classification algorithm are **validity** and **robustness** (or **stability**). Usually, the validity of a clustering procedure is proved by applying it to data sets of a well-known group structure. Such data sets can be generated by Monte Carlo-simulation or real data are used like R.A. Fisher's famous iris data published in 1936 in the "ANNALS OF EUGENICS" ([123]). The percentage of data classified correctly by the procedure is a measure of validity. (see [27], [122], [187], [247], [268], [279], [280], [398]). Robustness of classification procedures usually is judged by one of the following methods, after applying a procedure twice. The first method is to apply a cluster detecting algorithm at first to the whole data set and then to a slightly modified sample where either some objects have been left aside ([14], [269]), or some outliers have been added ([278]). In the second method, the original data set is disturbed by small random effects, before the procedure is applied a second time, or data have been raised twice (at different times) from the same objects ([13]). In both cases, the percentage of objects attached to the same clusters can serve as a measure of robustness. The methods named above, however, possibly may **not** evaluate the validity or robustness of a procedure but just the quality and efficacy of the variables chosen.

Obviously, researchers are interested in robust classification procedures. Especially when the data are not complete, as is often the case in medical research. Missing data, however, can cause severe problems to applying cluster detecting algorithms to data (see Subsection 2.2.2 and [427]). The techniques for validating clusters, as named above, only can provide hints, and illustrate some special aspects of classification. By no means do they provide an assurance for the quality of the result of a clustering of a **new** data set. This especially holds if groups of distinct shapes must be supposed in a set of objects as J. Krauth emphasizes in [240]. Thus, for real-life situations the researcher must be supplied with criteria to classify an object set as either being homogeneous — and thus a clustering as being artificial — or as composed of distinct groups ([101]).

As a contribution to this fact, today, adequate criteria to judge the quality of a classification are searched for with priority. While, at the beginning of the fifties, developing efficient algorithms to outline clusters of every shape was the main purpose of scientists (in some cases coupled with developing and defining cluster models), nowaday the main interest therefore is focussed in integrating cluster models into a relevant mathematical-statistical theory. Thus, apart from the more intuitively based methods named in the last paragraph, the task of judging the relevance of a classification can be performed by using probabilistic models and suitable statistical significance tests. Statistical theory, however, cannot provide a complete theory of classification. We cannot say how similarities should be judged, although we can give technical assistance in constructing distances. Different classifications are right for different purposes, so we cannot say any one classification method or classification to be best. Statistical theory in clustering provides a testing ground for various clustering methods as J.A. Hartigan emphasizes in [187]: We discover how well the methods work for various idealized forms of data, and reject those methods that fail, at least for application to similar types of real data.

In Chapters 2 and 3, we outlined how to detect clusters of level d (or $\vec{d}^T$, respectively) in a sample S by using a suitable distance matrix or tensor. The same tools can be used to build hierarchies as we have seen, too. We especially have been interested in k-clusters. Every group suggested by a numerical procedure, however, at this stage is nothing more than a cluster **candidate**. Its validity and relevance must be judged by the researcher as well as by the statistician. While the researcher judges the relevance from a possibly more qualitative point of view, the statistician has to rely on appropriate significance tests or other methods of data analysis.

In this chapter, we discuss a quantitative approach to cluster analysis. The set of objects, upon which measurements have been taken, is considered as a sample drawn from a (fictive) population; not the sample itself but the whole population is the true object of interest (see Subsection 2.2.1). The measurements x_{il} of the i-th object O_i and l-th attribute M_l are now interpreted as realizations of random variables X_{il} for $i = 1, \ldots, n$, and $l = 1, \ldots, t$. Thus, to discriminate between the hypothesis, on the one hand, that the observations are sampled from a "homogeneous" population, and an alternative involving "heterogeneity" or a "clustering structure", on the other hand, we need assumptions about the probability distribution of these random variables or of the similarities or distances derived. With some knowledge of probability models which are valid under several null hypotheses of "random clusterings", that is, of homogeneous populations, we then can compute test statistics to test the relevance of these hypotheses at a given significance level α. If, for a given α, such a test statistic exceeds the corresponding critical value $c = c(\alpha)$, then the hypothesis of homogeneity is rejected in favour of a clustering structure.

For such purposes, the basic ideas and concepts of researchers in different sciences about hypotheses, which they may call "structured, heterogeneous data" or "homogeneity", must be cast into mathematical models, which should be acceptable for an analyst. Probability models to detect those clusters from a set of classes suggested by an algorithm, which reflect a true group structure and are most likely not random, must be very general, too. This contributes to the fact that methods of cluster analysis are used mainly in exploratory statistics, when information about the possible structure is small. Rarely, information from the outset is available and can be used to some extend, as the examples of the last chapter of this booklet show.

Obviously, under ambiguous conditions like these, it cannot be easy to develop probability models and test statistics. Because of the diversity of probability models that can give rise to the dissimilarity matrix, the probabilistic approach to classification proves to be rather cumbersome: It is difficult to choose a model that is realistic over a wide variety of problems, and few of these models are at all tractable in theory. Moreover, depending on the genesis of the data vectors, dissimilarity matrices, respectively, certain assumptions of every probability model may be hurt more or less thus limiting the applyability of statistical tests.

With this chapter, we want to give a short overview of three different test statistics (based on the largest nearest neighbour distance, the mean similarity, and the minimum within-class sum of squares, respectively, see [45], [48]). This overview is followed by a more elaborate discussion of a rather general classification model of R.F. Ling ([259], [261]). Then we propose a probability model which is based on our concept of multigraphs and is a generalization of Ling's model (see Section 3.2). Its relevance in testing hypotheses of homogeneity is discussed in detail.

4.1 CURRENT PROBABILITY MODELS
IN CLUSTER ANALYSIS

The main difficulty in deriving test statistics for testing hypotheses of the structure of a data set lies in finding a suitable mathematical definition of the term "homogeneity". This definition should be plausible to be applied to research, and, at the same time, should provide a theoretical basis for statisticians to derive probability models: It should be **realistic** and **mathematical tractable**.

The first statistical procedures in classification theory have been derived for partitioning clustering algorithms (see also Subsubsection 2.2.4.1). They assumed multidimensional normal distributions of n t-dimensional observation points in Euclidean space. Here, some authors propose to interpret "homogeneity" as unimodal distribution of the data. "Heterogeneity" then means significantly different mean vectors for different clusters. For continuous data, these authors derive test statistics which are based on ideas analysis of variance (see [21], [134], [183], [229], [256], [352]). These significance tests, however, are based on considerations of analogy only. They can be regarded as heuristic recommendations, but are not valid for judging the quality of a cluster analysis performed. This is because in classification problems, we usually have no external criterion. The clusters are defined by the data vectors in a way that maximises the between-groups sum of variances. Thus, for classifications based on optimality criteria, we cannot accept a central F-distribution even if the null hypothesis of homogeneity is true. In this case the distribution of $\max |T|/|W|$ is more appropriate under the null hypothesis than that of Wilks' λ (by T we denote the total variance of the data vectors, and by W the within-groups sum of variances, see Subsubsection 2.2.4.1 and [117]).

This is regarded by L. Engelman and J. Hartigan in [109], [184], [185] (see also the survey papers [187] and [188]). Their test statistic $k_n(\mathcal{C})$ of a disjoint partition $\mathcal{C} = \{\mathcal{C}_1, \dots, \mathcal{C}_m\}$ is the **maximum F-statistic**, known from analysis of variance. That is, the statistic $k_n(\mathcal{C})$ is the F-value which is given by the "most significant" partition of the sample $\mathcal{S}$, the maximum F-value. The results of simulation studies are discussed in [109]. The asymptotic distribution of $k_n(\mathcal{C})$ is derived in [185]. However, the results are proved for univariate normally distributed observations only. H.H. Bock generalizes this Bayes procedure in [37] to random vectors.

In [45] and [48], the same author discusses three different types of test criteria which can be used to discriminate between the hypothesis of "homogeneity" and the alternative of "heterogeneity". Like the methods before, these tests are valid for quantitative data only. The hypothesis of homogeneously distributed data vectors can be mathematically formulated in two different ways, as can be the alternative (see also [40], [42], [47]):

(a) H_0: The data vectors $\vec{x}_{1*}, \dots, \vec{x}_{n*}$ are realizations of pairwise independent random variables $\vec{X}_{1*}, \dots, \vec{X}_{n*}$, which are identically uniformly distributed in a given bounded, open and connected region $\mathcal{G} \subset \mathcal{R}^t$ (**uniformity hypothesis**).

H_1: The common density $f(\vec{x})$ is multimodal, that is, a finite number $m \geq 2$ of distinct points $\vec{\mu}_1, \dots, \vec{\mu}_m$ exist, where f attains a strict relative maximum (the modes of f).

(b) H_0': The random variables are independent, and identically distributed with some unimodal density (**unimodality hypothesis**).

H_1': The common distribution is a translation mixture of densities of the same type (**translation mixture model**).

The alternative H_1' is a special case of H_1 (see Figure 4-1). It is also a special case of (2-12) in Subsubsection 2.2.4.2, namely $f_k(\vec{x}) = f(\vec{x} - \vec{\mu}_k)$. In the papers named above, H.H. Bock uses the so-called "gap" statistic to test H_0 against H_1. Both a statistic derived from a "mean similarity" and the maximum-F test can be used to test H_0' against H_1'.

The gap test is used to test the uniformity hypothesis. It is based on a modified nearest-neighbour distance. The test statistic D_n is the largest open ball, which can be centered at some data vector $\vec{x}_{i*}$ without containing in its interior some other point of $\mathcal{S}$ or the boundary $\partial\mathcal{G}$ of $\mathcal{G}$,

$$(4\text{-}1) \qquad D_n = \max_i \{\min_{j \neq i} \{\min_j d_{ij}, d(\vec{x}_{i*}, \partial\mathcal{G})\}\}.$$

The hypothesis H_0 is rejected if $D_n > c$. The threshold $c = c_n(\alpha)$ is to be calculated from $P(D_n > c) = \alpha$, the given error probability of the first kind. Since the exact distribution of D_n is very intricate to compute for finite sample size n, usually the asymptotic distribution — which is, up to a linear transformation, Gumbel's extreme value distribution — is used for calculating c if n is large enough (see [28], [45], [91], [104], [194], [195], [344], [345], [409], [410], [417]).

The asymptotic power of the gap test can be estimated. However, the use of this test is restrained because, for more than three-dimensional data, D_n and the asymptotic statistic both depend strongly on the type and volume of the domain $\mathcal{G}$ which rarely

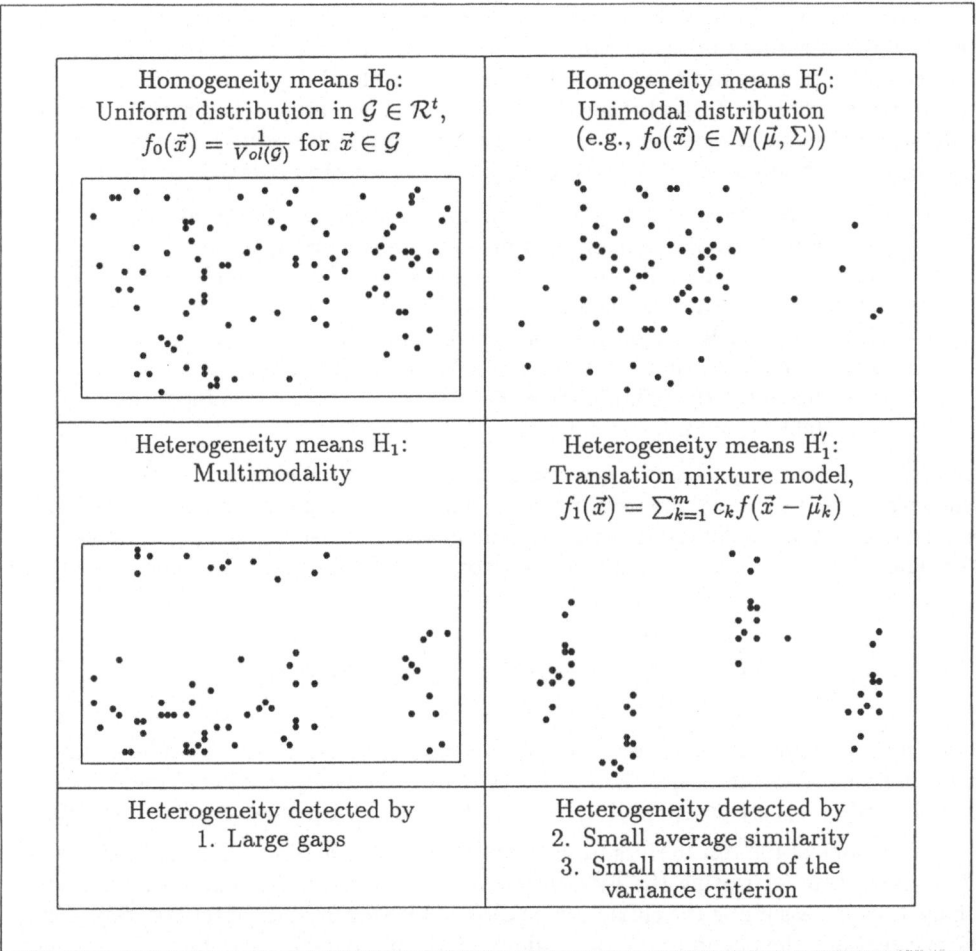

Figure 4-1 *Probability models (from [45], [48]).*

is known in practical situations. Furthermore, the convergence of the sequence of distributions of D_n towards its limit distribution is slow, namely $1/\log n$. Thus, the asymptotic threshold may be an inaccurate estimate of $c_n(\alpha)$. Several modifications and generalizations of the test statistic D_n and the corresponding gap test exist (see [48]).

For gap tests, the ranking of pairwise distances is the predominant feature. For mixture models, H.H. Bock uses the sum of mean similarities within the groups as test statistic. Under the mixture alternative H_1', we can expect that the random variable

$$(4\text{-}2) \qquad T_n = \frac{1}{\binom{n}{2}} \sum_{i=1}^{n-1} \sum_{j=i+1}^{n} S_{ij}$$

of the sum of mean similarities will be smaller than under H_0'. (The realizations s_{ij} can be calculated from pairwise distances using (2-5).) The unimodality hypothesis H_0' is rejected if $T_n > c$ holds for the sample S. Since only the asymptotic distribution of T_n is known, the threshold c just can be calculated approximatively (see [48]).

Since the tests given above neither use nor presuppose a clustering of the data, they can be applied without or before applying a clustering algorithm. However, in practice, the researcher is more interested in finding a suitable classification of his data. He only subsequently wants to test whether this classification is more marked than for random homogeneous data. Therefore he needs methods for simultaneously constructing and testing a classification of data. With H.H. Bock's words (from [48]):

> "It is expected that methods of this type will extract more information from the data concerning a prospective clustering structure and that the corresponding tests will exhibit a better power performance than a gap or a mean similarity test which is tailored more to the hypothesis of homogeneity than to the clustering alternative. But since their performance depends on both the clustering model and the clustering algorithm, it is evident that both must be adapted in a suitable sence from the outset."

In the framework of partition type classifications, this problem can be treated using the within-groups sum of squares criterion. This leads to the maximum-F test which is a generalization of the results of L. Engelman and J.A. Hartigan to continuous multivariate, not necessarily normally distributed random variables. By maximising the criterion

$$(4\text{-}3) \qquad k_n(\mathcal{C}) = \sum_{j=1}^{m} n_j \, d(\bar{\bar{x}}_{\mathcal{C}_j}, \bar{\bar{x}})^2 \Big/ \sum_{j=1}^{m} \sum_{O_j \in \mathcal{C}_j} d(\vec{x}_{i*}, \bar{\bar{x}}_{\mathcal{C}_j})^2$$

(see Subsubsection 2.2.4.1), we get an optimal partition of the n objects into m nonempty and disjoint classes $\mathcal{C}_1, \ldots, \mathcal{C}_m$. The hypothesis H_0 is rejected and the partition $\mathcal{C} = \{\mathcal{C}_1, \ldots, \mathcal{C}_m\}$ accepted if $k_n^* := \max_{\mathcal{C}} k_n(\mathcal{C}) > c$ with some critical value $c(\alpha)$.

The name of the test derives from the fact that $((n-m)/(m-1)) \cdot k_n(\mathcal{C})$ is just the F-ratio statistic for testing the hypothesis $\vec{\mu}_1 = \ldots = \vec{\mu}_m$ in the multivariate variance analysis model assuming the classification $\mathcal{C}$ to be **known** from an outer criterion. Thus, k_n^* is essentially the maximum F-ratio which can be obtained by searching for the "most significant" partition $\mathcal{C}$. This test has been derived correctly, avoiding incorrectnesses which have been made by other authors who relied on heuristically based, variance analytical considerations. The finite distribution of k_n^* is not known. However, some results on the asymptotic behaviour have been proved which allow an application of this test under rather general conditions ([48], [62], [63], [314], [315]). Under normal distribution assumptions, the test is optimal in some Bayesian sense ([37], [38]). Its major disadvantage is that the number m of clusters must be known in advance.

In [61], P. Bryant derived a homogeneity test which is based on the sample variance. This test can be applied if data are normally distributed and classification is performed by analysing the point density (Subsubsection 2.2.4.3). The hypotheses to be tested are H_0' against H_1', the unimodality model against the translation mixture model. This test, like the mean similarity test, is based on the conjecture that, unter the mixture alternative, certain functionals will be smaller than under the unimodality hypothesis. For P. Bryant's test, results concerning the error probability β of the second kind have been derived. In [397], D.J. Strauss developed a stochastic cluster model on the basis of Markov processes which was generalized by F.P. Kelly and B.D. Ripley in [227]. M. Westcott used Poisson processes to derive asymptotic results for cluster processes

([418]). For further reading on tests in cluster analysis, we refer to [19], [104], [105], [253], [313], [370], [371], [372].

Often structures and similarity relations can be described very well by graphs or digraphs. Thus, another basis for simultaneous constructing and testing a classification is provided by the graph-theoretic model. As discussed in the previous chapter, in this model of cluster analysis, objects are represented by vertices and those pairs of objects satisfying a particular similarity relation (for example, with a distance smaller than a given threshold d) are termed adjacent and constitute the edges of the graph. Clusters are then characterised by appropriately defined subgraphs. If we accept that "homogeneity" means a completely random assignment of the $N = N(d)$ edges to all $\binom{n}{2}$ pairs of objects (that is, a completely random choice of the N edges of our graph $\Gamma(d)$ from $\binom{n}{2}$ possible ones) then we are supplied with a rather general **combinatorial** cluster analysis model which is far simpler to use than the numeric-matrix oriented cluster analysis procedures: Accepting this model of **uniform distribution** of edges, we can adopt results from the theory of **random graphs** to derive test statistics for classification problems as has been proposed by C. Abraham in 1962 (which was only three years after the theory of random graphs had been founded by P. Erdős and A. Rényi, and N. Gilbert, respectively, see [1], [111], [112], [113], [142]).

For partitions, we can use the number of components (single-linkage clusters), or the size of the largest component as test statistic. For overlapping clusters, the size of the largest clique can be used. Their distribution is known under the null hypothesis. In [259], [261], [263], [264], [265], [272], this concept has been studied in detail. The authors derived exact — finite and asymptotic — results for $(1, d)$-clusters of Definition 3-4 and performed Monte-Carlo studies for $(2, d)$- and $(3, d)$-clusters. Tables which can be used for significance tests, can be found in these papers as well. In [148], we proved theorems on the exact finite distribution of the number of vertices in $(2, d)$-clusters. They are of theoretic interest only because they are very intricate. However, they could provide a basis for developing asymptotic results. Further papers which are of practical relevance, are [81], [108], [200], [203], [204], [224], [350], [351]. Another graph-theoretic approach to testing the relevance of a classification is described in [133]; an application of this model can be found in [169]. In [358], D.R. Shier used the number of possible trees and minimum spanning trees to construct distribution-free (nonparametric) tests of homogeneity. All these graph-theoretic models have a certain advantage when used in classification theory: They do not need any a-priori knowledge of the **number** of classes.

Relevance and validity of several significance tests for hierarchical classifications and dendrograms have been studied, using tools of simulation studies which were, of course, very time-consuming (see [15], [16], [168], [183], [184], [201], [257], [366], [369], [431]). But no theoretic models did exist until the beginning of the eighties; thus only approximative test statistics have been recommended (and that primarily for agglomerative procedures). In [288] however, F. Murtagh proposes a probability model of random dendrograms, which enables researchers to test the randomness of hierarchical structures in data set; and further steps have been done into this direction.

The random graph model also supplies the scientist with significance tests for testing the randomness of clusters in dendrograms without needing a special model of randomness of dendrograms. The preposition of a complete random choice of the N edges is equivalent to the assumption that the ranking of distances or similarities is completely random. Thus, R.F. Ling defined an isolation index for clusters on the basis of the hierarchical method described in Subsection 3.1.3 (see Definition 3-8 and Remark 3-2). For the hypothesis of random ranking of distances, he could derive the exact distribution of isolation index of $(1, d)$-clusters which follows from the hypergeometric distribution. For (k, d)-clusters, he gives asymptotic and approximate results ([38], [42], [148], [259], [261]).

4.2 GRAPH-THEORETIC MODELS OF CLASSIFICATION

4.2.1 THE MODEL OF R.F. LING

In chemistry and physics, scientists used gaph-theoretic models very early for describing structures ([69], [70], [71], [328], [405]). Obviously, not only chemical structures, but also relations between objects of other data sets can be described by graphs in the same way. This gave way to several cluster definitions as we described in Subsubsection 2.2.4.4 and Chapter 3.

As a consequence, the potential of the theory of random graphs as a tool to develop statistical tests for the randomness of clusters had been discovered soon after random graphs had been introduced in mathematics. C. Abraham and R.F. Ling used the analogy of single-linkage clusters or k-clusters, respectively, and certain subgraphs of the graph $\Gamma(d)$, to derive such test statistics. These tests are based on the probability models of random graphs proposed by N. Gilbert, and P. Erdős and A. Rényi. Ling's k-clusters from Definition 3-4 result from a distance matrix $\mathbf{D}$ or from the matrix Δ of ranks of distances: Every triple $(\mathcal{S}, \mathbf{D}, k)$ or $(\mathcal{S}, \Delta, k)$ determines a hierarchy of k-clusters, each quadruple $(\mathcal{S}, \mathbf{D}, d, k)$ or $(\mathcal{S}, \Delta, N(d), k)$ determines a class of k-clusters of level d (see [259], [261]). Hence we are not interested in the probability model of the data vectors $\vec{x}_{1*}, \ldots, \vec{x}_{n*}$, we merely need assumptions about the probability space $(\Omega_n, \mathcal{A}_n, P)$ of the distance matrices or rank matrices, respectively.

R.F. Ling makes presupposes, that all $\binom{n}{2}$ distances between the objects of a sample to be clustered are realizations of random variables and that homogeneity within a data set can be described by the following model:

$$(4\text{-}4)\text{-} \qquad D_{ij} = a + \epsilon_{ij} \qquad (1 \le i < j \le n).$$

The random variables D_{ij} — with realizations $d_{ij} = d(\vec{x}_{i*}, \vec{x}_{j*})$ from the data — consist of a positive constant a and continuous, indepent, identically distributed random variables ϵ_{ij} with expectation $E\,\epsilon_{ij} = 0$. Because of the assumption of continuity, $P(D_{ij} = D_{kl}) = 0$ for $i \ne k$ or $j \ne l$. Thus, with probability 1, we get $\binom{n}{2}$ different distances between the n objects, which can be arranged in $\binom{n}{2}!$ different ways above the main diagonal of the distance matrix. Under Assumption (4-4), $\binom{n}{2}!$ different rank matrices are possible, each of them with the same probability. Thus, for homogeneous data,

$$(4\text{-}5) \qquad P(\Delta) = \frac{1}{|\Omega_n|} = \frac{1}{\binom{n}{2}!}$$

is the probability to arrange all distances to a special symmetric distance matrix (or to get a special rank matrix Δ of the distances). Formula (4-5) remains true if the claim of different distances is abandoned and the ϵ_{ij} are not continuous random variables. (In this case, successive ranks are attached to distances with equal values by randomisation.) R.F. Ling justifies his model in [259] and [261] as follows:

(a) It is the simplest and therefore the most practicable probability model.

(b) The order of distances is invariant to monotonic transformations of the distances.

(c) Virtually no a priori information about the data to be classified is needed.

(d) Under the assumption of this model, it is possible to transform the problem of classification to a problem about the shape of random graphs.

According to [111], [112], [113], a random graph Γ_{nN} is a graph with n labelled vertices and N edges which have been drawn randomly from all $\binom{n}{2}$ possible ones. Now, the association to random graphs is as follows. There are $\binom{n}{2}!$ different ways to draw the $\binom{n}{2}$ edges in a graph one after the other. The probability to get a special graph with N given edges then is

$$(4\text{-}6) \qquad P(\Gamma_{nN}) = \binom{\binom{n}{2}}{N}^{-1}.$$

This is the same as the probability to get a certain rank matrix Δ of distances of the data to be classified, where the N smallest ranks δ_{ij} are attached to N given pairs (O_i, O_j) of objects, that is, to N given places (i, j) in Δ. (Equivalently we can say: The probability to get a certain graph $\Gamma_{n,N(d)} = \Gamma(d)$ is the same as the probability to get a distance matrix $\mathbf{D}$ with the $N(d)$ distances d_{ij}, for which $d_{ij} \leq d$ holds, located at given places.) Let us cite R.F. Ling:

> "In my formulation of the k-clustering problem, if S is the set of vertices $V_1, V_2, \ldots, V_n$ and the ranks $(1, 2, \ldots, N)$ in Δ correspond to the N edges to be drawn, then the probabilities associated with the properties of a k-cluster in (Ω_n, Σ_n, P) are identical to those probabilities associated with the subgraph in Γ_{nN} that corresponds to the k-cluster. The correspondence of the probability models can be seen as follows: In constructing clusters, the links between objects are considered in order of their ranks; thus assuming the ranks are randomly permuted is equivalent to adding edges randomly to the graph."

Many results on random graphs can be applied directly to prove hypotheses on complete-linkage clusters, and k-clusters if Assumptions (4-4) and (4-5) can be accepted; especially the results from [111], [112], [113], [264], [265], [271], [273], [294], [351] can be immediately applied to this theory of cluster analysis (see [38], [42], [224]). In fact, R.F. Ling proved exact or asymptotic results of random graphs only for single-linkage clusters — his 1-clusters or $(1, d)$-clusters for thresholds d of similarity. For $k > 2$, he used Monte Carlo simulations to derive approximative results. In [148], we computed exact results for the case of 2-clusters, where any point must be connected to at least two other points of a cluster in order to belong to that group.

The following example shows how statistics to test the hypothesis of randomness of clusters can be derived. Under the assumption of real-valued, continuous measurements, all $\binom{n}{2}$ distances d_{ij} are different $(1 \leq i < j \leq n)$. Let d_N be the N-smallest distance $(1 \leq N \leq \binom{n}{2})$. A classification into single-linkage clusters of level d_N, which results from drawing the first N edges, may contain one-element classes (or isolated vertices) — especially if N is small. Let X be the number of one-element classes in a classification. Then $Y = n - X$ is the number of objects lying in the union of all single-linkage

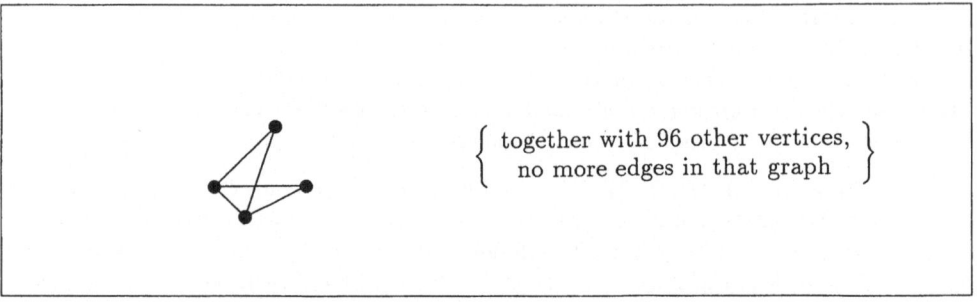

Figure 4-2a $P_{100,5}(Y = 4)$: *The first five edges connect four vertices (one single "cluster" immediately at the beginning of the edge-drawing process).*

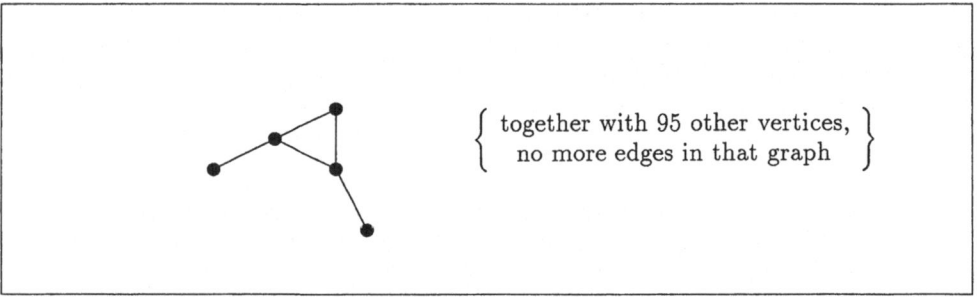

Figure 4-2b $P_{100,5}(Y = 5)$: *The first five edges connect five vertices (one single "cluster" immediately at the beginning of the edge-drawing process).*

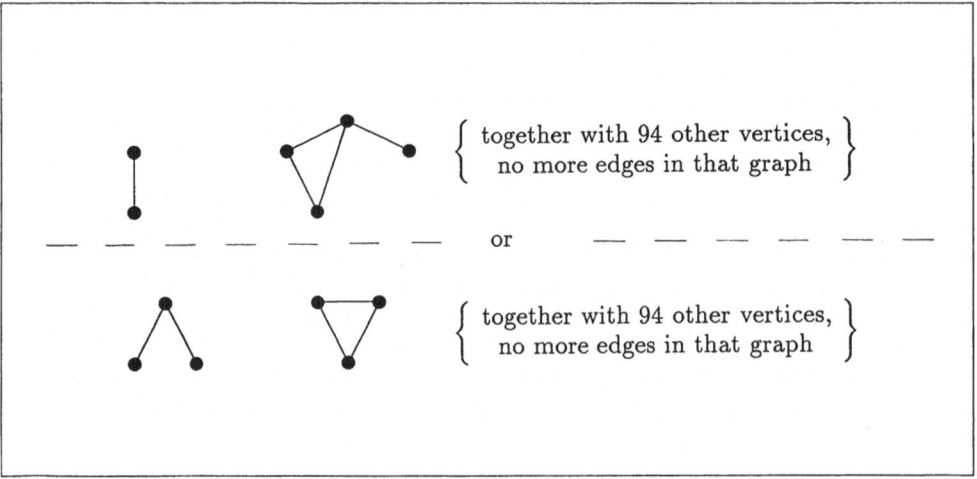

Figure 4-2c $P_{100,5}(Y = 6)$: *The first five edges are distributed between six vertices (two "clusters" immediately at the beginning of the edge-drawing process).*

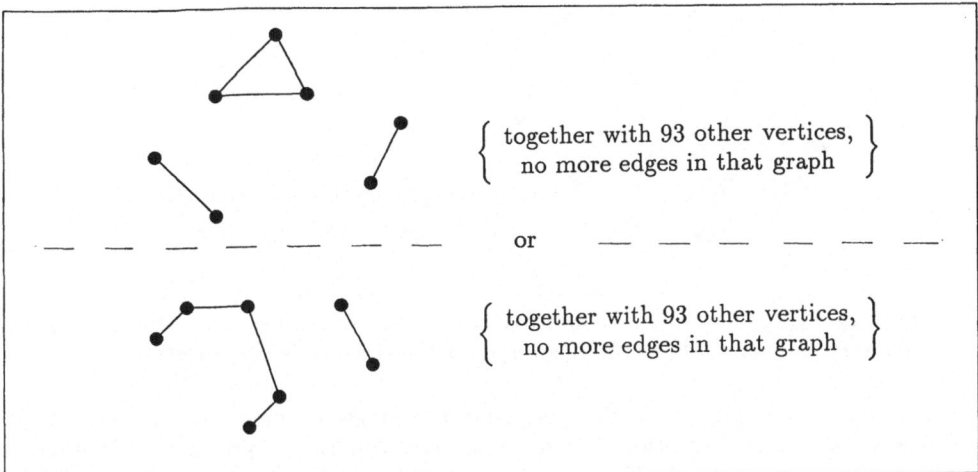

Figure 4-2d $P_{100,5}(Y = 7)$: *The first five edges are distributed between seven vertices (two or three "clusters" immediately at the beginning of the edge-drawing process).*

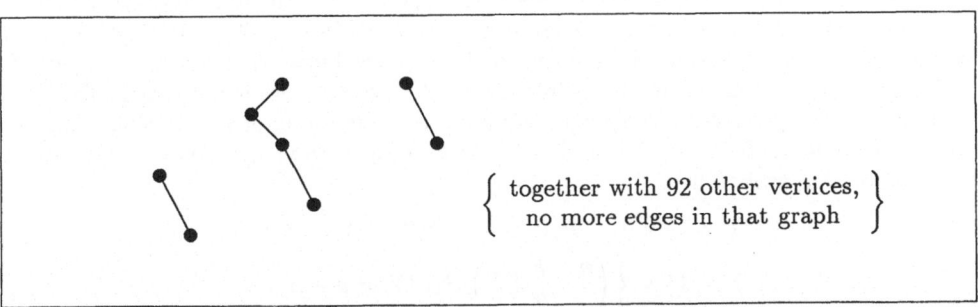

Figure 4-2e $P_{100,5}(Y = 8)$: *The first five edges are distributed between eight vertices (three "clusters" immediately at the beginning of the edge-drawing process).*

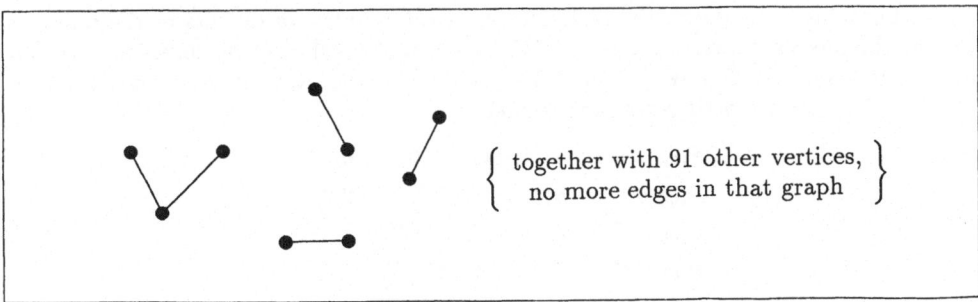

Figure 4-2f $P_{100,5}(Y = 9)$: *The first five edges connect nine vertices to three pairs and one three-point tree (four "clusters" immediately at the beginning of the edge-drawing process).*

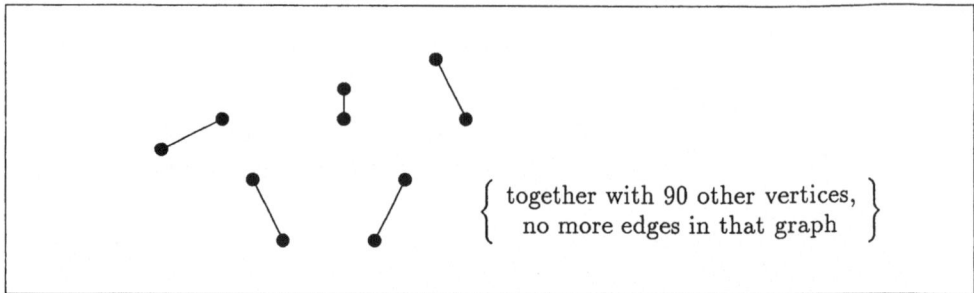

Figure 4-2g $P_{100,5}(Y = 10)$: *The first five edges connect ten vertices to five pairs (five "clusters" immediately at the beginning of the edge-drawing process).*

clusters of level d_N which contain more than one single element (which are $(1, d_N)$-clusters according to Definition 3-3; we sometimes call them "proper" single-linkage clusters). In other words: By Y we denote the number of non-isolated vertices in $\Gamma(d)$. Then for Y, we have

$$l_N := \min\left\{ l \in \mathcal{N}_0 : \binom{l}{2} \geq N \right\} \leq Y \leq 2N$$

(see [259], [261]). Under (4-4), the values of the $\binom{n}{2}$ different distances d_{ij} are attached at random as weights to all pairs of objects from $S = \{O_1, \ldots, O_n\}$. That is, the first N smallest distances are distributed randomly as edges between the objects. Thus, X and Y are integer-valued random variables. By $P_{nN}(Y = k)$, we denote the probability that exactly k objects of an n-element sample are in the union of proper single-linkage clusters of a level d_N. In [259] and [261], R.F. Ling proved the following recursion formula for this probability:

$$P_{n1}(Y = 2) = 1, \quad P_{n1}(Y = k) = 0 \quad \text{for} \quad k \neq 2,$$

$$P_{n,N+1}(Y = k) = \left\{ \binom{n - k + 2}{2} P_{nN}(Y = k - 2) \right.$$

$$+ (k - 1)(n - k + 1) P_{nN}(Y = k - 1)$$

$$+ \left. \left(\binom{k}{2} - N \right) P_{nN}(Y = k) \right\} \left\{ \binom{n}{2} - N \right\}^{-1}.$$

Other formulas to compute the distribution of the number of objects in the union of proper clusters have been derived in [111], [148] and [149]. Let us consider a sample of $n = 100$ objects. For $N = 5$, $4 \leq Y \leq 10$ holds true, and the distribution of the number Y of non-isolated vertices is defined by

$$P_{100,5}(Y = 4) = 0.000\,000\,000\,951\,931,$$
$$P_{100,5}(Y = 5) = 0.000\,000\,676\,252,$$
$$P_{100,5}(Y = 6) = 0.000\,076\,2535,$$
$$P_{100,5}(Y = 7) = 0.002\,856\,25,$$
$$P_{100,5}(Y = 8) = 0.044\,271\,9,$$
$$P_{100,5}(Y = 9) = 0.290\,929,$$
$$P_{100,5}(Y = 10) = 0.661\,865.$$

Attaching the five smallest distances at random as weights to $\binom{100}{2} = 4\,950$ pairs of objects most probably results in 90 isolated objects and five two-element "groups". Now the distributions of X or Y can be used to test the null hypothesis of homogeneity, that is, to test the relevance of (4-4). The idea proposed by R.F. Ling is as follows. In an n-element sample with the first N smallest distances drawn, let k objects be in the union of 1-clusters (and the remaining $n - k$ objects form as many one-element groups). If now $P_{nN}(Y \leq k) \leq \alpha$ holds for a given error probability α of the first kind, then the null hypothesis (4-4) of a random attachment of distances to pairs of objects can be rejected at this level α.

Since every clustering algorithm suggests clusters, R.F. Ling calls the groups resulting from a classification procedure "cluster candidates". If, however, the null hypothesis of homogeneity of the data can be rejected by a test and a class structure of the data is accepted, then at least the larger cluster candidates are considered as real groups. These groups are sometimes called "significantly compact". From the probabilities above, we get $P_{100,5}(Y \leq 8) < 0.05$. This means that at a level $\alpha = 0.05$, the null hypothesis of homogeneous data can be rejected if in a sample of 100 elements the five smalles distances are distributed between at most seven objects as shown in Figures 4-2a–4-2e. Only Figures 4-2f–4-2g show clusters for which the null hypothesis cannot be rejected.

The formulas to compute the exact values of $P_{nN}(Y = k)$ are computationally tractable for small values of N only. For larger ones, they become very intricate and time-consuming. If the sample size n is not too small, however, then we can rely on well-known asymptotic results from graph theory. If we choose $N \approx n$, then X and Y both are asymptotically normally distributed; for $2N/(n \log n) \to 1$, the number X of isolated vertices is asymptotically Poisson distributed ([111], [112], [149]). These results can be used in the same way as shown above to derive test statistics for testing (4-4). Obviously, the **power** of such tests is small for very small values of N as well as for very large ones. Unfortunately, until now nothing has been published on the power of such tests, and especially on how to choose the number N to get maximal power. Therefore, R.F. Ling suggests to use the distribution of the isolation indices $i(\mathcal{C})$ of 1-clusters to test the hypothesis of homogeneous data. This distribution can be derived from the hypergeometric distribution (see [148], [259]).

For weak 2-linkage or 3-linkage clusters, few is known about the distributions of $i(\mathcal{C})$, X and Y. For 2-clusters, we gave in [148] the exact distributions of X and Y, but these formulas are too clumsy to be of practical value. (Nevertheless, they may be of useful to derive asymptotic results.) Further investigations into the distributions of random variables, which are relevant for 2- and 3-clusters, have been made by R.F. Ling using Monte Carlo simulations ([259]).

4.2.2 A Probability Model Based on Random Multigraphs

As mentioned in Chapter 2, it is troublesome to calculate similarities s_{ij} or distances d_{ij} for mixed data (see Subsection 2.2.2). For methodological reasons, it may be useful to combine some attributes to blocks (layers) and to calculate a matrix $\mathbf{D}_l = (d_{ijl})$ of "local" dissimilarities (or a matrix $\mathbf{S}_l = (s_{ijl})$ of "local" similarities, respectively) for every such block ($l = 1, \ldots, t$). Doing this, we can merge all binary attributes of the raw data into one block and compute a local distance matrix for this block using the matching coefficient. All continuous, interval scaled measurements form another block from which the Euclidean distances between the objects are computed. Every attribute can form a block by its own if it is intended. The symmetric matrices $\mathbf{D}_1, \ldots, \mathbf{D}_t$ then are arranged to a **distance tensor** $\mathbf{D} = (d_{ijl})$. In our classification model, for a given random sample $\mathcal{S}$ and dissimilarity tensor $\mathbf{D}$, the class of weak k-linkage clusters of level $(\vec{d}^T; s)$ is uniquely determined by a quintuple $(\mathcal{S}, \mathbf{D}, \vec{d}^T, k; s)$, and $(k, \vec{d}^T; s)$-clusters can be interpreted as subgraphs of a multigraph Γ_{tnN} with $N = N(\vec{d}^T)$ edges, as Definitions 3-8-3-12 show — an edge κ_{ijl} is in Γ_{tnN} if and only if $d_{ijl} \leq d_l$.

For the following, we assume that homogeneity of a sample $\mathcal{S}$ can be described by a random order of the N smallest distances out of a total of $t\binom{n}{2}$ distances in the distance tensor $\mathbf{D}$. This corresponds to the assumption, that in the multigraph Γ_{tnN}, the $N(\vec{d}^T)$ edges are drawn at random. This is a generalization of Ling's model, and arguments for our assumption are similar to those of the previous subsection. (Sometimes it may be better to normalise the matrices $\mathbf{D}_1, \ldots, \mathbf{D}_t$ to get a threshold $(d, \ldots, d)$ instead of $(d_1, \ldots, d_t)$.) Hence, under the assumption of homogeneity of the data, the probability to get a multigraph with the vertices $\xi_1, \ldots, \xi_n$ and $N(\vec{d}^T)$ given edges is

$$(4\text{-}7) \qquad P(\Gamma_{tnN}) = \binom{t\binom{n}{2}}{N}^{-1},$$

because now N edges out of a total of $t\binom{n}{2}$ edges can be chosen.

By this assumption, the random graph of the previous subsection is generalized as follows: A **random multigraph** Γ_{tnN} is a multigraph with n labelled vertices — each consisting of t layers, thus allowing t distinct edges to link two vertices — in which the edges are chosen at random from the $t\binom{n}{2}$ possible ones. We have $(t\binom{n}{2})!$ different ways to draw the edges one after the other. From this, (4-7) immediately can be derived as the probability to have N given edges at the first N places (or to get a distance tensor $\mathbf{D} = (d_{ijl})$ with the $N = N(\vec{d}^T)$ smallest distances located at given places, see below). Hence, if we use Definitions 3-8-3-12 as cluster definitions, then we can adopt results on random multigraphs in cluster analysis (see [152], [154], [155]). Such results have been derived in [149] and [153]. Further theorems will be proved in the following chapter.

A justification for (4-7) is that, in case of homogeneous data, the computed distances or dissimilarities are realizations of random variables D_{ijl}, for which the condition

$$(4\text{-}8) \qquad D_{ijl} = a + \epsilon_{ijl} \qquad (1 \leq i < j \leq n, 1 \leq l \leq t)$$

holds true. (Here, a is a positive constant and the ϵ_{ijl} are independent, continuous, identically distributed random variables with $\mathrm{E}\epsilon_{ijl} = 0$.) In this case we get $(t\binom{n}{2})!$ different symmetric rank tensors Δ in Ω_n, with the same probability to be chosen, namely

$$(4\text{-}9) \qquad P(\Delta) = \frac{1}{|\Omega_n|} = \frac{1}{(t\binom{n}{2})!}.$$

The probability that the N lowest ranks take N given places regardless of their order is the same as in (4-7). (Formulas (4-7)–(4-9) are generalizations of (4-4)–(4-6).)

At first glance, (4-8) does not seem to be very realistic: The global constant a for all blocks makes this model even less acceptable than Ling's model (4-4). However, we can substitute this condition with

$$D_{ijl} = a_l + \epsilon_{ijl} \qquad (1 \leq i < j \leq n, 1 \leq l \leq t)$$

for the original data (with positive constants $a_1, \ldots, a_t$ and independent, continuous random variables ϵ_{ijl}, which are from the same family of distributions). This substitution follows from a simple transformation of the elements of every distance matrices $\mathbf{D}_l$. Again the preposition of continuous random variables D_{ijl}, or ϵ_{ijl}, respectively, is needed to get $t\binom{n}{2}$ different distances d_{ijl} with probability 1 and thus to keep (4-9) to be true. Again, we can abandon this preposition if equal distances are arranged in an ascending order by randomisation.

For the related multigraph, the occurrence of such **ties**, that is, of several distances of equal size, means that the edges corresponding to distances of equal value can be drawn simultaneously or in a random order. For the purpose of exploratory data analysis, it makes no difference whether one chooses the first or the second procedure if ties occur. For our probability model, however, all ranks must be different in order to get equiprobable rank matrices, or rank tensors of distances. Therefore, if ties occur, then the ranks must be attached randomly to those distances which have the same numeric values. In Figure 2-11, for example, the distances between the three objects are $d_{12} = 2.5$, and $d_{13} = d_{23} = 2.8$. Thus, the rank of d_{12} is $\delta_{12} = 1$. By some random mechanism we have to decide, whether d_{13} or d_{23} should get rank no. 2. The other distance will get rank no. 3.

For continuous measurements the probability of ties always is zero — at least theoretically. Nevertheless, the procedure of attaching ranks to ties randomly is common practice for most nonparametric procedures and no specific disadvantage of our graph theoretic model.

4.3 DISCUSSION OF THE GRAPH-THEORETIC PROBABILITY MODELS

Graph-theoretical models have certain advantages, when they are used in classification theory (see Section 2.3): Clusters have well-defined properties; probability models for testing null hypotheses of randomness of clusters are simpler than other models, which are based on numeric-matrix orientated models; results are invariant to monotonic transformations of distance measures; the simplicity of combinatorial models — especially of the multigraph model — allow a fairly good approach to the classification of mixed data; no a priori information on the classes is required. However, in most real situations, R.F. Ling's model (4-5) is not satisfied to some extent. The assumption, that is most often violated, is the equiprobability of the rank matrices Δ. The lack of uniformity of the distribution could be implied by the fact that the random variables D_{ij} $(1 \leq i < j \leq n)$ were statistically dependent. If the d_{ij} are actual metric distances instead of dissimilarities, then small values of d_{ij} and d_{jk} imply small values of d_{ik} through the triangle inequality. (For metric data, in most cases the Euclidean or L^2-distance is preferred, sometimes cluster analyses are based on the city-block or on the Chebyshev distance, rarely dissimilarity functions are used.) Another discrepancy between the model assumptions and real data is that certain matrices of Ω_n cannot occur, a known phenomenon if d_{ij} is a distance measure on points in a low dimensional space. In contrast to (4-5), for distances only those rank matrices Δ have positive probabilities, which are derived from matrices $\mathbf{D} = (d_{ij})$ with

$$d_{ij} + d_{jk} \geq d_{ik}.$$

Thus, we should choose the set Ω_n' of **admissible** rank matrices as the universe of a probability model instead of Ω_n. The same holds if we generate the rank matrices from similarities s_{ij} instead of distances d_{ij}. Here, (2-1d) is the analogon of the triangle inequality (2-2d). In [108], M. Eigener showed that Ω_n' has far less elements than Ω_n. Moreover, he proved that the assumption of equiprobability on Ω_n' is wrong at least if Euclidean distance is used for classification. Especially for classification theory, the following three results of his thesis are of interest. They have been proved, however, only for real, one-dimensional random variables $X_1, \ldots, X_n$.

Theorem 4-1 *Let S be an n-element set of points and d the Euclidean distance. Furthermore, let Ω_n' be the set of all admissible permutations of the elements of a distance matrix, that is, the set of all admissible rank matrices. Then, for $m_n = |\Omega_n'|$, we have*

$$(4\text{-}10a) \qquad\qquad m_3 = 6 = \binom{3}{2}!,$$

$$(4\text{-}10b) \qquad\qquad m_4 = 120 < 720 = \binom{4}{2}!,$$

and generally,

$$(4\text{-}10c) \qquad\qquad m_n \leq 2n \left(\frac{\binom{n-1}{2} + n - 4}{n - 3} \right) m_{n-1} < \binom{n}{2}!. \; \bullet$$

Theorem 4-2 *The cardinality of the set of rank matrices, which are admissible under the Euclidean distance, is very small as compared to the number of all rank matrices. We have*

$$\lim_{n \to \infty} \frac{m_n}{\binom{n}{2}!} = 0. \; \bullet$$

Theorem 4-3 *Let $a_n = m_n / \binom{n}{2}!$. Then the convergence rate of the sequence $(a_n)_{n \to \infty}$ is given by*

$$\frac{a_{n+1}}{a_n} < \frac{1}{(n-1)!}. \; \bullet$$

From Theorem 4-3, we can expect an extremely fast convergence towards zero. Thus, the number of rank matrices or random graphs admissible under the triangle inequality, grows very slowly in n as compared to the number of all possible rank matrices. Moreover, M. Eigener shows in [108] that the assumption of equiprobability is wrong on the spaces (Ω_n', Σ_n) of admissible rank matrices at least for continuous, one-dimensional data. The computation is rather cumbersome even for sample sizes n as small as $n = 6$. No closed formula for this distribution is known today — even if we simplify the model.

Thus, M. Eigener's results cannot be used to improve Ling's model or to derive exact statistical tests on the structure of a data set in a metric space. This must hold even more because no investigations have been made whether the inequality (4-10c) of the number of admissible rank matrices holds for vector-valued random variables $\vec{X}_{1*}, \ldots, \vec{X}_{n*}$ or not. It may be possible then that more rank matrices are allowed.

Thus, the disadvantages of R.F. Ling's graph-theoretic probability model seem to be serious at least for continuous, one-dimensional data. It is hasty, however, to reject graph-theoretic procedures as being not suitable to derive tests of homogeneity, since for multidimensional data no analogon of Theorems 4-1 and 4-2 is known. (Apart from this, for one-dimensional data better classification methods than graph-theoretic ones exist. Graph-theoretic based methods are used to preferably describe the structure of a sample of multidimensional data.) Computational problems and incorrect results are known also for other classification models. For the gap test, for example, the probability distributions strongly depend on assumptions on the domain $\mathcal{G}$, where the data are drawn from, and are expected to be uniformly distributed under the null hypothesis. Moreover, nearly all parametric test procedures are not robust. They are very sensitive against violations of assumptions on the distribution of the data under the null hypothesis. The gap test cannot be applied if the data are not uniformly, but normally distributed. (For a normal distribution, we expect large gaps in regions which are apart from the mean; thus large gaps can occur in homogeneous data sets under the unimodality hypothesis.) This property of parametric tests is detrimental if we are involved in judging the randomness of clusters of probably ramified shapes.

The disadvantages of Ling's model hold true also for each block and each matrix of local distances in our classification model of multigraphs. The differences between reality and model assumptions become smaller, however, if we consider all blocks at once. The $(k, \vec{d}^T; s)$-clusters as well as the strong k-linkage clusters, the k-overlap clusters or the complete-linkage clusters of a given level $(\vec{d}^T; s)$ are defined as certain subgraphs of the s-projection of the multigraph Γ_{tnN} with $N = N(\vec{d}^T)$ edges. For the s-projection, however, the triangle inequality will not hold true for $s < t$ (see Section 3.3). Thus, for a matrix $\mathbf{D}_{(s)}$ the triangle inequality will not hold even if it is valid for all matrices $\mathbf{D}_1, \ldots, \mathbf{D}_t$; for $s < t$, $d_{(s)}$ generally is a dissimilarity function even if all d_l are distance functions. Thus it may be possible for two pairs (O_i, O_j), and (O_j, O_k), respectively, to be very similar in s or more blocks of attributes, and the pair (O_i, O_k) being similar in less than s blocks. That means that O_i and O_j are connected by an s-fold connection as well as O_j and O_k are; O_i and O_k, however, are **not** connected by an s-fold connection (see Figure 3-6). Thus, for the s-projection $\tilde{\Gamma}_{nv}$ of the original multigraph $\Gamma_{t,n,N(\vec{d}^T)}$,

$$(4\text{-}11) \qquad P(\tilde{\Gamma}_{nv}) = \left(\binom{n}{2} \atop v \right)^{-1},$$

that is, the randomness of drawing edges can be easier adopted as null hypothesis of randomness of clusters than (4-6). Thus, our multigraph model of $(k; s)$-clusters provides more advantages than Ling's simple graph model: Results for ordinary random graphs can be applied to cluster analysis even if the similarity functions for the t blocks are based on (local) distances, that is, if the triangle inequality holds. They now yield conditional tests for the randomness of clusters under the condition that exactly v s-fold connections have been drawn in $\Gamma_{t,n,N(\vec{d}^T)}$.

Compared with the simple graph model, the multigraph model provides greater flexibility, that is to say that single (local) distance thresholds for each dimension or for blocks are easier defined than one global distance threshold. Moreover, our model tolerates that objects from the same cluster may differ in some variables; they only have to be similar in at least s blocks or variables. (This situation occurs in medical inquiries.) We recommend the construction of $(k, \vec{d}^T; s)$-clusters by s-projections especially for mixed data. In this case, attributes of the same scale level can be merged into blocks or layers of the multigraph. This yields finer gradings in the dissimilarities of each block, and thus gets closer to the continuity assumption (4-8) for distances.

Remark 4-1 The value of v in Formula (4-11) is the realization of a random variable V, the number of s-saturated connections in a random multigraph Γ_{tnN}. (4-11) gives the conditional probability to get a certain s-projection $\tilde{\Gamma}_{nv}$ with v s-fold connections by distributing the N edges. •

Remark 4-2 For given threshold vector $\vec{d}^T$, the number $N = N(\vec{d}^T)$ of distances (or dissimilarities) $d_{ij1} \leq d_1, \ldots, d_{ijt} \leq d_t$ also is the realization of a random variable. •

Remark 4-3 Especially for hierarchical procedures like those of Section 3.3, for which some components of the threshold vector remain fixed or are not changed simultaneously, another probability model of random multigraphs appears to be more suitable. In this model, not only the total number of edges to be drawn is given but also the number $N_l = N_l(d_l)$ of edges in every layer, $N_1 + \cdots + N_t = N$. Some investigations in this model showed that the results do not asymptotically differ from those derived in the following chapter if all N_l tend to ∞ with the same rate for growing vertex numbers n. The problems of applying graph-theoretical results to distance matrices, however, now occur for each matrix of local distances like in Ling's model. Therefore, we did not study this model to a deeper extent. •

CHAPTER 5

PROBABILITY THEORY OF COMPLETELY LABELLED RANDOM MULTIGRAPHS

The roots of graph theory are obscure. The famous eighteenth-century Swiss mathematician Leonard Euler was perhaps the first to solve a problem using graphs when he was asked to consider the problem of the Königsberg bridges (in the 1730s). Problems in (finite) graph theory are often enumeration problems, and thus can become rather intricate to be solved. However, in the late 1950s and early 1960s the Hungarian mathematicians Paul Erdős and Alfred Rényi founded the theory of random graphs and used probabilistic methods (limit theorems) to by-pass enumeration problems. These problems also became secondary with the emergence of powerful computers. Thus, perhaps no topic in mathematics has enjoyed such explosive growth in recent years as graph theory. This stepchild of combinatorics and topology has emerged as a fascinating topic for research in its own right. Moreover, during the last two decades, calculus of graph theory has proved to be a valuable tool in applied mathematics and life sciences as well. Using graph-theoretic concepts, scientists study properties of real systems by modelling and simulation. The aim of graph-theoretic investigations is, in fact, the simplest topological structure after that of isolated points: The structure of a graph is that of "points" or "vertices", and "edges" or "lines".

There are, however, different approaches to this mathematical topic. This led to different ways to define and to generalize the term "graph". That is, terms like "graph", "multigraph" or "pseudograph" may bear different meanings, depending on the author's intentions. Therefore, we will define here what *we* mean by these terms. A **graph** Γ is a finite nonempty set $\mathcal{G} = \{\xi_1, \ldots, \xi_n\}$ together with a (possibly empty) set $\mathcal{H}$ of two-element subsets of distinct elements of $\mathcal{G}$, that is, $\Gamma = (\mathcal{G}, \mathcal{H})$ with $\mathcal{H} \subseteq \mathcal{K} = \{\kappa_{ij} = \{\xi_i, \xi_j\} : 1 \leq i \leq n, 1 \leq j \leq n, i \neq j\} = \{\kappa_{ij} = (\xi_i, \xi_j) : 1 \leq i < j \leq n\}$. The elements of $\mathcal{G}$ are called "vertices", and the elements of $\mathcal{H}$ are called "edges" or "lines"; $\mathcal{G}$ is the "vertex set" and $\mathcal{H}$ is called "edge set" (see also Section 1.3 and Chapter 3). A graph can be conveniently depicted as a diagram where the vertices appear as small circular dots and the edges are indicated with line segments joining two appropriate dots.

In a directed graph or **digraph** $\Delta = (\mathcal{G}, \mathcal{H})$, the edge set $\mathcal{H}$ is defined as $\mathcal{H} \subseteq \mathcal{D} = \{\delta_{ij} = (\xi_i, \xi_j) : 1 \leq i \leq n, 1 \leq j \leq n, i \neq j\}$. In contrast to an (undirected) graph, the elements of $\mathcal{H}$ here are **ordered** pairs; they are called "arrows" or "arcs". The arc δ_{ij} shows from ξ_j to ξ_i (to indicate the order of the vertices of an edge or its direction, arrows are used instead of line segments when a digraph is pictured as diagram).

In both types of graphs, edges connecting a vertex with itself, so-called **loops**, or **slings**, are prohibited, and only one edge can link two different vertices ([180], [254]). Thus,

$$0 \leq N = |\mathcal{H}| \leq n(n-1) \qquad \text{in a digraph,}$$

$$0 \leq N = |\mathcal{H}| \leq \binom{n}{2} = \frac{1}{2}n(n-1) \qquad \text{in a graph}$$

must hold true. We get **pseudographs** if we admit loops. In **multigraphs**, more than one line can connect the same pair of vertices. Graphs, di-, pseudo- and multigraphs are **labelled** if all vertices and edges are labelled uniquely. In graphs, di- and pseudographs, the edges usually are labelled as pairs $\{\xi_i, \xi_j\}$ or (ξ_i, ξ_j), or by double indices (like κ_{ij} or δ_{ij} as is done above). In undirected graphs, the lower index is put in front of the higher one without restriction (see Chapters 0 and 1). In multigraphs, authors often do not distinguish between the edges connecting the same pair of vertices. These edges, however, can be labelled by a third index, as is done in our model.

Graphs and digraphs can be used to model many real systems. This usually works if the vertices can be identified as physical objects, and the edges can be identified as relations or equations between the objects. In fact, every binary structural relation can be described by a graph (thus it was a natural consequence to use graphs in classification theory; here, the similarity between the objects defines the relation). The advantage of using graphs when describing and modelling structures is based on the fact that researchers can rely on a plethora of mathematical theorems and results to gain insight into the structure of a real system.

Graph theory provides researchers of different topics with a single, mathematical language. Thus, it promotes the exchange of ideas to a degree which probably would be impossible if scientists would only rely on the technical terms of their own science. This is mirrored by the fact that many theoretical results in graph theory have not been found by pure mathematicians, but by scientists working in applied mathematics and research (see [1], [231], [259], [261], [263], [264], [265], [270], [272], [292], [322], [328], [390], [405], [420]). The problems and results for random multigraphs presented in this chapter, also have all been motivated by problems from classification theory — especially from medical research. We wanted to represent similarity structures between objects more subtle than by one single distance measure. We furthermore wanted to generate hypotheses on the homogeneity of samples which are suitable for mixed data, too ([152], [154], [155], [158]).

5.1 DEFINITIONS AND NOTATION

In Subsection 3.2.1, we defined a certain type of multigraph which can be used in cluster analysis. The notation in that section has been introduced by us originally in 1980 (see [149], [150], [152], [153], [154], [155]); until now, however, this concept is not commonly used. Therefore — and to prevent the reader from being confused with other definitions of multigraphs — the notation is repeated in this section as a series of definitions, together with other terms needed. If we speak of **graphs** or **multigraphs** in this chapter, then we always mean **undirected graphs** or **undirected multigraphs**. Let us start with the Definition of the undirected, completely labelled multigraph from Subsection 3.2.1.

Definition 5-1 For $t, n \in \mathcal{N}$, let $\mathcal{G} := \{\xi_1, \ldots, \xi_n\}$ be a nonempty set and $\mathcal{K} := \{\kappa_{ij} = (\xi_i, \xi_j) : 1 \le i < j \le n\}$ be the set of all two-element subsets (ordered pairs) from $\mathcal{G}$. Furthermore let $\mathcal{K}_t := \mathcal{K} \times \{1, \ldots, t\} = \{\kappa_{ijl} = ((\xi_i, \xi_j), l) : 1 \le i < j \le n, 1 \le l \le t\}$. For every subset $\mathcal{H} \subseteq \mathcal{K}_t$, the tuple $\Gamma_t := (\mathcal{G}, \mathcal{H})$ is called an **undirected, completely labelled multigraph** (of order n). The elements of $\mathcal{G}$ are the **vertices**, and the elememts of $\mathcal{H}$ are the **edges** of this multigraph Γ_t. Every nonempty set $\mathcal{E}_{ij} := \{\kappa_{ij1}, \ldots, \kappa_{ijt}\} \cap \mathcal{H}$ is called a **connection** between ξ_i and ξ_j. •

This definition does not allow loops. Furthermore, for undirected multigraphs, $\kappa_{ijl} = \kappa_{jil}$ and $\mathcal{E}_{ij} = \mathcal{E}_{ji}$ hold true. The simple undirected, labelled graph Γ — as defined in Section 1.3 and at the very beginning of this chapter — is equivalent to the case $t = 1$. Thus, for the remainder of this booklet we do not differ between Γ and Γ_1. For $t > 1$, two vertices can be directly conncted by at most t distinct edges. We can picture an undirected, completely labelled multigraph Γ_t as t undirected labelled graphs Γ, each with the same vertex set, stacked in layers one upon the other, and N_l edges per layer, $N_1 + \cdots + N_t = N$ (see Figures 3-4 and 5-1). Every layer of Γ_t can contain at most $\binom{n}{2}$ edges (if the graph of that layer is complete); thus, $0 \le N := |\mathcal{H}| \le t\binom{n}{2}$ must hold.

We repeated our definition of a multigraph because Definition 5-1 is different from multigraphs, as defined by other authors (see [12], [180]). Usually, neither they label edges, which connect the same pair of vertices, to distingiush between them, nor do they bound the number of these edges by a constant, say t. Moreover, the concept of layers as a labelling scheme is new. It is obvious to define the degree of connectedness of two vertices by the number of edges connecting them. This gives the following definition.

Definition 5-2 Two vertices ξ_i and ξ_j of a multigraph $\Gamma_t = (\mathcal{G}, \mathcal{H})$ are called **s-fold connected** if and only if $|\mathcal{E}_{ij} \cap \mathcal{H}| \ge s$ holds true, that is, if and only if at least s edges connect ξ_i and ξ_j directly. In this case, the connection $\mathcal{E}_{ij}$ is called **s-saturated** (or an **s-fold connection**). •

Obviously, the number of connections in a multigraph Γ_t is bounded by $\binom{n}{2}$. For $t = 1$, a 1-fold connection is the same as an edge. Since the term "s-connected vertices" already has a special meaning for digraphs, we use the more complicated term "s-fold connected vertices" for multigraphs and hope to avoid confusion.

Definition 5-3 The **s-degree** of a vertex ξ_i is the number of vertices of Γ_t which are s-fold connected with ξ_i. Every vertex with an s-degree 0 is called **s-isolated**. •

Definition 5-4 An alternating sequence $\xi_{i_1}, \mathcal{E}_{i_1, i_2}, \xi_{i_2}, \ldots, \xi_{i_{m-1}}, \mathcal{E}_{i_{m-1}, i_m}, \xi_{i_m}$ of m ($m \ge 2$) different vertices and $m - 1$ s-fold connections is an **s-path** (of length $m - 1$) in Γ_t. If additionally for $m \ge 3$, ξ_{i_m} and ξ_{i_1} are s-fold connected, then we speak of an **s-cycle** (of length m). Two different s-paths between two vertices ξ_{i_m} and ξ_{i_1} are **connection disjoint** if they have no s-fold connection in common; they are **vertex disjoint** if they share no common vertices (with the exception of the end points). •

The previous definitions specified properties of vertices of a multigraph Γ_t. Now we define properties of **submultigraphs** of Γ_t. Submultigraphs of Γ_t can be obtained in two different ways, either by removing edges of Γ_t or by removing some vertices together with all those edges having one of these vertices as an endpoint (see Subsection 3.2.1). Submultigraphs obtained by the first method are denoted by a "prime " as exponent, submultigraphs of the second type by a "star". Let $\Gamma_t = (\mathcal{G}, \mathcal{H})$ with $|\mathcal{G}| = n$ and $|\mathcal{H}| = N$. Then for a submultigraph $\Gamma_t' = (\mathcal{G}', \mathcal{H}')$ with m vertices and M edges, $\mathcal{G}' = \mathcal{G}$

and $\mathcal{H}' \subseteq \mathcal{H}$ hold true. A submultigraph $\Gamma_t^* = (\mathcal{G}^*, \mathcal{H}^*)$, however, is given by $\mathcal{G}^* \subseteq \mathcal{G}$ and $\mathcal{H}^* = \mathcal{H} \cap (\mathcal{G}^* \times \mathcal{G}^* \times \{1, \ldots, t\})$. If $M < N$ or $m < n$, that is, if $\mathcal{H}' \subset \mathcal{H}$ or $\mathcal{G}^* \subset \mathcal{G}$, then we speak of **proper** submultigraphs. It is important to note that in our multigraph model, the number of layers, that is, the number of edges which maximally can connect every pair of vertices, remains unchanged in submultigraphs.

The following definitions state properties of submultigraphs. They are stated here for submultigraphs $\Gamma_t^* = (\mathcal{G}^*, \mathcal{H}^*)$ only but they similarly hold true for $\Gamma_t' = (\mathcal{G}, \mathcal{H}')$. As usual, sets with a given property are called **maximal** with respect to this property if no proper superset has this property, too.

Definition 5-5 *Let $\Gamma_t^* = (\mathcal{G}^*, \mathcal{H}^*)$ be a submultigraph of $\Gamma_t = (\mathcal{G}, \mathcal{H})$. Let $|\mathcal{G}^*| = m$ and $|\mathcal{H}^*| = M$. Γ_t^* is s-**fold connected** if either it consists of only one vertex ($m = 1$ and $M = 0$), or any two vertices ξ_i and ξ_j from $\mathcal{G}^*$ are connected by an s-path in Γ_t^*.* •

Definition 5-6 *A maximal, s-fold connected submultigraph Γ_t^* with m vertices and M edges is called s-**component** (of order or size m). An s-component, in which every two vertices are connected by just one s-path, is called s-**tree**. An s-cycle which is an s-component at the same time, is called an **isolated s-cycle**.* •

Remark 5-1 According to Definitions 5-5 and 5-6, s-isolated vertices are s-components and s-trees of a multigraph Γ_t as well. They are called **degenerated** s-components to distinguish them from other — proper — s-components. From Definition 5-6 can be seen that s-trees contain no s-cycles (else two vertices would been connected by more than one s-path). Thus, an s-tree of size m has exactly $m - 1$ s-fold connections. •

Definition 5-7 *An s-fold connected submultigraph Γ_t^* which is maximal with respect to the fact that every vertex is s-fold connected with k other vertices in Γ_t^*, is called a **weak s-component of degree k** or simply a **weak $(k; s)$-component** of Γ_t. A **strong s-component of degree k** (or more simply, a **(strong) $(k; s)$-component**) of a multigraph Γ_t is an s-fold connected submultigraph, which is maximal with respect to the fact that every two vertices in Γ_t^* are connected by at least k different s-paths in Γ_t which are connection disjoint. A $(k; s)$-**block** Γ_t^* is a maximal, s-fold connected subgraph such that for every pair of vertices in Γ_t^* at least k vertex disjoint s-paths exist (and lie completely in Γ_t^*) which connect every two vertices of this submultigraph. An s-**clique** is a maximal submultigraph Γ_t^* such that every two vertices are directly linked together by an s-fold connection.* •

From Definition 5-7 follows that weak and strong $(k; s)$-components of degree k must contain at least $k + 1$ vertices. This also holds true for $(k; s)$-blocks. For $k = 1$, we get in all three cases the proper s-components from Definition 5-6. The vertex sets of different s-components, weak or strong $(k; s)$-components are disjoint. The cut of the vertex sets of two different $(k; s)$-blocks or s-cliques, however, must not necessary be empty. An s-clique of size m has at least $s\binom{m}{2}$ edges because all $\binom{m}{2}$ connections must be s-saturated.

Definitions 5-2–5-7 make sense for natural numbers s with $1 \leq s \leq t$. They generalize the terms "connected", "isolated", "component" etc. which, for $s = 1$, are well known terms in graph theory. We also could have used the s-**projection** of Subsection 3.2.1 to derive the definitions above from those for simple graphs; a submultigraph, for example, is an s-component of a multigraph if and only if its s-projection is a component, see Figures 3-5 and 5-1. (For given s, the s-projection maps an arbitrary multigraph $\Gamma_t = (\mathcal{G}, \mathcal{H})$ onto a simple graph $\tilde{\Gamma} = (\tilde{\mathcal{G}}, \tilde{\mathcal{H}})$ with $\tilde{\mathcal{G}} = \mathcal{G}$ and $\tilde{\mathcal{H}} = \{\kappa_{ij} : |\mathcal{E}_{ij}| \geq s \text{ in } \Gamma_t\}$ see Section 3.2). A simple graph Γ of size n is called

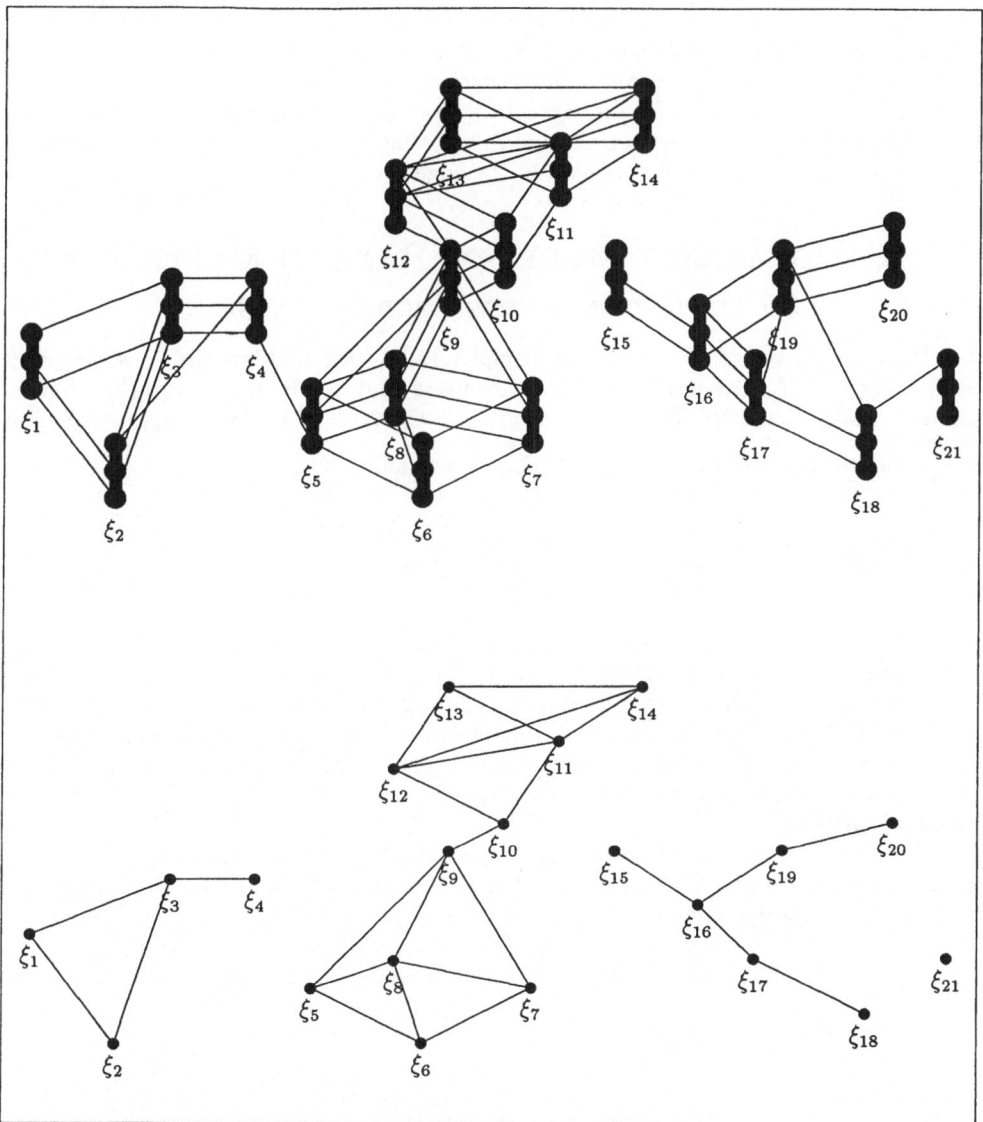

Figure 5-1 *Multigraph* Γ_3 *and 2-projection* $\tilde{\Gamma}$.

complete if it contains all $\binom{n}{2}$ edges. This term can be generalized to multigraphs Γ_t in two different ways: Γ_t can be called complete if it contains all $t\binom{n}{2}$ edges or if it has $\binom{n}{2}$ s-fold connections; multigraphs of the second type are called **s-complete** now.

For a given multigraph, every submultigraph $\Gamma_t^* = (\mathcal{G}^*, \mathcal{H}^*)$ is defined uniquely by its vertex set $\mathcal{G}^*$. Therefore, we will identify these submultigraphs by their vertex sets — as is common practice throughout in the literature. Thus, the set $\{\xi_5, \ldots, \xi_9\}$ in Figure 5-1 is called a strong $(3;2)$-component, and $\{\xi_{10}, \xi_{11}, \xi_{12}\}$ and $\{\xi_{11}, \ldots, \xi_{14}\}$ are two nondisjoint 2-cliques of Γ_3. (Note, however, that submultigraphs Γ_t' are defined by their edge set because the vertex set remains unchanged.) Because of the method to picture simple graphs as diagrams with dots and lines, we furthermore are customed to say that "the edge κ_{ij} lies between the vertices ξ_i and ξ_j" or that "it has been drawn" if $\kappa_{ij} \in \mathcal{H}$ holds true. Now, for multigraphs Γ_t we say that "an edge κ_{ijl} is drawn from the (possible) connection between ξ_i and ξ_j" or that "it is put in the respective layer of this connetion" if κ_{ijl} is an element of the edge set of Γ_t.

5.2 A Probability Model of Random Multigraphs

5.2.1 Definition of the Probability Space

Let Ω_{tnN} be the set of all undirected, completely labelled multigraphs Γ_t with vertex set $\mathcal{G} = \{\xi_1, \ldots, \xi_n\}$, together with an N-element edge set $\mathcal{H}$, $0 \le N \le t\binom{n}{2}$, that is, $\Omega_{tnN} := \{\Gamma_{tnN} = (\mathcal{G}, \mathcal{H}) : \mathcal{G} = \{\xi_1, \ldots, \xi_n\}, \mathcal{H} \subseteq \mathcal{K}_t, |\mathcal{H}| = N\}$. Thus, Ω_{tnN} consists of $\binom{t\binom{n}{2}}{N}$ elements. Hence, on $(\Omega_{tnN}, \mathcal{P}(\Omega_{tnN}))$ we define

(5-1)
$$P(\Gamma_{tnN}) = \binom{t\binom{n}{2}}{N}^{-1}$$

as probability to draw a certain multigraph at random from Ω_{tnN} (**uniform model**). It therefore is the same to draw an element Γ_{tnN} at random from Ω_{tnN} or to draw N edges from the $t\binom{n}{2}$ possible ones in $\mathcal{K}_t$ at random and without replacement. The probability to be chosen as the next edge is equal for all elements not yet be chosen from $\mathcal{K}_t$. We speak of an "urn model without replacement" and without regard to the order of the edges drawn. Every element Γ_{tnN} drawn at random from Ω_{tnN} is called a **random multigraph**.

Let $\mathcal{A}$ be the set of all multigraphs from Ω_{tnN} with a certain property. Then, by $P_{tnN}(\mathcal{A})$ we denote the probability that a multigraph Γ_{tnN} drawn at random will possess this property; that is,

$$P_{tnN}(\mathcal{A}) := P(\{\Gamma_{tnN} : \Gamma_{tnN} \text{ has the property } \mathcal{A}\}).$$

For our probability model (5-1),

(5-2)
$$P_{tnN}(\mathcal{A}) = a_{tnN} \binom{t\binom{n}{2}}{N}^{-1}.$$

holds true. Here, by $a_{tnN} := |\mathcal{A}|$ we denote the number of elements in Ω_{tnN} which possess the property in question. Because of the equiprobability we supposed, to compute the probability $P_{tnN}(\mathcal{A})$ is the same as to calculate the number of multigraphs Γ_{tnN} with this property.

5.2.2 Definition of the Random Variables

The following random variables are defined on $(\Omega_{tnN}, \mathcal{P}(\Omega_{tnN}))$. Their probability measure P_{tnN} is induced by (5-1). As usual, we denote by $P_{tnN}(X_{s.1} = k)$ the probability $P(\{\Gamma_{tnN} \in \Omega_{tnN} : X_{s.1}(\Gamma_{tnN}) = k\})$. This notation is helpful in the formulation of limit theorems, that is, for the case $n \to \infty$ and $N = N(n)$. The positive natural numbers t, n and N determine Ω_{tnN}, that is, the number and properties of the admitted multigraphs. Now, let i, j, l, m and s be from $\mathcal{N}$ with $1 \leq i \leq n$, $1 \leq j \leq n$, $1 \leq m \leq n$, $1 \leq l \leq t$ and $1 \leq s \leq t$, respectively. For all $\Gamma_{tnN} = (\mathcal{G}, \mathcal{H})$, let the random variable T_{ijl} be defined by

$$(5\text{-}3\text{a}) \qquad T_{ijl}(\Gamma_{tnN}) := \begin{cases} 1 & \text{if } \kappa_{ijl} \in \mathcal{H} \text{ or } \kappa_{jil} \in \mathcal{H}, \\ 0 & \text{else.} \end{cases}$$

The random variable T_{ijl} takes the value 1 for those multigraphs Γ_{tnN} with an edge in the l-th layer, connecting the vertices ξ_i and ξ_j. According to this definition, $T_{ijl}(\Gamma_{tnN}) = T_{jil}(\Gamma_{tnN})$ and $T_{iil}(\Gamma_{tnN}) = 0$ both hold always true for all Γ_{tnN}. Furthermore, the realization of the random variable $T_{ij.} := T_{ij1} + \cdots + T_{ijt}$ gives the number of edges between ξ_i and ξ_j in a random multigraph Γ_{tnN}.

Let

$$(5\text{-}3\text{b}) \qquad U_{sij}(\Gamma_{tnN}) := \begin{cases} 1 & \text{if } T_{ij.}(\Gamma_{tnN}) \geq s, \\ 0 & \text{else} \end{cases}$$

be another indicator variable. U_{sij} takes the value 1 for those multigraphs with s-fold connected vertices ξ_i and ξ_j. From $U_{si.}$, we get the s-degree of the vertex ξ_i, and $V_s := \frac{1}{2} U_{s..} = \sum_{i=2}^{n} \sum_{j=1}^{i-1} U_{sij}$ counts the number of s-fold connections. From a third indicator variable,

$$(5\text{-}3\text{c}) \qquad W_{sim}(\Gamma_{tnN}) := \begin{cases} 1 & \text{if } U_{si.}(\Gamma_{tnN}) = m, \\ 0 & \text{else,} \end{cases}$$

we get the number of vertices with an s-degree of m in a random multigraph Γ_{tnN} by summation over i: It is given by $W_{s.m}(\Gamma_{tnN})$.

Remark 5-2 For the edge set $\tilde{\mathcal{H}}$ of the s-projection $\tilde{\Gamma} = \tilde{\Gamma}_{nv} = (\tilde{\mathcal{G}}, \tilde{\mathcal{H}})$ of a random multigraph Γ_{tnN},

$$|\tilde{\mathcal{H}}| = V_s(\Gamma_{tnN})$$

always must hold; the number v of edges in the s-projection Γ_{nv} is the realization of a random variable, $V_s(\Gamma_{tnN})$. •

Let X_{sim} be defined by

$$(5\text{-}3\text{d}) \qquad X_{sim}(\Gamma_{tnN}) := \begin{cases} \dfrac{1}{m} & \text{if } \xi_i \text{ lies in an } s\text{-component of size } m, \\ 0 & \text{else.} \end{cases}$$

Then, the sum $X_{s.m}(\Gamma_{tnN})$ is the number of s-components of size m in this multigraph; another sum, $Z_s := X_{s..}$ counts the number of all s-components in Γ_{tnN}. From (5-3d),

$$X_{si1}(\Gamma_{tnN}) = \begin{cases} 1 & \text{if } U_{si.}(\Gamma_{tnN}) = 0, \\ 0 & \text{else} \end{cases}$$

follows. Thus, $X_{s.1}$ is the number of s-isolated vertices. And $X_{s.1}(\Gamma_{tnN}) = W_{s.0}(\Gamma_{tnN})$ holds true for all random multigraphs Γ_{tnN} since $W_{s.0}$ also counts the number of s-isolated vertices. Furthermore, by

$$(5\text{-}3\text{e}) \qquad \tilde{Z}_s := \sum_{m=2}^{n} X_{s.m} = Z_s - X_{s.1},$$

we get the number of proper s-components.

Finally, be the four random variables $X_{sim}^{(T)}$, $\tilde{X}_{sim}^{(C)}$, $X_{sim}^{(C)}$ and $X_{sim}^{(m,m)}$, respectively, as follows. $X_{sim}^{(T)} = 1/m$ if and only if the vertex ξ_i is in an s-tree of size m, $\tilde{X}_{sim}^{(C)} = 1/m$ if and only if ξ_i is in an s-cycle of size m, $X_{sim}^{(C)} = 1/m$ if and only if ξ_i is in an isolated s-cycle of size m, $X_{sim}^{(m,m)} = 1/m$ if and only if ξ_i is in an s-component with m vertices and m edges, that is, if it is in an s-component which contains exactly one s-cycle. Let all four variables take the value 0 in all other cases. The variables $X_{s.m}^{(T)}$, $\tilde{X}_{s.m}^{(C)}$, $X_{s.m}^{(C)}$ and $X_{s.m}^{(m,m)}$ then give the numbers of the respective s-components of size m or of the s-cycles of length m.

To draw a multigraph with l edges between the vertices ξ_i and ξ_j is the same as to choose l of the t layers between these vertices as places for the l edges and to choose places from the $\binom{t\binom{n}{2} - t}{N - l}$ possibilities outside the connection between ξ_i and ξ_j for the remaining $N - l$ edges. (Or in the language of the urn model without replacement: We have $\binom{t\binom{n}{2}}{N}$ balls in an urn — t of them are red — and we choose l of the t red balls and $N - l$ of the other balls without replacement.) By this combinatorial argument, it is easy to see that the number of edges between two vertices ξ_i and ξ_j follows a hypergeometric distribution, $\mathrm{H}(t\binom{n}{2}, t, N)$, that is,

$$(5\text{-}4) \qquad P_{tnN}(T_{ij.} = l) = \binom{t}{l}\binom{t\binom{n}{2} - t}{N - l}\binom{t\binom{n}{2}}{N}^{-1};$$

and the expected number of vertices between any pair of vertices is $\mathrm{E}_{tnN}T_{ij.} = N/\binom{n}{2}$. From this, for $i \neq j$ we get

$$p := P_{tnN}(U_{sij} = 1) = P_{tnN}(T_{ij.} \geq s)$$
$$(5\text{-}5a) \qquad = \sum_{l=s}^{t} \binom{t}{l}\binom{t\binom{n}{2} - t}{N - l}\binom{t\binom{n}{2}}{N}^{-1}$$

as the probability that two vertices ξ_i and ξ_j are s-fold connected, and

$$p_{12} := P_{tnN}(U_{si_1j_1} = 1, U_{si_2j_2} = 1)$$
$$(5\text{-}5b) \qquad = \sum_{l=s}^{t} \sum_{m=s}^{t} \binom{t}{l}\binom{t}{m}\binom{t\binom{n}{2} - 2t}{N - l - m}\binom{t\binom{n}{2}}{N}^{-1}$$

as the probability that two different connections $\mathcal{E}_{i_1j_1}$ and $\mathcal{E}_{i_2j_2}$ are both s-connections ($i_1 \neq j_1$, $i_2 \neq j_2$; $i_1 \neq i_2$ or $j_1 \neq j_2$). Because of the negative quadrant dependency, we have $p_{12} < p^2$. By elementary calculations, the following formulas for the expectations and variances of the numbers of s-fold connections in a random multigraph Γ_{tnN}, respectively, can be derived:

$$(5\text{-}6a) \qquad \mathrm{E}_{tnN}V_s = \binom{n}{2}p = 2^{s-1}\frac{\binom{t}{s}}{t^s}\frac{N^s}{n^{2s-2}}\left\{1 + O\left(\frac{N^2 + Nn + n^2}{Nn^2}\right)\right\},$$

$$(5\text{-}6b) \qquad \mathrm{Var}_{tnN}V_s = \binom{n}{2}p(1-p) + \binom{n}{2}\left(\binom{n}{2} - 1\right)(p_{12} - p^2),$$

$$(5\text{-}6c) \qquad \mathrm{E}_{tnN}V_s > \mathrm{Var}_{tnN}V_s, \qquad \mathrm{E}_{tnN}V_s \sim \mathrm{Var}_{tnN}V_s.$$

For the s-degree of an arbitary vertex ξ_i, we get the following results by virtually the same calculations:

(5-7a) $\qquad \mathrm{E}_{tnN} U_{si.} = (n-1)p = 2^s \dfrac{\binom{t}{s}}{t^s} \dfrac{N^s}{n^{2s-1}} \left\{ 1 + O\left(\dfrac{N^2 + Nn + n^2}{Nn^2} \right) \right\},$

(5-7b) $\qquad \mathrm{Var}_{tnN} U_{si.} = (n-1)p(1-p) + (n-1)(n-2)(p_{12} - p^2),$

(5-7c) $\qquad \mathrm{E}_{tnN} U_{si.} > \mathrm{Var}_{tnN} U_{si.}, \qquad \mathrm{E}_{tnN} U_{si.} \sim \mathrm{Var}_{tnN} U_{si.}.$

These formulas are the main tools to prove the limit theorems for random multigraphs which are given in this chapter; for a broader discussion of (5-6a)–(5-6c) and (5-7a)–(5-7c), we refer to [149].

5.2.3 RELATIONS TO CURRENT PROBABILITY MODELS

In this subsection, we want to discuss shortly the three main probability models which provide the bases of the bulk of the papers on random graphs and random multigraphs. Moreover, we explain the relations between these models and our model. All three probability models have in common that n vertices are given and edges are distributed "at random" between them.

Model 1 is defined by E.N. Gilbert in [141], [142] as follows:

> "Let n points, numbered $1, 2, \ldots, n$, be given. There are $n(n-1)/2$ lines which can be drawn joining pairs of these points. Choosing a subset of these lines to draw, one obtains a graph; there are $2^{n(n-1)/2}$ possible graphs in total. Pick one of these graphs by the following random process. For pairs of points make a random choice, independent of each other, whether or not to join the points of the pair by a line. Let the common probability of joining be p. Equivalently, one may erase lines, with common probability $q = 1 - p$, from the complete graph."

Random graphs resulting from this process usually are named G_{np}. The number N of edges in such a random graph is a random variable. The probability to get a certain random graph G_{np} with exactly N edges is

$$ P(G_{np}) = p^N (1-p)^{\binom{n}{2} - N}. $$

Thus, the probability to get a graph with N edges is $\dbinom{\binom{n}{2}}{N} P(G_{np})$. This means that N is $\mathrm{B}(\binom{n}{2}, p)$-distributed with expectation $\mathrm{E}N = \binom{n}{2}p$. This model is known as the **binomial model**. Properties of this model are discussed in [55], [64], [228], [234], [303], [395].

In [12], T.L. Austin, R.E. Fagen, W.F. Penney and J. Riordan studied the behaviour of random multigraphs. They defined Model 2 (as we call it) as follows.

> "Given n distinct points, N selections of pairs are made independently and at random, each of the $\binom{n}{2}$ possible pairs having the same chance $1/\binom{n}{2}$ of selection at each trial. Once selected, a pair is connected by a line joining its two points, labelled by the order number of its

selection; thus, after N selections, a linear graph with n distinct (labelled) points and N distinct (labelled) lines connecting pairs of points is formed. (Note that the rule of formation implies the graph contains no slings, but may contain lines in parallel.)"

This model differs significantly from our model of Subsection 5.2.1: Here we have no "layers of simple graphs", the next edge beteen two vertices is put beyond the preceeding one; the number of edges between a pair of vertices is not bounded (all N edges may link the same two vertices together). In this model, the number of possible multigraphs with n vertices and N edges is $\binom{n}{2}^N$. These are considered as equiprobable. We thus can speak of an "urn model with replacement".

Most results in random graph theory have been proved on the following Models 3 and 3*. These two models have been proposed — and at the same time studied extensively — by P. Erdős and A. Rényi who virtually founded the theory of random graphs with their famous paper "ON THE EVOLUTION OF RANDOM GRAPHS" (see [111], [112], [113]).

"Our aim is to study the probable structure of a random graph Γ_{nN} which has n given labelled vertices $P_1, P_2, \ldots, P_n$ and N edges; we suppose that these N edges are chosen at random among the $\binom{n}{2}$ possible edges, so that all $\binom{\binom{n}{2}}{N} = c_{nN}$ possible choices are supposed to be equiprobable. Thus if G_{nN} denotes any of the c_{nN} graphs formed from n given labelled points and having N edges, the probability that the random graph Γ_{nN} is identical with G_{nN} is $1/c_{nN}$. If A is a property which a graph may or may not possess, we denote by $P_{nN}(A)$ the probability that the random graph Γ_{nN} possesses the property A, that is, we put $P_{nN}(A) = a_{nN}/c_{nN}$ where a_{nN} denotes the number of those G_{nN} which have the property A. Another equivalent formulation is the following: Let us suppose that n labelled vertices $P_1, P_2, \ldots, P_n$ are given. Let us choose at random an edge among the $\binom{n}{2}$ possible edges, so that all these edges are equiprobable. After this let us choose at random an edge among the remaining $\binom{n}{2} - 1$ edges, and continue this process so that if already k edges are fixed, any of the remaining $\binom{n}{2} - k$ edges have equal probability to be chosen as the next one."

These models are "urn models without replacement". Both models, the model where a graph with N edges is chosen directly, that is, where the N edges are drawn at once, and the model where the edges are drawn one after the other, are equivalent and result in the same probabilities for all properties of graphs. We therefore do not distinguish between them.

Models 2 and 3 have in common that the number N of edges drawn is fixed. The difference between the probable properties of graphs or multigraphs from these models are in most (but not in all) cases negligible. The corresponding probabilities are in general (if the number of edges is not too large) asymptotically equal. This is due to the fact that for fairly small N, we don't expect many edges in parallel between the same pair of vertices. By the random process in Model 1, N is a random variable with expectation $EN = \binom{n}{2}p$, the decisions concerning the different edges being completely independent. In many (though not all) problems there is no essential difference in the results, however, for random graphs Γ_{nN} or G_{np} if only $p = N / \binom{n}{2} \sim 2N/n^2$ — p denoting the probability of an individual edge to be chosen — converges "moderately" to 0, neither too fast nor too slowly. In this case, we get the same limit distributions for all random graph models ([54], [112], [141], [142], [174], [223], [228], [303]). This is due to the fact that, for large n, the number of edges in G_{np} does not differ significantly from its expectation N with probability 1.

In our multigraph model, as defined in Subsections 5.2.1 and 5.2.2, the number N of edges is fixed, too. This model contains Models 3 and 3* as a special case, $t = s = 1$. The multigraph model of T.L. Austin, R.E. Fagen, W.F. Penney and J. Riordan — that is, Model 2 — can be considered as a limit case of our model: It is given by $s = 1$ and $t \to \infty$ (see the discussion in Section 5.5). Moreover, in our model, the equiprobability in drawing the edges leads to

(5-8a) $$P_{tnN}(X_{s.1} = k | V_s = v) = P_{nv}(X_{.1} = k),$$

(5-8b) $$E_{tnN}(X_{s.1} | V_s = v) = E_{nv}X_{.1},$$

(5-8c) $$\mathrm{Var}_{tnN}(X_{s.1} | V_s = v) = \mathrm{Var}_{nv}X_{.1}.$$

That means that the conditional distribution of the number of s-isolated vertices in Γ_{tnN} under the condition that Γ_{tnN} has exactly v s-saturated connections, is the same as the distribution of the number of isolated vertices in random graphs Γ_{nv} with v edges (and the same will hold for other random variables, too). By the s-projection of Γ_{tnN}, we get a random graph G_{np} with p given by (5-6a), which is in its probabilities for nearly all properties asymptotically equal to a random graph $\Gamma_{n, \lfloor EV \rfloor}$ with $\lfloor EV \rfloor = \lfloor E_{tnN} V_s \rfloor$ edges if N is not too small or not too large (see also Remark 5-2, Figure 5-1, and the discussion of the results).

For simple random graphs, we omit the leading indices $t = 1$ and $s = 1$: We write $E_{nv}X_{.1}$ instead of $E_{1nv}X_{1.1}$. This is done for the sake of brevity. If the indices are obvious from the context, then we will omit them completely, as said in Chapter 0. We write $\Gamma_{n, \lfloor EV \rfloor}$ instead of $\Gamma_{n, \lfloor E_{tnN} V_s \rfloor}$. A second reason for the shorter nomenclature is to use the notations which are known for simple graphs from literature.

5.3 SOME RESULTS FOR RANDOM GRAPHS Γ_{nN} AND G_{np}

For most mathematicians involved in graph theory, the results of this section are well known. They are repeated here, however, since they provide the basis for the proofs of the limit theorems of Section 5.4. They also can be used for numerical computations of numbers (or probabilities, respectively) of random graphs Γ_{nN} or G_{np} with given properties. Let us call a graph Γ_{nN} of type C_k if it consists of exactly one component with $n - k$ vertices, together with k isolated vertices ($k = 0, 1, 2, \ldots$). By C_{nN}, we denote the number of connected graphs Γ_{nN}, $C_{nN} = |C_0|$. The following theorem states formulas for C_{nN} for arbitrary edge numbers N as well as for the two special cases $N = n - 1$ and $N = n$. (The formulas for these special cases have to be proved on their own; they cannot be derived from the formula for the general case.)

Theorem 5-1 *For $n - 1 \leq N \leq \binom{n}{2}$, the number C_{nN} of nonempty, connected graphs Γ_{nN} with n vertices and N edges is given by*

(5-9a) $$C_{nN} = \sum_{m=1}^{n} \frac{(-1)^m}{m} \sum_{n_1=1}^{n} \cdots \sum_{n_m=1}^{n} \binom{n}{n_1, \ldots, n_m} \binom{\binom{n_1}{2} + \cdots + \binom{n_m}{2}}{N}.$$

For any N outside the range named above, $C_{nN} = 0$ holds true. The number $C_{n,n-1}$ of (nonempty) trees with n vertices is

$$(5\text{-}9\text{b}) \qquad\qquad C_{n,n-1} = n^{n-2}.$$

From n labelled vertices and n edges, $\frac{1}{2}(n-1)!$ different cycles can be built. Furthermore, the number of different connected graphs Γ_{nn} is

$$(5\text{-}9\text{c}) \qquad C_{nn} = \frac{1}{2}\,(n-1)!\left(1 + n + \frac{n^2}{2!} + \cdots + \frac{n^{n-3}}{(n-3)!}\right) \sim \sqrt{\pi/8}\,n^{n-1/2}. \;\bullet$$

R.J. Riddell and G.E. Uhlenbeck proved (5-9a) in [328]. They applied graph-theoretic results like this one to the thermodynamics of ideal gases. The validity of (5-9b) had been proved by A. Cayley in 1857. He also applied graph theory to chemistry ([69], [70], [71]). The asymptotic equation in (5-9c) had been proved in [323], and the left part of (5-9c) can be found in [12]. R.F. Ling and G.G. Killough used (5-9a) in [265] to compute tables for the probabilities to get connected graphs with n vertices and N edges. They compared their results with the empirical formulas in [351], and the asymptotic ones from [111], [112].

Remark 5-3 For the development of numeric algorithms, it is useful to allow the vertex set of a graph to be empty, and to put the number of such an "empty graph" Γ_{00} with given properties either to 1 or to 0 (to get an initial value especially in recursion formulas). This also prevents us from always being forced to either include or exclude special cases during the statement of theorems. Here, we define

$$(5\text{-}9\text{d}) \qquad\qquad C_{00} := 1.$$

By this definition, the empty graph is always a connected graph. Furthermore, graphs with one vertex and one edge, and graphs with two vertices and two edges cannot exist. Thus, we define

$$(5\text{-}9\text{e}) \qquad\qquad C_{11} := C_{22} := 0. \;\bullet$$

With these formulas, the expected number of components of size m in random graphs Γ_{nN} can be calculated as

$$(5\text{-}10) \qquad \mathrm{E}_{nN}X_{.m} = \binom{n}{m}\sum_{\mu} C_{m\mu}\left(\binom{\binom{n-m}{2}}{N-\mu}\right)\binom{\binom{n}{2}}{N}^{-1}.$$

The expected number of all (proper or degenerated) components of any size then is given by $\sum_{m=1}^{n}\mathrm{E}_{nN}X_{.m}$ and the expected number of the proper components only is $\sum_{m=2}^{n}\mathrm{E}_{nN}X_{.m}$. These exact formulas are too intricate to be used for practical computations. Thus, we are bound to approximate the exact results. In [263], it is shown that for "medium sized" values of N, that is, for $N = O(n)$,

$$\mathrm{E}_{nN}X_{.m} \approx \binom{n}{m}\left\{C_{m,m-1}\binom{\binom{n-m}{2}}{N-m+1} + C_{mm}\binom{\binom{n-m}{2}}{N-m}\right\}\binom{\binom{n}{2}}{N}^{-1}$$

is a fairly good approximation: Most components in a random graph are trees or components with exactly one cycle, and only a negligibly small number of components has a more complicated structure if N is small enough. More approximations are given in. [57], [112], and in [263]. A formula to compute the variance of $E_{nN}X_{.m}$ is given by J.I. Naus and L. Rabinowitz in [294]. Since the exact (finite) distribution of the number of components of size m, however, is known if the components are isolated vertices (that is, for $m = 1$), O. Frank uses a kind of "bootstrapping" to estimate it for $m > 1$: He draws subgraphs of Γ_{nN} at random ([132]).

The number L_{nN} of graphs Γ_{nN} without any isolated vertices can be computed using the formula of Sylvester and Poincaré (the formula of inclusion and exclusion or sieve formula). This number is given in the next theorem.

Theorem 5-2 *The number L_{nN} of graphs Γ_{nN} without isolated vertices is*

$$(5\text{-}11) \qquad L_{nN} = \sum_{j=0}^{n} (-1)^j \binom{n}{j} \left(\binom{\binom{n-j}{2}}{N} \right). \quad \bullet$$

Note that the graphs may be composed of more than one component. By this formula, the distribution of isolated vertices in random graphs Γ_{nN} can be calculated as

$$(5\text{-}12\text{a}) \qquad P_{nN}(X_{.1} = k) = \binom{n}{k} L_{n-k,N} \left(\binom{\binom{n}{2}}{N} \right)^{-1}.$$

Substituting the index $n - k$ in (5-12a) by k, gives the distribution of the number of nonisolated vertices, that is, the number of vertices lying in proper components of a random graph Γ_{nN}. Formula (5-10) contains the expected number of isolated vertices as special case $m = 1$. Independent of this, the expextation can be found by simple combinatorial arguments and elementary calculations: To keep a vertex — say ξ_i — isolated, all N edges have to be distributed between the remaining $n - 1$ vertices. We get

$$P_{nN}(\xi_i \text{ is isolated}) = \left(\binom{\binom{n-1}{2}}{N} \right) \left(\binom{\binom{n}{2}}{N} \right)^{-1}.$$

Thus, the expected number of isolated vertices is

$$E_{nN}X_{.1} = n\, P_{nN}(\xi_i \text{ is isolated})$$
$$(5\text{-}12\text{b}) \qquad = n \left(\binom{\binom{n-1}{2}}{N} \right) \left(\binom{\binom{n}{2}}{N} \right)^{-1} \sim n\, e^{-2N/n},$$

where the last part holds if $N = o(n^{3/2})$. (The asymptotical equivalence in (5-12b) can easily be proved by using the well known inequality

$$(5\text{-}13\text{a}) \qquad e^{-bd/(a-d-b+1)} \leq \binom{a-d}{b} \binom{a}{b}^{-1} \leq e^{-bd/a}$$

for integers a, b, d with $0 \leq b \leq a - d$; for all these formulas see also [111], [149], [259]). The variance of the number of isolated vertices follows from the paper of J.I. Naus and L. Rabinowitz which we cited above.

In [149], we derived the distribution of the number of vertices in weak components of degree 2 in random graphs Γ_{nN}. This exact formula, however, is far too intricate to be used for numeric calculations. An asymptotic formula, on the other hand, is not known. Only Monte Carlo simulations have been performed to get some idea on the distribution of vertices in weak components of degrees 2 and 3 ([259], [261]).

The distribution of the degree of an arbitrary vertex ξ_i in Γ_{nN} again follows from simple combinatorial arguments: The degree of ξ_i is k if k other vertices are linked to ξ_i by k edges; the remaining $N - k$ other edges can be distributed in $\binom{n-1}{2}$ different ways between the other $n-1$ vertices. Thus, the degree follows a $H(\binom{n}{2}, n-1, N)$-distribution,

$$(5\text{-}14a) \qquad P_{nN}(U_{i\cdot} = k) = \binom{n-1}{k}\binom{\binom{n-1}{2}}{N-k}\binom{\binom{n}{2}}{N}^{-1} \sim \frac{1}{k!}\left(\frac{2N}{n}\right)^k e^{-2N/n}.$$

Again, the asymptotic relation holds if $N = o(n^{3/2})$. The expectation is given by

$$(5\text{-}14b) \qquad \mathrm{E}_{nN}U_{i\cdot} = (n-1)\binom{\binom{n}{2}-1}{N-1}\binom{\binom{n}{2}}{N}^{-1} = \frac{N(n-1)}{\binom{n}{2}} = \frac{2N}{n}.$$

Until now, we considered random graphs Γ_{nN}, graphs given by Model 3 or Model 2 and only few edges (such that the probability to get more than one edge connecting the same pair of vertices remains negligibly small). From the theory of random graphs G_{np}, the following theorem from Z. Palka is important ([303]).

Theorem 5-3 *Let $X_{\cdot m}^{(T)}$ count the number of isolated trees of size m in random graphs G_{np}. Then*

$$(5\text{-}15) \qquad P_{np}(X_{\cdot m}^{(T)} = k) = \sum_{j=0}^{l-k}(-1)^j\binom{k+j}{k}S_{k+j} \qquad (k = 0, 1, \ldots, l),$$

where $l = \lfloor n/m \rfloor$, and the numbers S_r are given by $S_0 = 1$ and

$$S_r = \frac{1}{r!}\,n_{(rm)}\left(\frac{t_m}{m!}\right)^r (1-p)^s \qquad \text{with} \qquad s = rmn - \binom{r+1}{2}m^2$$

for $r = 1, 2, \ldots, m$. In S_r, $t_m = m^{m-2}p^{m-1}(1-p)^{(m-1)(m-2)/2}$, that is, t_m is the probability that a set of m labelled vertices spans a tree. •

We will study the "evolution" of random graphs and multigraphs if N is increased (starting with $N = 0$) or decreased (then starting with $N = \binom{n}{2}$, or $N = t\binom{n}{2}$). We want to uncover the "typical" structures of random graphs Γ_{nN} or G_{np} (or random multigraphs Γ_{tnN}, respectively) at a given stage, that is, for special values of N. This typical structure generally can be better described by so-called **limit theorems** than by formulas of finite distributions or probabilities, that is, as $n \to \infty$ and N is equal, or asymptotically equal, to a given function $N(n)$ of n. Thus, by a "typical" structure of a graph or multigraph, we mean a property the probability of which tends to 1 as $n \to \infty$ and $N = N(n)$. We then say that almost all graphs possess this property.

For the purpose of studying the structure of graphs, let us consider a sequence of probability spaces, $\left((\Omega_{nN}, \mathcal{P}(\Omega_{nN}), P_{nN})\right)_{n\to\infty, N=N(n)}$, the elements of this sequence be defined according to Subsection 5.2.1. From every Ω_{nN}, we choose a graph Γ_{nN} "at random". Doing this, we get a sequence $(\Gamma_{nN})_{n\to\infty}$ of random graphs together with the related sequences of distributions and parameters of all random variables defined on the Ω_{nN}. Thus, for different values of n and N, the arguments Γ_{nN} are from different probability spaces; the random variables, however, bear the same meaning for all spaces. The situation differs from the case, which is usually considered in limit theorems (usually, sequences of random variables on the same probability space are studied. The case, however, is the same as in Poisson's limit theorem for binomial distributions. Nevertheless, we also speak — not correctly — of "sequences of random variables".

If we speak of "convergence of random variables", then we shall deal mainly with **convergence in probability**. That is to say, the assertion "$X_n \to l$" means $X_n \to l$ in probability as $n \to \infty$; in other words, for any $\varepsilon > 0$, $P(|X_n - l| > \varepsilon) \to 0$ as $n \to \infty$. Convergence in distribution also plays a major role in random graphs. The statement "The number of isolated vertices in random graphs Γ_{nN} is asymptotically Poisson distributed" means that, as $n \to \infty$,

$$P_{nN}(X_{.1} = k) := P(\{\Gamma_{nN} : X_{.1}(\Gamma_{nN}) = k\}) \to e^{-\lambda}\frac{\lambda^k}{k!} \quad (k = 0, 1, 2, \ldots),$$

for some $\lambda > 0$.

If $\mathcal{A}$ is a property of graphs, then an assertion such as "A graph Γ_{nN} (or G_{np}, respectively) has property $\mathcal{A}$ almost surely" means

$$P(\{\Gamma_{nN} : \Gamma_{nN} \text{ has property } \mathcal{A}\}) = P_{nN}(\mathcal{A}) \to 1 \quad (n \to \infty).$$

Probabilists may insist that this usage is at odds with the notion of "almost sure convergence" of random variables. Sometimes such statements will be generalizations of certain earlier enumeration theorems; results of the latter type are commonly stated as "almost all graphs on n vertices have property $\mathcal{A}$" as we said above. All limits are taken as $n \to \infty$.

To study the typical behaviour of such sequences of random graphs (concerning certain properties), usually the edge numbers $N(n)$ are chosen as monotonic increasing functions of n. The best known results on the asymptotic behaviour of random graphs have been published by the Hungarian mathematicians B. Bollobás, P. Erdős and A. Rényi (also as co-authors, together or with others); the latter two persons being the founders of random graph theory. (This was due to their ideas to use probabilistic methods in their proofs instead of the enumeration methods known before. Thus, today graph theory is as well a part of probability theory as of combinatorics.) In [111] and [112], they proved the following assertions (see also [160]).

Theorem 5-4 *Let $(\Gamma_{nN})_{n\to\infty, N=N(n)}$ be a sequence of random graphs. For*

(5-16) $$N(n) = \left\lfloor \frac{1}{2} n \left(\log n + c + o(1)\right)\right\rfloor,$$

the expected number of isolated vertices remains bounded: We get $\mathrm{E}_{nN}X_{.1} \to e^{-c}$. In the limit, the number of isolated vertices follows a $\mathrm{P}(\lambda)$-distribution with $\lambda = e^{-c}$. The same holds for the number of vertices, lying not in the largest component of Γ_{nN}, and for the number of components, diminished by 1. The probability of a random graph Γ_{nN} being completely connected, tends to $\exp(-e^{-c})$. •

Remark 5-4 The fact that $E_{nN}X_{.1} \to e^{-c}$ holds true can easy be verified by inserting (5-16) into (5-12b) and using (5-13a). (For $b^2d = o(a^2)$ and $bd^2 = o(a^2)$, the left side and the right side in (5-13a) are asymptotically equivalent as $a \to \infty$, $b \to \infty$, $d \to \infty$. This is satisfied for $a = \binom{n}{2}$, $d = n-1$ and $b = N(n)$ according to (5-16), see also [149].) As a consequence of this theorem, almost every random graph Γ_{nN} consists of one single — very large — proper component and some (or none) isolated vertices besides it if n is not too small and the numbers $N = N(n)$ of edges satisfy (5-16), $P_{nN}(\tilde{Z}_s = 1) \to 1$. Moreover, edge sequences given by (5-16) serve as **threshold functions**. Substituting the converging sequence $(c_n)_{n\to\infty}$ — with $c_n = c + o(1)$ — in (5-16) by a sequence $c_n \to \infty$ gives sequences of random graphs Γ_{nN} wich are with probability 1 completely connected for large n. Choosing a sequence $c_n \to -\infty$, however, results in sequences of graphs Γ_{nN} which with probability 1 are composed of more than one proper component. The probability of being completely connected tends either to 1 (for $c_n \to \infty$) or to 0 (for $c_n \to -\infty$). Only for edge sequences given by (5-16), we get a nondegenerated distribution function for this probability, depending only on the parameter c (see [112] for a discussion of threshold functions). With other words, let us start with the complete graph and omit edges at random. Then the property of being connected is a typical property of random graphs as long as $N(n)$ increases faster than (5-16). If however $N(n)$ passes beyond this threshold function, then the probability of being connected tends to 0, that is, the typical structure of graphs with fewer edges is that of being composed of different components. •

In [303], Z. Palka proved the following limit theorems on the distributions of trees of size m in random graphs G_{np}.

Theorem 5-5 If the probability p for drawing an edge in random graphs G_{np} satisfies

$$(5\text{-}17a) \qquad np = \frac{1}{m}\log n + \frac{m-1}{m}\log\log n + c + o(1),$$

then for the number of trees of size m in G_{np}, $E_{np}X_{.m}^{(T)} \to \mu = \frac{e^{-mc}}{mm!}$ holds true. This number tends to a $P(\mu)$-distribution as $n \to \infty$,

$$(5\text{-}17b) \qquad P_{np}(X_{.m}^{(T)} = k) \to e^{-\mu}\frac{\mu^k}{k!} \qquad (k = 0,1,2,\dots). •$$

Theorem 5-6 If the probability p for drawing an edge in random graphs G_{np} satisfies

$$(5\text{-}18a) \qquad \lim_{n\to\infty} pn^{m/(m-1)} = c \qquad (c > 0),$$

then for the number of trees of size m in G_{np}, $E_{np}X_{.m}^{(T)} \to \nu = c^{m-1}\frac{m^{m-2}}{m!}$ holds true. The number of trees of size m tends to a $P(\nu)$-distribution as $n \to \infty$,

$$(5\text{-}18b) \qquad P_{np}(X_{.m}^{(T)} = k) \to e^{-\nu}\frac{\nu^k}{k!} \qquad (k = 0,1,2,\dots). •$$

Inserting $N = \binom{n}{2}p$ in (5-17a) and (5-18a) results in the following two edge sequences:

(5-17c) $\qquad N(n) = \left\lfloor \dfrac{1}{2}\, n\Big(\dfrac{1}{m}\, \log n + \dfrac{m-1}{m}\, \log\log n + c + o(1)\Big)\right\rfloor,$

and

(5-18c) $\qquad N(n) = \left\lfloor \dfrac{1}{2}\, n^{(m-2)/(m-1)}\big(c + o(1)\big)\right\rfloor,$

respectively. It now is easy to show that Theorems 5-5 and 5-6 similarly hold for random graphs Γ_{nN} if the edge sequences be chosen according to (5-17c) and (5-18c), respectively. (In this form, they have been stated and proved in [112].) (5-16) can be derived as special case $m = 1$ from (5-17c). Further results on the distribution of the number of trees in Γ_{nN} are given in [110], [225] and [226].

In both cases, that is, for edge sequences satisfying (5-17c) or (5-18c), the number of isolated trees of size m in random graphs Γ_{nN} is in the limit Poisson distributed. The fundamental difference for the two sequences is that if $N(n)$ fulfills (5-17c) (which is increasing more slowly than (5-16) for $m > 1$), then trees of size m are almost surely the second largest components besides a "giant"-component which contains nearly all vertices and edges. (Thus, this giant-comonent is of a rather complcated structure.) For edge sequences of the form (5-18c), however, trees of size m are, with probability tending to 1, the largest components themselves in random graphs. (The structure of such random graphs is rather simple.) For edge sequences which grow faster than (5-17c) or slower than (5-18c), we don't expect any tree of size m. For edge sequences between these threshold functions, the number of trees in random graphs is asymptotically normally distributed with $\mathrm{E}_{nN} X_{.m}^{(T)} \sim \mathrm{Var}_{nN} X_{.m}^{(T)}$ (see [112], [303]).

In [55] and [304], B. Bollobás and Z. Palka derive the distribution of the number of vertices which are connected with exactly $m - 1$ other vertices, that is, which have degree $m - 1$. This is given as follows.

Theorem 5-7 *For random graphs Γ_{nN} or G_{np}, let the respective edge sequences as given by (5-18c) or the probabilities of independently drawing edges as defined by (5-18a). Then the distribution of the random variable $W_{.,m-1}$, the number of vertices with degree $m - 1$, is asymptotically Poisson distributed,*

(5-18d) $\qquad P_{nN}(W_{.,m-1} = k) \to e^{-\zeta}\, \dfrac{\zeta^k}{k!} \qquad (k = 0, 1, 2, \dots)$

with expectation $\zeta = n \binom{n}{m-1} p^{m-1} (1-p)^{n-m} = \dfrac{c^{m-1}}{(m-1)!} \cdot$ •

In [55], B. Bollobás proves assertions on the distributions of the numbers of vertices with maximum or minimum degree, respectively, in random graphs G_{np}.

For edge sequences $N(n)$ being of the order of magnitude of n, the expected degree of an arbitrary vertex ξ_i in Γ_{nN} — as given by (5-14b) — remains positive and bounded. The distribution of this number tends to a Poisson distribution as $n \to \infty$. The same holds for the random variables $X_{.m}^{(C)}$, $X_{.m}^{(m,m)}$ and $\tilde{X}_{.m}^{(C)}$, that is, the number of isolated cycles, components with exactly one cycle, and not necessarily isolated cycles. This is shown in the following theorem.

Theorem 5-8 *In sequences $(\Gamma_{nN})_{n\to\infty, N=N(n)}$ or $(G_{np})_{n\to\infty, p=p(n)}$ of random graphs, where $p = N/\binom{n}{2}$ and*

(5-19a)
$$N(n) = \left\lfloor \frac{1}{2} n \left(c + o(1)\right) \right\rfloor \qquad (c > 0),$$

the distribution of the degree of an arbitrary vertex ξ_i is given by

(5-19b)
$$P_{nN}(U_{i.} = k) \to e^{-c} \frac{c^k}{k!} \qquad (k = 0, 1, 2, \dots).$$

The numbers of isolated cycles, components with exactly one cycle, and (not necessarily isolated) cycles, respectively, of any size m $(m \geq 3)$ are also Poisson distributed as $n \to \infty$. We get

(5-19c)
$$P_{nN}(X_{.m}^{(C)} = k) \to e^{-\kappa} \frac{\kappa^k}{k!} \qquad (k = 0, 1, 2, \dots),$$

(5-19d)
$$P_{nN}(X_{.m}^{(m,m)} = k) \to e^{-\eta} \frac{\eta^k}{k!} \qquad (k = 0, 1, 2, \dots),$$

(5-19e)
$$P_{nN}(\tilde{X}_{.m}^{(C)} = k) \to e^{-\theta} \frac{\theta^k}{k!} \qquad (k = 0, 1, 2, \dots),$$

respectively. The asymptotic expectations of all these numbers are $\mathbf{E}_{nN} U_{i.} \to c$, $\mathbf{E}_{nN} X_{.m}^{(C)} \to \kappa = \frac{c^m}{2m} e^{-cm}$, $\mathbf{E}_{nN} X_{.m}^{(m,m)} \to \eta = \frac{c^m}{2m} e^{-cm} \left(1 + m + \frac{m^2}{2!} + \dots + \frac{m^{m-3}}{(m-3)!}\right)$ and $\mathbf{E}_{nN} \tilde{X}_{.m}^{(C)} \to \theta = \frac{c^m}{2m}$, respectively. •

Proof (Sketch) Inserting (5-19a) in (5-14b) and (5-14a) immediately leads to $\mathbf{E}_{nN} U_{i.} \to c$ and $P_{nN}(U_{i.} = k) \to e^{-c} \frac{c^k}{k!}$ $(k = 0, 1, 2, \dots)$, by straightforward calculations. This proves (5-19b) (a detailed proof can be found in [153]). All the proofs of (5-19c)–(5-19e) are given in [112]. •

The results of this theorem are different from those of the preceeding theorems. In all those theorems, the size m of the components to be under consideration always occurs also in the edge sequences. We thus expect from Theorem 5-6 that with probability 1 no trees of either size $m - 1$ (or smaller) or size $m + 1$ (or larger) occur. Only for trees of size m, the probability is between 0 and 1. In contrast, the size of the cycles does **not** occur in the edge sequences. The probabilities of getting cycles or isolated cycles of any size m, or components with m vertices and m edges are in the open interval $(0; 1)$, independent of the value of m. Furthermore, the expected number of isolated cycles in random graphs Γ_{nN} takes its maximum for $c = 1$, that is, for $N \sim n/2$ in (5-19a); the probability that a graph contains an isolated cycle of a given size m never reaches 1 (as, in contrast, happens for trees of size m).

For $c = 1$ in (5-19a), the structure of random graphs changes dramatically as we know. For $c < 1$, it almost surely consists of isolated vertices together with trees and components containing exactly one cycle, the largest component being a tree with probability 1. (This justifies the use of the approximation formula for the expected number of components of size m as given after Remark 5-3.) For $c > 1$, however, the largest component is of a fairly complex structure; besides it only isolated vertices, trees, and some — far less — isolated cycles or components with exactly on cycle do emerge (for the possible structures of random graphs for different edge sequences $N(n)$ see also [18], [22], [54], [56], [57], [112], [113], [174], [223], [234], [434], [435], [436]).

D.W. Matula shows in [270], [272] that the expected number of cliques in random graphs G_{np} is of order of magnitude of $2\log(1/p)^n$, as $p = const$. We cannot transfer this result to random graphs Γ_{nN} since the probability p of drawing edges remains constant as $n \to \infty$ (and thus $p = o(n^{-1/2})$ doesn't hold). The same author also derives the distribution of the vertices in the largest clique ([271]).

5.4 LIMIT THEOREMS FOR RANDOM MULTIGRAPHS

In this chapter, we will study the asymptotic behaviour of sequences of undirected, completely labelled random multigraphs. We especially shall generalize the theorems of the previous chapter to multigraphs. For any fixed integer t, let us consider sequences $\left(\Omega_{tnN}, \mathcal{P}(\Omega_{tnN}), P_{tnN}\right)_{n\to\infty, N=N(n)}$ of probability spaces as defined in Subsection 5.2.1. for different edge functions $N(n)$, we are interested in the "typical" behaviour of sequences of random multigraphs from these probability spaces. We derive limit theorems for the distributions (and their parameters) of the sequences of the random variables T_{ijl}, U_{sij}, U_{si}. etc., defined on these sequences of probability spaces (see Subsection 5.2.2). As in the preceeding section, for different n and $N(n)$, the arguments of the random variables (now being multigraphs Γ_{tnN}) are from different spaces while the random variables bear the same meaning.

Remark 5-5 The number t of layers of the multigraph, and the integer s, by which the degree of connectedness between two vertices is controlled, are fixed parameters of this model. The edge sequences $N(n)$ are assumed to be monotonic nondescending for all $n \geq n_0$ — as it is usually proposed for random graphs, too. •

Let us start with a complete multigraph, omitting edges. In [149], we answered the following question: How many edges can we remove such that the multigraph remains almost surely s-fold connected? To be more precisely (see also [150], [153]):

(A) Do edge sequences exist wich, at one hand, grow fast enough such that for $N = N(n)$, the sequence $(\mathrm{E}_{tnN}X_{s.1})_{n\to\infty}$ of expected numbers of s-isolated vertices in random multigraphs Γ_{tnN} remains bounded, but which, on the other hand, are slow enough such that for this expectation, $(\mathrm{E}_{tnN}X_{s.1}) > 0$ holds true for all n? If such sequences $N(n)$ exist, what can we say about the limit distribution of $X_{s.1}$ and other random variables?

To answer this question, we proved the following theorem. For this theorem, we denote by $\mathcal{C}_k^s$ the event that a multigraph Γ_{tnN} has exactly k s-isolated vertices and $n - k$ vertices in one single, proper s-component. ($\mathcal{C}_k^s$ is a generalization of the event $\mathcal{C}_k$ in the previous section.)

Theorem 5-9 *In sequences* $(\Gamma_{tnN})_{n\to\infty, N=N(n)}$ *of random multigraphs with*

$$(5\text{-}20\mathrm{a}) \qquad N(n) = \left\lfloor \frac{t}{2\binom{t}{s}^{1/s}} \, n^{2-1/s} \left(\log n + c + o(1)\right)^{1/s} \right\rfloor,$$

the sequence of expected s-isolated vertices remains bounded; moreover, we get

$$(5\text{-}20\mathrm{b}) \qquad \mathrm{E}_{tnN}X_{s.1} \to \lambda := e^{-c}.$$

In the limit, the distribution of this random variable follows a $P(\lambda)$*-distribution with* $\lambda = e^{-c}$,

(5-20c) $$P_{tnN}(X_{s.1} = k) \to e^{-e^{-c}} \frac{e^{-ck}}{k!} \qquad (k = 0, 1, 2, \dots).$$

For large n*, and* N *as above, a random multigraph* Γ_{tnN} *almost surely consists of exactly one giant* s*-component and only few* s*-isolated vertices besides it,*

(5-20d) $$P_{tnN}\left(\bigcup_{k=0}^{n} C_k^s\right) = P_{tnN}(\tilde{Z}_s = 1) \to 1.$$

For the number of s*-components in random multigraphs* Γ_{tnN},

(5-20e) $$P_{tnN}(Z_s = k - 1) \to e^{-e^{-c}} \frac{e^{-ck}}{k!} \qquad (k = 0, 1, 2, \dots)$$

holds true. For the probability that a random multigraph is completely s*-fold connected,*

(5-20f) $$\lim_{n \to \infty} P_{tnN}(C_0^s) = e^{-e^{-c}}$$

holds true if $N = N(n)$ *satisfies condition* (5-20a)*. Furthermore, the expected number of* s*-saturated connections and the expected* s*-degree of an arbitrary vertex* ξ_i *in* Γ_{tnN} *satisfy*

(5-20g) $$E_{tnN} V_s = \frac{1}{2} n \left(\log n + c + o(1)\right),$$

(5-20h) $$E_{tnN} U_{si.} = \log n + c + o(1). \bullet$$

PROOF For $t = s = 1$, edge sequences given by (5-20a) degenerate to those given by (5-16), that is, $N(n) = \lfloor \frac{1}{2} n (\log n + c + o(1)) \rfloor$, and the results of Theorem 5-9 are the same as those of Theorem 5-4, which has been proved by P. Erdős and A. Rényi in [111] and [112]. (In this case, a 1-saturated connection is an edge.)

Let t and s be positive integers with $t \geq s \geq 1$. Inserting (5-20a) in (5-6a) and (5-7a) and performing only straightforward calculations yields (5-20g) and (5-20h), respectively. The correspondence between the right sides of (5-20g) and (5-16) gives the idea to prove (5-20b)–(5-20f): We show that

(5-20i) $$E_{tnN} X_{s.1} = E E_{tnN}(X_{s.1}|V_s) = E\left(n \binom{\binom{n-1}{2}}{V_s} \binom{\binom{n}{2}}{V_s}^{-1}\right)$$
$$\sim \binom{\binom{n-1}{2}}{\lfloor EV_s \rfloor} \binom{\binom{n}{2}}{\lfloor EV_s \rfloor}^{-1} = E_{n, \lfloor EV \rfloor} X_{.1} \sim n\, e^{-2\lfloor EV \rfloor / n}$$

and

(5-20j) $$P_{tnN}(X_{s.1} = k) = E P_{tnN}(X_{s.1} = k|V_s) \sim P_{n, \lfloor EV \rfloor}(X_{.1} = k)$$

both hold true (the right side of (5-20i) now for $\lfloor E_{tnN}V_s \rfloor = o(n^{3/2})$), and that we can say the same about the random variables Z_s, $\tilde{Z}_s$ and the property C_k^s. The behaviour of random multigraphs Γ_{tnN} with $N(n)$ edges — concerning the property of being completely s-fold connected, the distribution of the number of s-isolated vertices, etc. — is the same as the behaviour of random graphs $\Gamma_{n,\lfloor EV \rfloor}$ with $\lfloor EV \rfloor$ edges — now concerning the property of being completely connected, the the distribution of number of isolated vertices, etc. (which is intuitively clear because of our s-projection, see (5-8a)–(5-8c)). Inserting (5-20g) into the right side of (5-20i) gives $E_{tnN}X_{s.1} \sim E_{n,\lfloor EV \rfloor}X_{.1} \to e^{-c}$ and the other assertions of Theorem 5-9 because of (5-6a) and Theorem 5-4 (with $E_{tnN}V_s$ now playing the role of the edge sequence $N(n)$ in simple graphs).

For nonnegative integers n, let $s_n(v)$, $f_n(v)$ and $g_n(v)$, respectively, be functions defined by

$$s_n(v) = E_{nv}X_{.1} = n\left(\binom{n-1}{2}{v}\right)\left(\binom{n}{2}{v}\right)^{-1},$$

$$f_n(v) = n\,e^{-2v/n},$$

$$g_n(v) = n\,\exp\left(-\frac{(n-1)v}{\binom{n-1}{2}-v-1}\right) \qquad \text{for} \qquad 0 \le v \le \binom{n-1}{2}.$$

(For $s_n(v)$, v must be a positive integer, and for $f_n(v)$ and $g_n(v)$, let v be nonnegative and real-valued; for $v > \binom{n-1}{2}$, we put $g_n(v) = 0$.) For fixed n, $s_n(v)$, $f_n(v)$ and $g_n(v)$ are monotonic decreasing and convex. Because of inequality (5-13a),

$$g_n(v) \le s_n(v) = E_{nv}X_{.1} \le f_n(v)$$

holds true. This together with Jensen's inequality for convex functions gives

$$E_{tnN}X_{s.1} = EE_{tnN}(X_{.1}|V_s) = Es_n(V_s)$$
$$\ge s_n(\lfloor EV_s + 1 \rfloor) \ge g_n(\lfloor EV_s + 1 \rfloor) \ge g_n(EV_s + 2).$$

This yields immediately

$$\liminf_{n\to\infty} E_{tnN}X_{s.1} \ge \limsup_{n\to\infty} s_n(\lfloor EV_s + 1 \rfloor) = \limsup_{n\to\infty} E_{n,\lfloor EV_s + 1 \rfloor}X_{.1}$$
$$\ge \limsup_{n\to\infty} g_n(EV_s + 2).$$

On the other hand, as the functions are convex and monotonic decreasing we can write

$$E_{tnN}X_{s.1} = Es_n(V_s) = \sum_v s_n(v)\,P(V_s = v)$$
$$\le s_n(0)\,P(V_s \le k_n) + s_n(\lfloor k_n \rfloor)\,P(k_n < V_s \le EV_s - m_n)$$
$$+ s_n(\lfloor EV_s - m_n \rfloor)\,P(V_s > EV_s - m_n)$$
$$\le n\,P(V_s \le k_n) + f_n(k_n - 1)\,P(V_s \le EV_s - m_n) + s_n(\lfloor EV_s - m_n \rfloor)$$

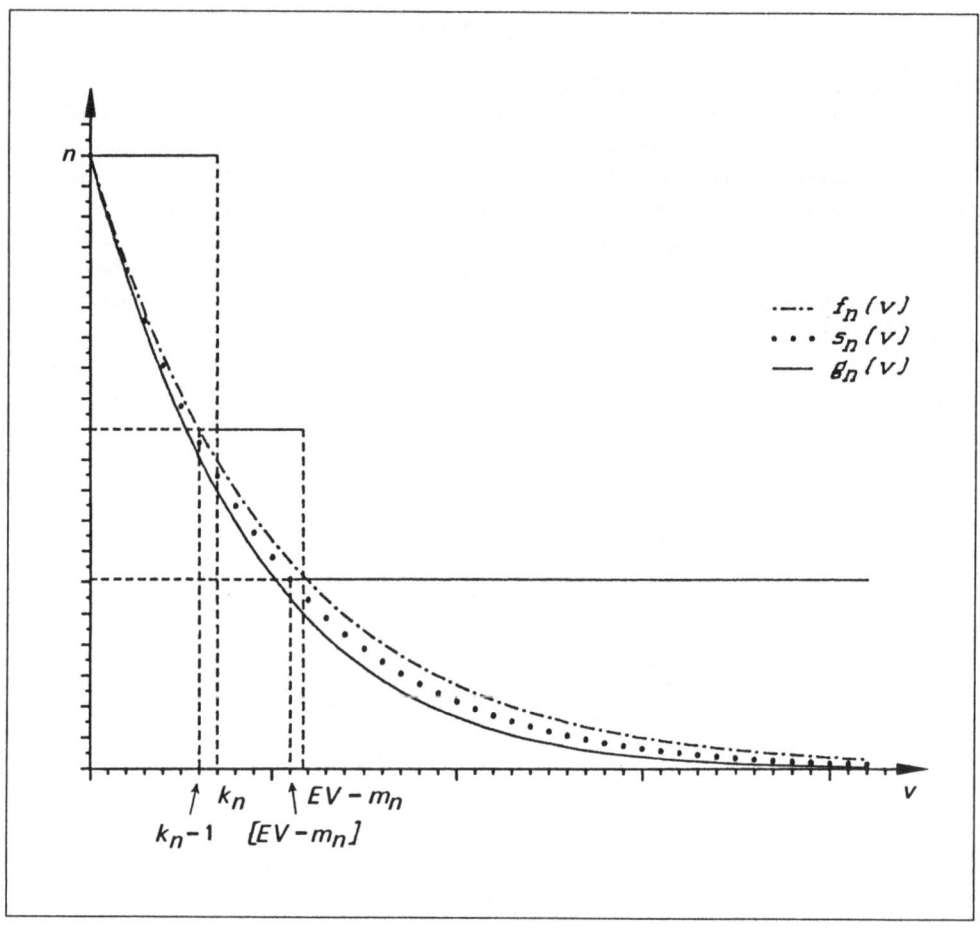

Figure 5-2 *Plot of the functions $s_n(v)$, $f_n(v)$ and $g_n(v)$, respectively, for $n = 20$, and $c = 0$, $o(1) = 0$, $\varepsilon = 0.8$. This yields $\mathrm{E}_{tnN}V_s = \frac{1}{2}n\log n = 29.96$, $m_n = 18.22$, $k_n = 6.99$, and $f_n(k_n - 1) = 20^{0.8} = 10.98$, $s_n(\lfloor EV_s - m_n\rfloor) = 20$, $\binom{\binom{19}{2}}{11}\binom{\binom{20}{2}}{11}^{-1} = 6.07$.*

(see Figure 5-2). From this we get, using Chebyshev's inequality together with the left part of (5-6c),

$$\mathrm{E}_{tnN}X_{s.1} \le n\,\frac{\mathrm{Var}V_s}{(EV_s - k_n)^2} + f_n(k_n - 1)\,\frac{\mathrm{Var}V_s}{m_n^2} + s_n\left(\lfloor EV_s - m_n\rfloor\right)$$

$$\le n\,\frac{EV_s}{(EV_s - k_n)^2} + f_n(k_n - 1)\,\frac{EV_s}{m_n^2} + s_n\left(\lfloor EV_s - m_n\rfloor\right).$$

Defining the sequences (k_n) and (m_n) as

$$k_n = \frac{1}{2}(1 - \varepsilon)\, n \log n + 1 \qquad (0 < \varepsilon < 1),$$

$$m_n = \frac{n}{\log \log n},$$

such that $f_n(k_n - 1) = n^\varepsilon$ and $m_n = o(n)$, we obtain

$$\mathrm{E}_{tnN} X_{s.1} \leq \frac{n\, \mathrm{EV}_s}{(\mathrm{EV}_s - \frac{1-\varepsilon}{2} n \log n - 1)^2} + \frac{(\log \log n)^2 \mathrm{EV}_s}{n^{2-\varepsilon}} + s_n \left(\lfloor \mathrm{EV}_s - n/\log \log n \rfloor\right).$$

Inserting $\mathrm{E}_{tnN} V_s = \frac{1}{2} n (\log n + c + o(1))$ gives

$$\limsup_{n \to \infty} \mathrm{E}_{tnN} X_{s.1} \leq \liminf_{n \to \infty} s_n \left(\lfloor EV_s - m_n \rfloor\right) = \liminf_{n \to \infty} \mathrm{E}_{n, \lfloor EV_s - m_n \rfloor} X_{.1}$$

$$\leq \liminf_{n \to \infty} f_n(EV_s - m_n - 1).$$

If $\mathrm{E}_{tnN} V_s$ satisfies (5-20g), then $\lfloor \mathrm{E}_{tnN} V_s + 1 \rfloor$ and $\lfloor \mathrm{E}_{tnN} V_s - m_n \rfloor$ do also. Thus, $\lim_{n \to \infty} s_n (\lfloor EV_s - m_n \rfloor) = \lim_{n \to \infty} s_n (\lfloor EV_s + 1 \rfloor) = \lim_{n \to \infty} s_n (EV_s) = e^{-c}$ can be proved by straightforward calculations using the definitions of $g_n(v)$, $s_n(v)$ and $f_n(v)$ and the fact that $g_n(v) \leq s_n(v) = E_{nv} X_{.1} \leq f_n(v)$, which is a consequence of (5-13a) (see also Remark 5-4). The assertion $\mathrm{E}_{tnN} X_{s.1} \sim \mathrm{E}_{n, \lfloor EV \rfloor} X_{.1} \to e^{-c}$, that is, (5-20b) then follows for sequences $(\mathrm{E}_{tnN} V_s)$ satisfying (5-20g), that is, for edge sequences $N(n)$ satisfying (5-20a) because of the two inequalitites for $\liminf_{n \to \infty} \mathrm{E}_{tnN} X_{s.1}$ and $\limsup_{n \to \infty} \mathrm{E}_{tnN} X_{s.1}$, respectively.

Let us prove (5-20c). Because of the equiprobability in drawing edges, (5-8a) holds. Using this together with (5-11) and (5-12a), we get

$$P_{tnN}(X_{s.1} = k) = E P_{tnN}(X_{s.1} = k | V_s) = \binom{n}{k} \mathrm{E}\left(L_{n-k, V_s} \binom{\binom{n}{2}}{V_s}^{-1}\right)$$

$$= \binom{n}{k} \sum_{j=0}^{n-k} (-1)^j \binom{n-k}{j} \mathrm{E}\left(\binom{\binom{n-k-j}{2}}{V_s} \binom{\binom{n}{2}}{V_s}^{-1}\right)$$

$$= \binom{n}{k} \sum_{j=0}^{n-k} (-1)^j \binom{n-k}{j} \sum_{v=0}^{\binom{n}{2}} \binom{\binom{n-k-j}{2}}{v} \binom{\binom{n}{2}}{v}^{-1} P_{tnN}(V_s = v).$$

Using Chebyshev's inequality and the fact that the ratio of the two binomial coefficients is monotonic decreasing, we can overestimate this probability as follows:

$$P_{tnN}(X_{s.1} = k) \leq \frac{\mathrm{Var}V_s}{m_n^2} + \sum_{|v - EV_s| \leq m_n} P_{tnN}(V_s = v)$$

$$\times \binom{n}{k} \sum_{\substack{j=0 \\ j \text{ even}}}^{2\mu} \binom{n-k}{j} \binom{\binom{n-k-j}{2}}{\lfloor EV_s - m_n \rfloor} \binom{\binom{n}{2}}{\lfloor EV_s - m_n \rfloor}^{-1}$$

$$- \sum_{|v - EV_s| \leq m_n} P_{tnN}(V_s = v)$$

$$\times \binom{n}{k} \sum_{\substack{j=0 \\ j \text{ odd}}}^{2\mu-1} \binom{n-k}{j} \binom{\binom{n-k-j}{2}}{\lfloor EV_s + m_n \rfloor} \binom{\binom{n}{2}}{\lfloor EV_s + m_n \rfloor}^{-1}$$

$$\leq \frac{\mathrm{Var}V_s}{m_n^2} + \binom{n}{k} \sum_{\substack{j=0 \\ j \text{ even}}}^{2\mu} \binom{n-k}{j} \binom{\binom{n-k-j}{2}}{\lfloor EV_s - m_n \rfloor} \binom{\binom{n}{2}}{\lfloor EV_s - m_n \rfloor}^{-1}$$

$$- \left(1 - \frac{\mathrm{Var}V_s}{m_n^2}\right) \binom{n}{k} \sum_{\substack{j=0 \\ j \text{ odd}}}^{2\mu-1} \binom{n-k}{j} \binom{\binom{n-k-j}{2}}{\lfloor EV_s + m_n \rfloor} \binom{\binom{n}{2}}{\lfloor EV_s + m_n \rfloor}^{-1}.$$

This holds true for every integer μ. Now, for $E_{tnN}V_s = \frac{1}{2}n\left(\log n + c + o(1)\right)$ we get immediately

$$\lim_{n \to \infty} \left\{ \binom{n}{k} \binom{n-k}{j} \binom{\binom{n-k-j}{2}}{\lfloor EV_s \rfloor} \binom{\binom{n}{2}}{\lfloor EV_s \rfloor}^{-1} \right\}$$

$$= \lim_{n \to \infty} \left\{ \binom{n}{k} \binom{\binom{n-k}{2}}{\lfloor EV_s \rfloor} \binom{\binom{n}{2}}{\lfloor EV_s \rfloor}^{-1} \binom{n-k}{j} \binom{\binom{n-k-j}{2}}{\lfloor EV_s \rfloor} \binom{\binom{n-k}{2}}{\lfloor EV_s \rfloor}^{-1} \right\}$$

$$= \frac{e^{-ck}}{k!} \frac{e^{-cj}}{j!},$$

and this limit remains unchanged if we replace EV_s by $EV_s - m_n$ or $EV_s + m_n$ with $m_n = n/\log\log n = o(1)$. Thus, we have proved

$$\limsup_{n \to \infty} P_{tnN}(X_{s.1} = k) \leq \lim_{n \to \infty} \left\{ \binom{n}{k} \sum_{j=0}^{n-k} (-1)^j \binom{n-k}{j} \binom{\binom{n-k-j}{2}}{\lfloor EV_s \rfloor} \binom{\binom{n}{2}}{\lfloor EV_s \rfloor}^{-1} \right\}$$

$$= e^{-e^{-c}} \frac{e^{-ck}}{k!}$$

since $\mathrm{Var}V_s / m_n \leq EV_s / m_n \to 0$ holds true and the sum above converges uniformly and absolutely. On the other hand, we can use the same technique as above to underestimate

$P_{tnN}(X_{s.1} = k)$ by

$$P_{tnN}(X_{s.1} = k) \geq \left(1 - \frac{\operatorname{Var} V_s}{m_n^2}\right)\binom{n}{k} \sum_{\substack{j=0 \\ j \text{ even}}}^{2\mu} \binom{n-k}{j}\left(\frac{\binom{n-k-j}{2}}{\lfloor EV_s + m_n\rfloor}\right)\left(\frac{\binom{n}{2}}{\lfloor EV_s + m_n\rfloor}\right)^{-1}$$

$$- \binom{n}{k} \sum_{\substack{j=0 \\ j \text{ odd}}}^{2\mu-1} \binom{n-k}{j}\left(\frac{\binom{n-k-j}{2}}{\lfloor EV_s - m_n\rfloor}\right)\left(\frac{\binom{n}{2}}{\lfloor EV_s - m_n\rfloor}\right)^{-1}.$$

This at the same time gives

$$\liminf_{n\to\infty} P_{tnN}(X_{s.1} = k) \geq e^{-e^{-c}}\frac{e^{-ck}}{k!}.$$

Combining both results, we have proved (5-20c). That is, $\lim_{n\to\infty} P_{tnN}(X_{s.1} = k) = \lim_{n\to\infty} P_{n,\lfloor E_{tnN}V_s\rfloor}(X_{.1} = k)$ for sequences of random multigraphs with edge sequences $N = N(n)$ satisfying (5-20a).

Using virtually the same ideas as in the proof of (5-20c), we can show that $\lim_{n\to\infty} P_{tnN}(\tilde{Z}_s = 1) \geq \lim_{n\to\infty} P_{n,\lfloor E_{tnN}V_s\rfloor}(\tilde{Z} = 1)$ holds true. That proves (5-20d) since for simple random graphs with $\lfloor E_{tnN}V_s\rfloor = \lfloor \frac{1}{2}n(\log n + c + o(1))\rfloor$ edges, the expression on the right hand of the equation tends to unity as has been said in Remark 5-4 (see also [149] and [153]).

Formula (5-20c) together with (5-20d) gives us (5-20e) and (5-20f) because of

$$P_{tnN}(Z_s = k+1) = P_{tnN}(\{Z_s = k+1\}\cap\{\tilde{Z}_s = 1\}) + P_{tnN}(\{Z_s = k+1\}\cap\{\tilde{Z}_s \geq 2\})$$
$$= P_{tnN}(\{X_{s.1} = k\}\cap\{\tilde{Z}_s = 1\}) + P_{tnN}(\{Z_s = k+1\}\cap\{\tilde{Z}_s \geq 2\})$$
$$\to P_{tnN}(\{X_{s.1} = k\}\cap\{\tilde{Z}_s = 1\})$$

and

$$P_{tnN}(X_{s.1} = 0) = P_{tnN}(\{X_{s.1} = 0\}\cap\{\tilde{Z}_s = 1\}) + P_{tnN}(\{X_{s.1} = 0\}\cap\{\tilde{Z}_s \geq 2\})$$
$$= P_{tnN}(\mathcal{C}_0^s) + P_{tnN}(\{X_{s.1} = 0\}\cap\{\tilde{Z}_s \geq 2\})$$
$$\to P_{tnN}(\mathcal{C}_0^s).$$

This completes the proof because of $P_{tnN}(\tilde{Z}_s \geq 2) \to 0$. ●

For large n, almost all random multigraphs Γ_{tnN} consist of one big s-component and some s-isolated vertices if $N(n)$ satisfies (5-20a). This is the reason for $P_{tnN}(Z_s = k+1) \to P_{tnN}(X_{s.1} = k)$. In the limit, the value of the random variable $X_{s.1}$ almost surely is identical with the number of vertices lying not in the biggest s-component. However, what happens if $c_n \neq c + o(1)$, that is, if $c_n \to \infty$ or $c_n \to -\infty$ holds or if this sequence does not converge? The results of the theorem above can be reversed as follows.

Theorem 5-10 Let $(\Gamma_{tnN})_{n\to\infty, N=N(n)}$ be sequences of random multigraphs with edge sequences given by

(5-21a)
$$N(n) = \left\lfloor \frac{t}{2\binom{t}{s}^{1/s}} n^{2-1/s}(\log n + c_n)^{1/2}\right\rfloor.$$

For diverging sequences $(c_n)_{n \to \infty}$, *the following assertions hold true:*

(a)   If $c_n \to \infty$, *then*

(5-21b) $$\mathrm{E}_{tnN} X_{s.1} \to 0,$$

(5-21c) $$P_{tnN}(X_{s.1} = 0) \to 1,$$

(5-21d) $$P_{tnN}(\tilde{Z}_s = 1) \to 1,$$

(5-21e) $$P_{tnN}(Z_s = 1) \to 1,$$

(5-21f) $$P_{tnN}(\mathcal{C}_0^s) \to 1.$$

(b)   If $c_n \to -\infty$, *then*

(5-21g) $$\mathrm{E}_{tnN} X_{s.1} \to \infty,$$

(5-21h) $$P_{tnN}(\mathcal{C}_0^s) \to 0.$$

(c)   If $\limsup_{n \to \infty} c_n > \liminf_{n \to \infty} c_n$, *then the sequences (5-20b), (5-20c), (5-20e) and (5-20f), respectively, have more than one limit point.* ●

PROOF (SKETCH) Let Γ_{t,n,N_1} and Γ_{t,n,N_2} be multigraphs with N_1 and N_2 edges, respectively. Let $N_1 < N_2$. Because of the monotoneity of probabilities, we get $P_{t,n,N_1}(\mathcal{C}_0^s) < P_{t,n,N_2}(\mathcal{C}_0^s)$ and $\mathrm{E}_{t,n,N_1} X_{s.1} > \mathrm{E}_{t,n,N_1} X_{s.1}$. From this, parts (a) and (b) follow immediately. (Formulas (5-21c)–(5-21e) are direct implications of (5-21f).)

To prove part (c), let us choose from $(c_n)_{n \to \infty}$ two convergent subsequences $(c_{n_1}) \to c_1^*$ and $(c_{n_2}) \to c_2^*$ with $c_1^* \neq c_2^*$, and the respective subsequences $N_1(n)$ and $N_2(n)$ from $N(n)$. Because of Theorem 5-9, $\mathrm{E}_{t,n,N_1(n)} X_{s.1} \to \exp(-c_1^*)$ must hold for (c_{n_1}) and $N_1(n)$, and $\mathrm{E}_{t,n,N_2(n)} X_{s.1} \to \exp(-c_2^*)$ holds true for (c_{n_2}) and $N_2(n)$. We get more than one limit point for the sequence of expectations. The same can be shown for the probabilities, too. ●

A more detailed proof of Theorem 5-10 has been published together with the proof of Theorem 5-9 in [149], parts of both in [150]. By repeating them here, we want to illustrate the principle of applying concepts from probability theory instead of enumeration concepts to prove combinatorial results. They also will lead to an easier understanding of the following theorems.

Both theorems show that edge sequences satisfying condition (5-20a) are threshold functions: If "too much" edges remain in a multigraph (if $c_n \to \infty$), then it is almost surely completely s-fold connected; if we omit too many edges (if $c_n \to -\infty$), then with probability 1 it will be split into more than one proper s-components. For edge sequences satisfying (5-20a), multigraphs have a rather complex structure (virtually the same as graphs with edge sequences fulfilling (5-16)). We again expect exactly one proper s-component which contains almost all vertices and edges, and some s-isolated vertices. In one point, however, random graphs and random multigraphs differ: In random graphs with (5-16), no edges are outside the giant component, in random multigraphs, however, some edges may link the s-isolated vertices together or to the proper s-component. (The s-degree of these vertices must be 0 but not neccessary the $(s-1)$-degree, ..., 1-degree.)

Every edge sequence for which the expected number of proper s-components tends to a value larger than unity, must grow slower in n than (5-20a) because of Theorem 5-10. In the following, we consider sequences of random multigraphs with such "slow growing" edge sequences. We especially want to answer the following questions:

(B) Do edge sequences $N(n)$ exist such that the related sequences of expected numbers of certain s-components — of s-trees of size m or of isolated s-cycles, for example — are almost surely bounded and greater than 0?

(C) For what kind of edge sequences $N(n)$ do the sequences of s-degrees of an arbitrary vertex ξ_i satisfy $\mathrm{E}_{tnN}U_{si.} \to c\ (c > 0)$?

(D) For what kind of edge sequences $N(n)$ do the sequences of expected numbers of vertices with given s-degree m converge to a positive constant?

(E) For what kind of edge sequences $N(n)$ do the expected numbers of s-saturated connections remain positive and bounded?

For all these problems, we are not only interested in the expectations but also in the limit distributions of the respective random variables. As a first answer to Question (B), the following theorem gives the distributions of the numbers of s-trees of size m.

Theorem 5-11 *In sequences* $(\Gamma_{tnN})_{n\to\infty,N=N(n)}$ *of random multigraphs with edge sequences satisfying*

$$(5\text{-}22\mathrm{a}) \qquad N(n) = \left\lfloor \frac{t}{2\binom{t}{s}^{1/s}}\, n^{2-1/s} \left(\frac{1}{m} \log n + \frac{m-1}{m} \log \log n + c + o(1) \right)^{1/s} \right\rfloor,$$

the expected numbers of s-trees of size m are given by

$$(5\text{-}22\mathrm{b}) \qquad \mathrm{E}_{tnN}X_{s.m}^{(T)} \to \mu := \frac{e^{-mc}}{mm!}.$$

Only for sequences $N(n)$, satisfying (5-22a), the sequence of expectations tends to a positive, finite limit. For this case, the limit distribution is

$$(5\text{-}22\mathrm{c}) \qquad P_{tnN}(X_{s.m}^{(T)} = k) \to e^{-\mu}\frac{\mu^k}{k!} \qquad (k = 0, 1, 2, \dots).$$

The expected numbers of s-saturated connections and s-isolated vertices and the expected s-degree of any vertex ξ_i are given by

$$(5\text{-}22\mathrm{d}) \qquad \mathrm{E}_{tnN}V_s = \frac{1}{2}\, n \left(\frac{1}{m} \log n + \frac{m-1}{m} \log \log n + c + o(1) \right),$$

$$(5\text{-}22\mathrm{e}) \qquad \mathrm{E}_{tnN}U_{si.} = \frac{1}{m} \log n + \frac{m-1}{m} \log \log n + c + o(1),$$

$$(5\text{-}22\mathrm{f}) \qquad \mathrm{E}_{tnN}X_{s.1} = \left(\frac{n}{\log n} \right)^{(m-1)/m} e^{-(c+o(1))}. \ \bullet$$

PROOF (SKETCH) The proof of $\mathrm{E}_{tnN}X_{s.m}^{(T)} \sim \mathrm{E}_{n,\lfloor EV\rfloor}X_{.m}^{(T)}$ can be performed in the same way as that of $\mathrm{E}_{tnN}X_{s.1} \sim \mathrm{E}_{n,\lfloor EV\rfloor}X_{.1}$ in Theorem 5-9. Assertions (5-22b) and (5-22c) both follow from (5-17c) and Theorem 5-5. Assertions (5-22d) and (5-22e) can be verified by inserting (5-22a) in (5-6a) and (5-7a). (5-22f) can be proved by inserting (5-22d) instead of $N(n)$ into the right side of (5-12b) which can be done because of (5-20i). $\bullet$

The results of Theorem 5-9 on the limit distribution of the s-isolated vertices, and of the expected numbers of s-saturated connections or of the s-degree of an arbitrary vertex, are contained as special case $m = 1$ in Theorem 5-11. This agrees with the definition that s-isolated vertices are s-trees of size 1, too. The expected number of s-trees with less than m vertices grows to infinity if $N(n)$ satisfies (5-22a). For large n however, we don't expect s-trees with more than m vertices.

Problem (C) can be rewritten as follows. $E_{tnN}U_{si.} = c + o(1)$ for $c > 0$. Because of (5-7a), the edge sequence then must be given by

$$N(n) = \left\lfloor \frac{t}{2\binom{t}{s}^{1/s}} n^{2-1/s} \left(c + o(1)\right)^{1/s} \right\rfloor \qquad (c > 0).$$

This gives $E_{tnN}V_s = \frac{1}{2} n\left(c + o(1)\right)$ for $c > 0$ because of (5-6a) or $E_{tnN}V_s = \frac{1}{2} n E_{tnN}U_{si.}$ (which follows from the definitions of both random variables). Because of (5-19b) from Theorem 5-8, we now can prove the following results.

Theorem 5-12 *Only in sequences* $(\Gamma_{tnN})_{n\to\infty, N=N(n)}$ *of random multigraphs with edge sequences* $N(n)$ *satisfying*

(5-23a) $$N(n) = \left\lfloor \frac{t}{2\binom{t}{s}^{1/s}} n^{2-1/s} \left(c + o(1)\right)^{1/s} \right\rfloor \qquad (c > 0),$$

the sequence $(E_{tnN}U_{si.})_{n\to\infty}$ *of the expected s-degrees of an arbitrary vertex ξ_i remains bounded and tends to a positive limit,*

(5-23b) $$E_{tnN}U_{si.} \to c.$$

In this case, the random variable $U_{si.}$ *is asymptotically* $P(c)$-*distributed,*

(5-23c) $$P_{tnN}(U_{si.} = k) \to e^{-c} \frac{c^k}{k!} \qquad (k = 0, 1, 2, \dots).$$

The expected number of s-saturated connections or of s-isolated vertices, respectively, is

(5-23d) $$E_{tnN}V_s = \frac{1}{2} n\left(c + o(1)\right),$$

(5-23e) $$E_{tnN}X_{s.1} = n e^{-(c + o(1))}. \; \bullet$$

PROOF Formulas (5-23a), (5-23b), (5-23d) and (5-23e) follow from each other by equivalent transformations because of (5-6a), (5-7a) and (5-20i). Using the idea of the s-projection, that is, using (5-8a), together with the left side of (5-14a), we get

$$P_{tnN}(U_{si.} = k) = E P_{tnN}(U_{si.} = k | V_s) = E\left(\binom{n-1}{k} \binom{\binom{n-1}{2}}{V_s - k} \binom{\binom{n}{2}}{V_s}^{-1} \right)$$

$$= \binom{n-1}{k} \sum_{v=0}^{\binom{n}{2}} P(V_s = v) \binom{\binom{n-1}{2}}{v - k} \binom{\binom{n}{2}}{v}^{-1}.$$

The ratio of the binomial coefficients is monotonic decreasing in v. This together with Chebyshev's inequality and the left part of (5-6c), yields

$$P_{tnN}(U_{si.} = k) \leq \binom{n-1}{k} \sum_{\substack{v \\ |v - EV_s| \leq m_n}} P(V_s = v) \left(\binom{\binom{n-1}{2}}{v-k}\right) \left(\binom{\binom{n}{2}}{v}\right)^{-1} + \frac{\operatorname{Var} V_s}{m_n^2}$$

$$\leq \binom{n-1}{k} \left(\binom{\binom{n-1}{2}}{\lfloor EV_s - m_n\rfloor - k}\right) \left(\binom{\binom{n}{2}}{\lfloor EV_s - m_n\rfloor}\right)^{-1} + \frac{EV_s}{m_n^2}.$$

Inserting $E_{tnN}V_s = \frac{1}{2} n(c + o(1))$ and $m_n = n/\log\log n = o(1)$ and using the right part of (5-14a), we get $\limsup_{n\to\infty} P_{tnN}(U_{si.} = k) \leq e^{-c} \frac{c^k}{k!}$ by elementary calculations. With similar arguments, we can show $\liminf_{n\to\infty} P_{tnN}(U_{si.} = k) \geq e^{-c} \frac{c^k}{k!}$. Both estimations together give $P_{tnN}(U_{si.} = k) \sim P_{n,\lfloor EV\rfloor}(U_{i.} = k) \sim e^{-c} \frac{c^k}{k!}$ for those edge sequences $N(n)$ satisfying (5-23a). •

In virtually the same manner, the other results of Theorem 5-8 can be generalized from random graphs Γ_{nN} to random multigraphs Γ_{tnN}. This holds for the results of Theorems 5-6 and 5-7 as well. We state these generalizations here as Theorems 5-13–5-15 without proof.

Theorem 5-13 *In sequences* $(\Gamma_{tnN})_{n\to\infty, N=N(n)}$ *of random multigraphs with edge sequences satisfying (5-23a), we have*

(5-23f) $\qquad E_{tnN}X_{s.m}^{(C)} \to \kappa = \dfrac{c^m}{2m} e^{-cm},$

(5-23g) $\qquad E_{tnN}X_{s.m}^{(m,m)} \to \eta = \dfrac{c^m}{2m} e^{-cm}\left(1 + m + \dfrac{m^2}{2!} + \cdots + \dfrac{m^{m-3}}{(m-3)!}\right),$

(5-23h) $\qquad E_{tnN}\tilde{X}_{s.m}^{(C)} \to \theta = \dfrac{c^m}{2m}.$

The expected numbers of isolated s-cycles, of s-components of size m, $m \geq 3$, with exactly one cycle, and of (not necessary isolated) s-cycles in such random multigraphs tend to positive limits. The limit distributions of these random variables exist,

(5-23i) $\qquad P_{tnN}(X_{s.m}^{(C)} = k) \to e^{-\kappa} \dfrac{\kappa^k}{k!} \qquad (k = 0, 1, 2, \dots),$

(5-23j) $\qquad P_{tnN}(X_{s.m}^{(m,m)} = k) \to e^{-\eta} \dfrac{\eta^k}{k!} \qquad (k = 0, 1, 2, \dots),$

(5-23k) $\qquad P_{tnN}(\tilde{X}_{s.m}^{(C)} = k) \to e^{-\theta} \dfrac{\theta^k}{k!} \qquad (k = 0, 1, 2, \dots).$ •

Theorem 5-14 *In sequences* $(\Gamma_{tnN})_{n\to\infty, N=N(n)}$ *of random multigraphs with edge sequences $N(n)$ satisfying*

(5-24a) $\qquad N(n) = \left\lfloor \dfrac{t}{2\binom{t}{s}^{1/s}} n^{2-1/s} \left(n^{-1/(m-1)}(c + o(1))\right)^{1/s} \right\rfloor \qquad (c > 0),$

the expected number of s-trees of size m, $m = 2, 3, 4, \ldots$, *converges to a positive limit,*

$$(5\text{-}24\text{b}) \qquad \mathrm{E}_{tnN} X_{s.m}^{(T)} \to \nu = c^{m-1} \frac{m^{m-2}}{m!} \, .$$

In the limit, this number is $\mathrm{P}(\nu)$-*distributed,*

$$(5\text{-}24\text{c}) \qquad P_{tnN}(X_{s.m}^{(T)} = k) \to e^{-\nu} \frac{\nu^k}{k!} \qquad (k = 0, 1, 2, \ldots).$$

The expected numbers of s-saturated connections, of the s-degree of a vertex ξ_i, *and of the s-isolated vertices, respectively, in* Γ_{tnN} *are*

$$(5\text{-}24\text{d}) \qquad \mathrm{E}_{tnN} V_s = \frac{1}{2} n^{(m-2)/(m-1)} \big(c + o(1)\big),$$

$$(5\text{-}24\text{e}) \qquad \mathrm{E}_{tnN} U_{si.} = n^{-1/(m-1)} \big(c + o(1)\big),$$

$$(5\text{-}24\text{f}) \qquad \mathrm{E}_{tnN} X_{s.1} = n \, \exp\!\Big(-n^{-1/(m-1)} \big(c + o(1)\big)\Big). \; \bullet$$

Theorem 5-15 *In sequences* $(\Gamma_{tnN})_{n \to \infty, N = N(n)}$ *of random multigraphs with edge sequences* $N(n)$ *satisfying (5-24a), the expectation of the random variable* $W_{s,.,m-1}$, *the number of vertices with s-degree* $m - 1$, *is*

$$(5\text{-}24\text{g}) \qquad \mathrm{E}_{tnN} W_{s,.,m-1} \to \zeta = \frac{c^{m-1}}{(m-1)!} \, ,$$

The expected number remains bounded. The distribution tends to a $\mathrm{P}(\zeta)$-*distribution,*

$$(5\text{-}24\text{h}) \qquad P_{tnN}(W_{s,.,m-1} = k) \to e^{-\zeta} \frac{\zeta^k}{k!} \qquad (k = 0, 1, 2, \ldots). \; \bullet$$

Theorems 5-11 and 5-14 show that two answers exist to Question (B): The first defines an edge sequence, where s-trees exist besides a big s-component; the second gives a — slowly growing — edge sequence, where s-trees of size m are the biggest expected s-components themselves in random multigraphs. The edge sequences from Theorem 5-14 are the same which are asked for in (D): The expected numbers of vertices with a given s-degree remains positive and finite.

For the special case $m = 2$, the number of edges in (5-24a) is so small that almost surely only few s-trees of size 2 (s-fold connected pairs of vertices) exist besides s-isolated vertices. We don't expect bigger proper s-components. In this case, the expected number of s-saturated connections also remains finite as the following theorem shows as an answer to Problem (E).

Theorem 5-16 *In sequences* $(\Gamma_{tnN})_{n \to \infty, N = N(n)}$ *of random multigraphs with edge sequences* $N(n)$ *satisfying*

$$(5\text{-}25\text{a}) \qquad N(n) = \left\lfloor \frac{t}{2\binom{t}{s}^{1/s}} \, n^{2 - 2/s} \big((c + o(1))\big)^{1/s} \right\rfloor \qquad (c > 0),$$

the expected number of *s*-saturated connections satisfies

(5-25b)
$$\mathrm{E}_{tnN} V_s \to \frac{c}{2}.$$

Only for edge sequences as above, this sequence of expectations remains bounded and tends to a positive limit. Furthermore,

(5-25c)
$$\mathrm{E}_{tnN} U_{si.} = \frac{1}{n}\left(c + o(1)\right),$$

(5-25d)
$$\mathrm{E}_{tnN} X_{s.1} = n \exp\left(-\frac{1}{n}\left(c + o(1)\right)\right). \bullet$$

From Theorem 5-15, we get for $m = 2$ that $\mathrm{E}_{tnN} W_{s.1} \to c$ holds true. We expect c vertices with *s*-degree 1. Together with Theorem 5-16 — which tells us to expect $c/2$ *s*-saturated connections for large n — this means that only *s*-trees of size 2 can occur with probability 1.

Remark 5-6 Meanwhile, detailed proofs of Theorems 5-10, 5-12 and 5-16 have been published in [153], the proofs of Theorems 5-11, 5-13 and 5-14 in [154] and [155]. To prove Theorem 5-15 we need Theorem 5-7 which is generalized to random multigraphs Γ_{tnN} using

$$P_{tnN}(W_{s,.,m-1} = k) \sim P_{n,\lfloor EV\rfloor}(W_{.,m-1} = k). \bullet$$

Remark 5-7 Because of $\mathrm{E}_{tnN} \sim \mathrm{Var}_{tnN} V_s$ (see (5-6d)), we suppose that V_s is asymptotically $P(c/2)$-distributed if $N(n)$ satisfies Condition (5-25a). However, we didn't prove this assertion. $\bullet$

5.5 Discussion of the Results

For the remainder of this chapter, we write $N_{ts}(n)$ to denote edge sequences which satisfy one of the conditions (5-20a), (5-22a)–(5-25a). This is done mainly to emphasize the dependency of $N_{ts}(n)$ of the parameters t and s, respectively. Using the left part of (5-6c) and the inequality of Chebyshev,

$$\lim_{n\to\infty} P_{tnN}(|V_s - \mathrm{E}_{tnN} V_s| \geq \sqrt{n}\,\log n) \to 0$$

can be verified easily. The density of the random variable V_s, the number of *s*-saturated connections in random multigraphs Γ_{tnN}, is concentrated in a range of size $O(\sqrt{n}\,\log n)$ round its expectation. This is of special interest for such edge sequences which satisfy at least (5-24a) or grow even faster in n (because then $\sqrt{n}\,\log n = o(\mathrm{E}_{tnN} V_s)$ holds true).

From (5-6a), we immediately get the following facts for the expected numbers of $(s + 1)$-fold connections.

(a) For those edge sequences $N = N_{ts}(n)$ satisfying either (5-20a) or even, more generally, (5-22a) for an arbitrary integer m, we get

$$\mathrm{E}_{tnN} V_{s+1} = O\left(n^{1-1/s}(\log n)^{1+1/s}\right) \qquad (s + 1 \leq t),$$
$$\mathrm{E}_{tnN} V_{s+2} = O\left(n^{1-2/s}(\log n)^{1+2/s}\right) \qquad (s + 2 \leq t)$$

etc. for $s + 3, \ldots, t$.

(b) For edge sequences $N_{ts}(n)$ satisfying (5-23a), we get

$$\mathbb{E}_{tnN}V_{s+1} = O(n^{1-1/s}) \qquad (s + 1 \leq t),$$
$$\mathbb{E}_{tnN}V_{s+2} = O(n^{1-2/s}) \qquad (s + 2 \leq t)$$

etc. for $s + 3, \ldots, t$.

From these estimations, we can derive that the expected number of *at least* $(s + 1)$-saturated connections — connections with $s+1$ *or more* edges — is very small compared to the expectation of s-saturated connections in multigraphs Γ_{tnN} with $N = N_{ts}(n)$ edges.

We further get from these formulas that the ratio of $(s+k+1)$-saturated connections to $(s + k + 1)$-saturated connections is given as follows:

(a) For those edge sequences $N = N_{ts}(n)$ satisfying either (5-20a) or (5-22a) for an arbitrary integer m, we get

$$\frac{\mathbb{E}_{tnN}V_{s+k+1}}{\mathbb{E}_{tnN}V_{s+k}} = O\left((\log n/n)^{1/s}\right).$$

(b) For edge sequences $N_{ts}(n)$ satisfying (5-23a), we get

$$\frac{\mathbb{E}_{tnN}V_{s+k+1}}{\mathbb{E}_{tnN}V_{s+k}} = O\left((1/n)^{1/s}\right).$$

This ratio depends only on s for any k, $k = -s, -s + 1, \ldots, t - s - 1$. Here, $\mathbb{E}_{tnN}V_0$ is defined as $\mathbb{E}_{tnN}V_0 := \binom{n}{2}$: Every connection obviously is 0-saturated which means that at least 0 edges connect two vertices. Similar ratios can be calculated for edge sequences satisfying other conditions than either (5-20a), (5-22a) or (5-23a).

Let us consider the special case $s = 1$. As mentioned earlier, for $t = s = 1$ a 1-saturated connection is the same as an edge, and $\mathbb{E}_{1nN}V_1 = N$ by definition. For $s = 1$ and arbitrary $t \in \mathcal{N}$, (5-6a) implies

$$\mathbb{E}_{t,n,N_{t1}(n)}V_1 \sim N_{t1}(n)$$

if $N_{t1}(n) = o(n^2)$. Moreover, for every edge sequence, the order of magnitude of the ratio of more than s-saturated connections to s-saturated ones is smallest for the case $s = 1$. Thus, results which have been proved for random graphs Γ_{nN} hold true without any modifications of the edge sequences for random multigraphs Γ_{tnN} for $s = 1$ and arbitrary $t \in \mathcal{N}$. The edge sequences can be chosen independently of t, and an extremely small portion of connections only is 2-, 3- or even higher connected. This also explains why the model for multigraphs from [12] — called Model 2 in Subsection 5.2.3 — leads to virtually the same results as Model 3 if $N(n)$ does not grow too fast in n: We simply do not expect more than one edge connecting two vertices for moderate values of N. Thus, Model 2 can be derived as special case $t \to \infty$ from our model (see [149] for more details).

Table 5-1 *Values t, n and N, respectively, for which $EX_{2.1} = e^0 = 1$ holds true (from [149]).*

$s =$	2	$n = 140$	$n = 200$
$t =$	2	$N = 3633$	$N = 6447$
$t =$	20	$N = 2878$	$N = 5044$
$t =$	100	$N = 2836$	$N = 4966$

The factor $t\left(2\binom{t}{s}^{1/s}\right)^{-1}$ is common to all edge sequences (5-20a)–(5-25a). It equals $1/2$ for $s = 1$ and all t. Now let $s > 1$. For $t_1 < t_2$, we get $t_1^s/\binom{t_1}{s} > t_2^s/\binom{t_2}{s}$. This leads to

$$\lim_{n \to \infty} \frac{N_{t_1,s}(n)}{N_{t_2,s}(n)} = \frac{t_1}{t_2}\left(\frac{\binom{t_2}{s}}{\binom{t_1}{s}}\right)^{1/s} > 1 \qquad (1 < s \le t_1 < t_2).$$

That means that edge sequences $N_{t_1,s}(n)$ increase faster than sequences $N_{t_2,s}(n)$. To say it in a somehow unmathematical way: The lower the number t of layers in a multigraph is, the greater must be the number N_t of edges in such a multigraph which is necessary to get the same expected number of s-isolated vertices (of s-trees of size m, the same expected s-degree of a vertex etc.) in that random multigraph. From Theorem 5-10, we get

$$E_{t_2,n,N_{t_1,s}(n)}X_{s.1} \to 0 \qquad (s > 1).$$

This is contrary to the result for $s = 1$ which gives the same expected number of 1-isolated vertices for every number t of layers in Γ_{tnN}. Table 5-1, a small abstract from the tables in [149], illustrates the fact that for fixed n the number N of edges in Γ_{tnN} decreases in t. For this table, we derived an exact formula for $EX_{2.1}$, the expected number of 2-isolated vertices (see the next section for computational formulas).

For $t = s$ and $t \to \infty$, the factor $t\left(2\binom{t}{s}^{1/s}\right)^{-1}$ is given by

$$\frac{t}{2\binom{t}{s}^{1/s}} = \begin{cases} \dfrac{s}{2} & \text{for } t = s, \\[2ex] \dfrac{(s!)^{1/s}}{2} \sim \dfrac{s}{2e} & \text{for } t \to \infty \end{cases}$$

(the last part follows from Stirling's formula). From this, we get

$$\lim_{s \to \infty} \lim_{t \to \infty} \lim_{n \to \infty} \frac{N_{ss}(n)}{N_{ts}(n)} = e,$$

no matter whether the edge sequence $N(n)$ satisfies (5-20a) or one of the other conditions, (5-22a)–(5-25a). Taking the limit of multigraphs with an "infinite" number of layers, we get different edge sequences depending on the fact whether we first let $t \to \infty$ for fixed s, followed by $s \to \infty$, or we choose $s = t \to \infty$ directly.

From the theorems and proofs of the previous section, it can be seen that the relation between edges in simple graphs and s-saturated connections in undirected, completely labelled multigraphs together with formulas like

$$E_{tnN}X_{s.m} \sim E_{n,\lfloor EV \rfloor}X_{.m} \qquad (m = 1, 2, \ldots),$$

$$E_{tnN}U_{si.} \sim E_{n,\lfloor EV \rfloor}U_{i.} \qquad (i = 1, \ldots, n),$$

$$P_{tnN}(X_{s.m} = k) \sim P_{n,\lfloor EV \rfloor}(X_{s.m} = k) \qquad (k = 0, 1, 2, \ldots; \ m = 1, 2, \ldots)$$

are the decisive points to generalize the results in Section 5.3 to multigraphs for arbitrary natural numbers s and t $(1 \leq s \leq t)$. Obviously, many other results on the structure of random graphs Γ_{nN} can be generalized to random multigraphs Γ_{tnN}, as defined in this monograph, in virtually the same way. This statement holds at least for edge sequences $N(n)$ or the respective expectations $E_{t,n,N(n)}V$ in multigraphs $\Gamma_{t,n,N(n)}$, for which the random graphs $\Gamma_{n,\lfloor EV \rfloor}$ of Model 3 and G_{np} of Model 1 have similar structures. For all the proofs in the previous section can be performed by using the fact that the number v of s-fold connections in Γ_{tnN} is the realization of a random variable, namely V_s. This leads to random graphs G_{np} with p, the probability that two vertices are s-fold connected, given by (5-5a) or (5-6a), respectively, which states the connection between random graphs $\Gamma_{n,\lfloor EV \rfloor}$ and G_{np}. The probability p of getting an s-fold connection in Γ_{tnN} — of drawing edges in a graph G_{np} — must be small enough such that it virtually does not depend on whether some other pairs of vertices are also s-fold connected or not (some kind of "quasi-independence"). This at least seems to be the case as long as $EV_s = o(n^{3/2})$ holds true.

Digraphs also can be handled by our model. We choose $t = 2$. The first layer of Γ_2 contains an edge κ_{ij1} connecting the vertices ξ_i and ξ_j if and only if the arc δ_{ij} $(i < j)$ is in the edge set of the digraph Δ. The second layer contains an edge κ_{ij1} between ξ_i and ξ_j if and only if the arc δ_{ji} $(i < j)$ is in the edge set of the digraph Δ, pointing from ξ_i to ξ_j (see Section 1.3). Thus, results like those proved in [68] or [302] can be verified using our model.

5.6 Hints for the Numerical Computation of the Expectations and Distributions

In the theory of random graphs and random multigraphs, one is interested mainly in the asymptotic behaviour of certain graph parameters on probability spaces of random graphs or random multigraphs. These parameters characterize the properties of classes of such graphs of size n as $n \to \infty$. This especially holds when properties of random graphs are studied for their own sake and without any intention of practical applications. By now, random graphs are studied mostly for their own sake, and calculus of random graphs sometimes is considered as a branch of probability theory and not of combinatorics. The probabilistic technique in random graph theory, however, enables one to prove the existence of graphs with special properties without being forced to construct such graphs directly. Thus in the late fifties when random graph theory originated, random graphs mainly were used when it was thought to be hard to construct a graph with a given property by pure enumeration, that is, by combinatorial arguments.

On the other hand, calculus of graph theory provides researchers of different topics with a model to simulate many systems — especially in applied mathematics and life sciences. Cluster analysis or numerical taxonomy is one example. As the assertions tend to hold only "if n is sufficiently large" (with the exception of Theorems 5-1–5-3), the question arises how large does n have to be to make the results "nearly true". Scientists who want to apply results of random graph theory to real data as in cluster analysis, however, normally deal with graphs of small sizes. The edge numbers rarely exceed $n = 500$. Exact, finite formulas — for example, for the calculation of the probabilities of Γ_{nN} having exactly k components of size m or of Γ_{tnN} having exactly k s-components of size m — usually are fairly intricate as (5-9a) and (5-11) show (see also [149], where some finite formulas for random multigraphs have been derived). Hence, asymptotic results are often applied for rather small values of n, and it is of interest to know how good these approximations are when n is not too large (see [57]). This knowledge is especially then valuable when the theory of random graphs is applied to statistical inference, for example, to test the hypothesis of random clusters in cluster analysis (see [154], [265], [350], [351]).

For the sake of brevity, we want to discuss this problem here for random graphs and for the distribution of the number of isolated vertices only. Other results simply are stated here because they can be proved in virtually the same manner. Thus, let us consider the two well-known theorems from [111] and give the generalization to random multigraphs afterwards.

Theorem 5-17 *Let the random variable $X_{.1}$ denote the number of isolated vertices in a random graph, and let $P_{nN}(X_{.1} = k)$ denote the probability that a random graph Γ_{nN} has exactly k isolated vertices. Then, for*

$$(5\text{-}16b) \qquad N = N(n) = \left\lfloor \frac{1}{2}\, n\,(\log n + c) \right\rfloor$$

the following holds true:

$$\lim_{n \to \infty} P_{nN}(X_{.1} = k) = e^{-\lambda}\, \frac{\lambda^k}{k!} \qquad (k = 0, 1, 2, \dots)$$

with $\lambda = e^{-c}$; that is, the number of isolated vertices in random graphs Γ_{nN}, in the limit, has a Poisson distribution with expectation $\mathrm{E}_{nN} X_{.1} = e^{-c}$. •

Theorem 5-18 *Let the edges of a random graph with n vertices be chosen successively among the possible edges in such a manner that at each stage every edge which has not yet been chosen has the same probability to be chosen as the next, and let us continue this process until the graph is connected. Let V denote the number of edges of the resulting connected graph Γ. Then we have for $|l| = O(n)$,*

$$(5\text{-}26) \qquad P\left(V = \left\lfloor \frac{1}{2}\, n \log n \right\rfloor + l \right) \sim \frac{2}{n} e^{-2l/n - e^{-2l/n}}.$$ •

The result of Theorem 5-17 can be easily generalized to edge sequences $N = N(n)$ with

$$(5\text{-}16a) \qquad N(n) = \left\lfloor \frac{1}{2}\, n\,(\log n + c + o(1)) \right\rfloor,$$

where $o(1)$ is an arbitrary null sequence. In this form, it has been already mentioned as part of Theorem 5-4. It yet has been stated first by P. Erdős and A. Rényi in the form as shown here.

Obviously, whether the expected value for the numbers of isolated vertices in a random graph $\Gamma_{n,N(n)}$ is close to the limit e^{-c} or not, depends heavily on the choice of the null sequence $o(1)$ in the edge sequences (5-16a) (that is, whether the asymptotic formula gives a good approximation to the exact (finite) values or not and hence can be used even for small n). In [149], we proved that chosing $o(1) \equiv 0$ as in Theorem 5-17 yields no sufficiently good approximation of e^{-c} for $n \leq 200$. Table 5-2 indicates this fact. This is in accord with two papers of J. Schultz and L. Hubert who judged the results of P. Erdős and A. Rényi cited here as Theorems 5-17 and 5-18 as somehow "poor" for random graphs of small order (see [350], [351], and our discussion in [156]).

A first guess of how to choose $o(1)$ in (5-16a) so that the asymptotic results give a good approximation to the exact — finite case — values even for small n can be made by using (5-12b) and the inequality (5-13a) of Section 5.3. From these formulas, we get for the expected number of isolated vertices in a random graph Γ_{nN},

$$\exp\left(\log n - \frac{(n-1)N}{\binom{n-1}{2} - N + 1}\right) \leq \mathrm{E}_{nN}X_{.1} = n\left(\binom{\binom{n-1}{2}}{N}\right)\left(\binom{\binom{n}{2}}{N}\right)^{-1} \leq \exp\left(\log n - \frac{2N}{n}\right).$$

Inserting e^{-c} for $\mathrm{E}_{nN}X_{.1}$ and rearranging yields
(5-27a)
$$\left\lfloor \frac{1}{2}(n-2)(\log n + c)\left(1 - \frac{(n-2)(\log n + c) - 2}{(n-2)((n-1) + \log n + c)}\right)\right\rfloor \leq N \leq \left\lfloor \frac{1}{2}n(\log n + c)\right\rfloor.$$

Thus, $o(1)$ should not be positive. Throughout the remainder of this paper, let us denote by N_1 the edge sequence at the right of inequality (5-27a), and by N_4 the lower bound in (5-27a).

For $2 \leq b \leq a - d$ and $0 \leq d \leq a$, the right side of (5-13a) can be improved. This improvement is easy to derive (see [149]), thus we state it here as

(5-13b) $$e^{-bd/(a-d-b+1)} \leq \binom{a-d}{b}\binom{a}{b}^{-1} \leq e^{-2bd/(2a-b+1)}.$$

This gives a sharper upper bound on $\mathrm{E}_{nN}X_{.1}$, namely

$$\mathrm{E}_{nN}X_{.1} \leq \exp\left(\log n - \frac{2(n-1)N}{2\binom{n-1}{2} - N + 1}\right).$$

Inserting e^{-c} for $\mathrm{E}_{nN}X_{.1}$ and rearranging now gives

(5-27b) $$N \leq \left\lfloor \frac{1}{2}n(\log n + c)\left(1 - \frac{n(\log n + c) - 2}{n(2(n-1) + \log n + c)}\right)\right\rfloor.$$

Let us denote the sequence at the right of (5-27b) by N_2. This, together with (5-27a) indicates that the null sequence in (5-16a) should be negative and of order of magnitude $O((\log n)^2/n)$. It also indicates that choosing $o(1) \equiv 0$ cannot be good if one is interested in a good approximation of the expected number of isolated vertices in random

Table 5-2 *Number N of edges and related expectation $\mathrm{E}_{nN}X_{.1}$ of isolated vertices in random graphs Γ_{nN} for $n = 10\,(10)\,100\,(20)\,200$ and different edge sequences as defined in the text (from [149]).*

n	N_1	$\mathrm{E}X$	N_2	$\mathrm{E}X$	N_3	$\mathrm{E}X$	N_4	$\mathrm{E}X$	N_b	$\mathrm{E}X$
10	11	0.5919	10	0.7968	9	1.0624	7	1.8394	9	1.0624
20	29	0.7221	27	0.9259	26	1.0473	23	1.5081	26	1.0473
30	51	0.7080	48	0.8950	46	1.0452	42	1.4220	47	0.9673
40	73	0.7827	70	0.9275	68	1.0382	64	1.2995	69	0.9813
50	97	0.8064	94	0.9212	92	1.0064	87	1.2546	92	1.0064
60	122	0.8249	118	0.9560	116	1.0261	111	1.2300	117	0.9895
70	148	0.8354	144	0.9452	142	1.0053	136	1.2092	142	1.0053
80	175	0.8378	170	0.9579	168	1.0106	161	1.2185	168	1.0106
90	202	0.8530	197	0.9602	195	1.0066	188	1.1876	195	1.0066
100	230	0.8577	225	0.9535	222	1.0160	215	1.1781	223	0.9947
120	287	0.8726	281	0.9692	279	1.0038	271	1.1545	279	1.0038
140	345	0.8931	339	0.9768	337	1.0063	329	1.1337	337	1.0063
160	406	0.8910	399	0.9758	397	1.0015	388	1.1256	397	1.0015
180	467	0.9027	460	0.9784	458	1.0012	449	1.1103	458	1.0012
200	529	0.9135	522	0.9819	520	1.0024	510	1.1114	520	1.0024

graphs of small size n. Both choices — N_2 or N_4 — however, yield fairly good results, as can be seen from Table 5-2.

Very good results, even for small n, can be achieved by choosing the edge sequence

$$(5\text{-}27c) \qquad N_3 = \left\lfloor \frac{1}{2}\,(n-1)\,(\log n + c)\left(1 - \frac{(n-1)(\log n + c) - 2}{(n-1)(2(n-1) + \log n + c)}\right)\right\rfloor$$

as a comparison with the last column of Table 5-2 shows. (This edge sequence, which looks somehow like a "mean" between N_2 and N_4, was found by experimenting with different edge sequences on the computer. It's quality as a sharp lower bound of the "best" edge sequence had been proved later on.)

For Table 5-2, we computed $\mathrm{E}_{nN}X_{.1} = n\dbinom{\binom{n-1}{2}}{N}\dbinom{\binom{n}{2}}{N}^{-1}$, inserting the four edge sequences N_i $(i = 1, 2, 3, 4)$ with $c = 0$. For the last column, we calculated the values $N_b = N_b(n)$ such that the expectation $\mathrm{E}_{n,N_b}X_{.1}$ was closest to $e^0 = 1$. In this table, only the case $c = 0$ is illustrated but similar computations have been done for the cases $e^{-c} = 2$, 4, and 8; and the approximations are all of the same quality. As can be seen from Table 5-2, the sequence N_3 generally gives the best approximation. It always equals N_b or $N_b + 1$. This may indicate that it is a sharp lower bound of the "best" edge sequence N_b. For $n \geq 60$, the relative deviation between $\mathrm{E}_{n,N_3}X_{.1}$ and e^{-c} did not exceed 5%, and for $n \geq 120$ it was always below 1%.

Trying to improve the quality of the result of Theorem 5-18, we inserted N_2, N_3 and N_4 instead of N_1 in (5-26), with $c = 0$ because l plays the role of $2c$ in this formula.

Table 5-3 *Asymptotic cumulative distribution function of V, the number of edges, at which a graph becomes connected, together with the exact distribution of V for n = 15 edges (exact values from [265]).*

V	Asymptotic cdf of V (n = 15)				Exact cdf of V
	$N_1 = 20$	$N_2 = 18$	$N_3 = 17$	$N_4 = 14$	
13	0.0924	0.1615	0.2028	0.3424	0.000
14	0.1245	0.2028	0.2475	0.3914	0.021
18	0.2948	0.3924	0.4410	0.5768	0.298
22	0.4885	0.5777	0.6187	0.7239	0.604
26	0.6569	0.7248	0.7545	0.8270	0.803
30	0.7816	0.8280	0.8477	0.8942	0.909
34	0.8654	0.8952	0.9076	0.9361	0.961
38	0.9187	0.9371	0.9447	0.9616	0.984
42	0.9515	0.9626	0.9672	0.9769	0.994
46	0.9712	0.9779	0.9806	0.9860	0.998
50	0.9830	0.9870	0.9886	0.9914	0.999
54	0.9900	0.9923	0.9933	0.9945	1.000
58	0.9941	0.9955	0.9960	0.9964	1.000

Table 5-4 *Asymptotic cumulative distribution function of V, the number of edges, at which a graph becomes connected, together with the exact distribution of V for n = 20 edges (exact values from [265]).*

V	Asymptotic cdf of V (n = 20)				Exact cdf of V
	$N_1 = 29$	$N_2 = 27$	$N_3 = 26$	$N_4 = 23$	
18	0.0573	0.0962	0.1203	0.2083	0.000
19	0.0752	0.1203	0.1472	0.2419	0.004
20	0.0962	0.1472	0.1766	0.2769	0.016
30	0.4229	0.4943	0.5286	0.6236	0.472
40	0.7287	0.7717	0.7910	0.8405	0.834
50	0.8901	0.9091	0.9174	0.9381	0.958
60	0.9581	0.9655	0.9688	0.9767	0.991
70	0.9844	0.9872	0.9884	0.9914	0.999
80	0.9942	0.9953	0.9957	0.9968	1.000
90	0.9979	0.9983	0.9984	0.9988	1.000

For $n \leq 20$, all four edge sequences N_i ($i = 1, 2, 3, 4$) give rather bad approximations of the exact finite case values of the cumulative distribution function of V (see Tables 5-3 and 5-4). However, what already can be seen from these tables is that for small values of V, using N_1 in (5-26) gives better results than using N_2, N_3 or N_4. For greater values of V, N_4 gives the best results (of course, they are not good enough to be used in practice; as noted in [350], [351]). A table of exact percentiles of V is given in [265] for $n = 20\,(1)\,30\,(5)\,80\,(10)\,100$. We compared these percentiles with those obtained from the four asymptotic distribution functions of V, which can be obtained by inserting each of the four edge sequences N_i $(i = 1, 2, 3, 4)$ in (5-26). Table 5-5 shows the result for $n = 80$.

Tables 5-3 and 5-4 show that it is not admissible to apply the asymptotic result of Theorem 5-18 to graphs of order $n \leq 20$. (The values of the exact cumulative distribution function of V in the last column of these tables are cited from [265]). Referring to our calculations, however, we can say that for $n \geq 40$ the approximation of the exact distribution of V given in Theorem 5-18 is not generally as poor as is stated in [350], [351], and it can be improved by switching between the different edge sequences N_i $(i = 1, 2, 3, 4)$ for the different intervals of percentiles p. Our tables indicate that for smaller values of V it is better to use N_1 (the original formula (5-26)), and replace it by N_4 for greater values of V (recall that $c = 0$ holds here by definition).

We have no table of the exact cumulative distribution of V for $n \geq 20$ but only a table of percentiles ([265]). We compared these percentiles with those one can obtain from (5-26). It again can be seen that, for percentiles p with $5\% < p < 25\%$, the asymptotic percentiles obtained with the edge sequence N_1 are nearest to the exact values, while for $25\% < p < 45\%$ the use of N_2 gave better approximations. For $45\% < p < 80\%$, it was best to use N_3 while for $80\% < p < 96\%$, N_4 gives the best approximation. These rules also hold true for other values of n if $n \geq 40$. Using these facts, (5-26) can be used to calculate the cumulative distribution function of V instead of the exact formula which is rather cumbersome for calculations (see [265] or [328]). We achieved a good congruence between the tables in [57], [170], [259], [261], [263], [264], [265], [434] and our results.

For $s = 1$ and arbitrary values of $t \in \mathcal{N}$, the expected number of 1-isolated vertices in random multigraphs Γ_{tnN} is $\mathrm{E}_{nN}X_{.1} = n \binom{t\binom{n-1}{2}}{N} \binom{t\binom{n}{2}}{N}^{-1}$. Hence for $s = 1$ and arbitrary integers t, the same calculations can be performed as for $t = s = 1$ to get an idea how the null sequence $o(1)$ should be chosen to give good approximative results for multigraphs Γ_{tnN} of small sizes. Thus, the choice of

$$(5\text{-}28) \qquad N_{t1}(n) = \left\lfloor \frac{1}{2}(n-1)(\log n + c)\left(1 - \frac{(n-1)(\log n + c) - 2}{(n-1)\big(2t(n-1) + \log n + c\big)}\right) \right\rfloor$$

yields the best possible approximation. Our investigations in [149] show that — as with N_3 for $s = 1$ — this choice of $N_{t1}(n)$ equals either the value of the edge number for which the expected number of 1-isolated vertices is closest to e^{-c} or this value plus 1. The sequence $N_{t1}(n)$, moreover, has an advantage over the value of the "best" number of edges: With (5-28), the expectations never become underestimated. This does not hold for N_b. For $n \geq 60$, the relative deviation between $\mathrm{E}_{t,n,N_{t1}(n)}X_{1.1}$ and e^{-c} did not exceed 5%, and for $n \geq 120$, it was always below 1.5%.

For $t = s > 1$, some elementary calculations give

$$(5\text{-}29) \qquad N_{tt}(n) = \left\lfloor \frac{1}{2}tn(n-1)^{1-1/t}(\log n + c)^{1/t}\left(1 - \frac{n(\log n + c) - 2}{n\big(2(n-1) + \log n + c\big)}\right)^{1/t} \right\rfloor$$

Table 5-5 *Asymptotic and exact percentiles of V for n = 80 (p in %; the exact values are from [265], a "*" indicates that this value was uncertain); all tabulated values are the largest l for which $P(V \leq l) \leq p/100$.*

p	Asymptotic percentiles of V (n = 80)				Exact perc. of V
	$N_1 = 175$	$N_2 = 170$	$N_3 = 168$	$N_4 = 161$	
1	113	108	106	99	*
2	119	114	112	105	124
3	124	119	117	110	127
4	127	122	120	113	130
5	130	125	123	116	132
6	133	128	126	119	135
7	135	130	128	121	136
8	137	132	130	123	138
9	139	134	132	125	140
10	141	136	134	127	141
15	148	143	141	134	148
20	155	150	148	141	154
25	161	156	154	147	159
30	167	162	160	153	164
35	172	167	165	158	169
40	178	173	171	164	173
45	183	178	176	169	178
50	189	184	182	175	183
55	195	190	188	181	189
60	201	196	194	187	194
65	208	203	201	194	200
70	215	210	208	201	207
75	224	219	217	210	215
80	234	229	227	220	224
85	247	242	240	233	236
90	264	259	257	250	251
91	268	263	261	254	255
92	273	268	266	259	260
93	279	274	272	265	265
94	285	280	278	271	270
95	293	288	286	279	277
96	302	297	295	288	285
97	314	309	307	300	296
98	330	325	323	316	310
99	358	353	351	344	335

as a fairly good choice to achieve useful approximations. Nevertheless, the results which we get by using (5-29) are not of the same quality as those which we can achieve using (5-28) for $s = 1$ (see [149] for details).

Our investigations have shown that for the cases $s = 1$ or $s = t$, respectively, the null sequences in (5-16a) and (5-20a) should be negative and of order of magnitude

$O((\log n)^2/n)$. The same also holds for the edge sequences given by (5-22a) and (5-23a). For the other edge sequences similar results for the orders of magnitude of $o(1)$ can be verified easily (obviously, the functions to be inserted for $o(1)$ all should be negative here). For all cases $1 < s < t$, however, the null sequence should be of order of magnitude of $o((\log n)^2/n)$ and positive! An explanation for this phenomenon is that, for $1 < s < t$, there are two ways to "waste" edges: Drawing them from connections which either are already s-fold connected or are not yet $(s-1)$-fold connected. Until now, we have not derived what we can call a good edge sequence useful for approximations of asymptotic results for $1 < s < t$.

The first programs for computing $E_{tnN}X_{s.1}$ originally have been written in FOR-TRAN IV and assembler on a Telefunken TR445 of the Computer Centre of the University of Düsseldorf. Mr. M. Schnell wrote these programs in the Institute of Medical Statistics and Biomathematics of the University of Düsseldorf. Our cooperation made it possible to show that N_3 according to (5-27c) and $N_{t1}(n)$ according to (5-28) are sharp lower bounds of the best edge sequences. Continuing in our investigations, we developed completely new programs to compute the expectations of the random variables $X_{s.m}$ (the number of s-components of size m), $X_{s.m}^{(T)}$ (the number of s-trees of size m), $X_{s.m}^{(C)}$ (the number of isolated s-cycles of size m), $X_{s.m}^{(m,m)}$ (the number of s-components with m vertices and m edges). We additionally developed algorithms to compute the distributions — and their parameters — of the random variables V_s (the number of s-saturated connections), and $U_{si.}$ (the s-degree of a vertex ξ_i) in random multigraphs Γ_{tnN}. (The expectations and variances of these two random variables are given as Formulas (5-6a), (5-6b), (5-7a) and (5-7b), respectively.)

Mr. H. Herrmann wrote these final programs in FORTRAN 77 and Pascal on the PR1ME 550/II of the Institute of Medical Dokumentation and Statistics of the University of Köln. In these programs, he used the exact combinatorial formulas in [149]. The number C_{tsnN} of s-fold connected multigraphs Γ_{tnN}, for example, is

$$C_{tsnN} = \sum_{v=n-1}^{\binom{n}{2}} C_{nv} \, \tilde{K}_{t,s,v,\binom{n}{2}-v,N}$$

with C_{nv} given by (5-9a). From this general formula, the number $C_{t,s,n,n-1}$ of s-trees with n vertices can be derived as

$$C_{t,s,n,n-1} = n^{n-2} \tilde{K}_{t,s,n-1,\binom{n}{2}-(n-1),N}.$$

Similarly, the number C_{tsnn} of s-fold connected multigraphs with n vertices and n s-saturated connections is

$$C_{tsnn} = \frac{1}{2}(n-1)! \left(1 + n + \frac{n^2}{2!} + \cdots + \frac{n^{n-3}}{(n-3)!}\right) \tilde{K}_{t,s,n,\binom{n}{2}-n,N}.$$

By $\tilde{K}_{t,s,v,\binom{n}{2}-v,N}$ we denote the number of ways to distribute N edges between $\binom{n}{2}$ connections such that exactly v connections are s-saturated and the remaining $\binom{n}{2} - v$ are not. This is the number of ways to get s-projections with v edges. In [149], we gave some formulas of how to compute $\tilde{K}_{t,s,v,\binom{n}{2}-v,N}$, including recursive ones.

Moreover, the distribution of the number of s-isolated vertices in random multigraphs Γ_{tnN} is computed using

$$P_{tnN}(X_{s.1} = k) = \binom{n}{k} \sum_{\lceil (n-k)/2 \rceil}^{\binom{n-k}{2}} L_{n-k,v} \tilde{K}_{t,s,v,\binom{n}{2}-v,N} \binom{t\binom{n}{2}}{N}^{-1}$$

with $L_{n-k,v}$ given by (5-11). (All these formulas show the idea of how the results for random graphs Γ_{nN} can be generalized to random multigraphs Γ_{tnN}: We determine the number of s-projections Γ_{nv} with the desired property, together with the number $\tilde{K}_{t,s,v,\binom{n}{2}-v,N}$ of ways to get such an s-projection of Γ_{tnN} with N edges. The last formula also shows the difficulties for the programmer: The numbers $L_{n-k,v}$ are given by an alternating series. Thus we have to obey small differences of large numbers.

At the moment of writing, we are busy to implement the programs for computing the formulas mentioned above, and programs to uncover clusters by graph-theoretic means on IBM compatible microcomputers. As a first step, the program for the computation of the values of Tables 5-3–5-5 has been rewritten by Mr. J. Kunert in Turbo Pascal 3.0 and will be transferred to Microsoft C.

Chapter 6

Classifications by Multigraphs: Three Examples from Medicine

Calculus of graph theory provides researchers of different topics with a model to simulate many systems — especially in applied mathematics and life sciences. Cluster analysis or numerical taxonomy is one example. Though usually, in taxonomy, clusters are nothing more than the result of an algorithm, graph theory makes it possible to define clusters properly in mathematical terms. One convenient definition — as shown in Subsections 3.1.2 and 3.2.2 — is to speak of (single-linkage) clusters as components of a graph $\Gamma(d)$ or as s-components of a multigraph $\Gamma(\vec{d}^T)$. Beyond the possibility to define clusters mathematically, graph theory provides the scientist with tools to formulate and test null hypotheses on the structures of data sets. Thus, cluster analysis is raised from a tool of exploratory statistics to a helpful method of inference statistics.

From Chapter 4, we know that every cluster detecting algorithm suggests a classification of the data, even if the sample is homogeneous. If we agree to interpret possible "randomness" of clusters as "random distribution of distances" between the objects, that is, as "random choice of edges", then we can derive conditional tests on the basis of random graphs for testing the null hypothesis that a sample S has been drawn from a population with unimodal (or uniform) density and is partitioned into clusters randomly (terminated by the algorithm or by a poorly chosen threshold $\vec{d}^T$ only). For this kind of conditional test, the test is performed under the condition that $N = N(d)$ edges in $\Gamma(d)$ have been drawn. Thus, R.F. Ling used Formulas (5-9a)–(5-9e) and (5-10) in [259], [261], [265] to test the randomness of single-linkage clusters of level d. Additionally, he used the distribution of the isolation index $i(C)$, that is, the distribution of the number of edges which must be added to a graph $\Gamma(d)$ such that a given component becomes part of a larger component (see Definition 3-7 and Remark 3-1).

Ling's method has some disadvantages which we discussed in Chapter 4. Often, especially for medical data, the scale levels vary considerably between the different items, that is, between the dimensions of the data vectors as Example 2.1 has shown. It then is questionable if not impossible to compute overall (or global) similarities s_{ij} or distances d_{ij}, respectively, between the elements of a data set S. The structure of a data set consisting of n t-dimensional vectors can be described better by a multigraph as we discussed in Chapter 3. In [149], we derived exact, finite results on random multigraphs, which can be used to construct conditional tests for testing the randomness of single-linkage clusters if these are defined as s-components of a multigraph $\Gamma(\vec{d}^T)$. (Moreover, using multigraphs $\Gamma_{t,n,N(\vec{d}^T)}$ together with the s-projection onto simple graphs Γ_{nv} has the advantage that some weak points of Ling's method could be avoided. The triangle inequality of metric data, for example, must not hold for the s-projection; thus the equiprobability model seems to be more plausible for this projection as we discussed in Section 4.3.)

If the number $N(d)$ of edges drawn at a threshold d is not very small, however, then the formulas to calculate such exact probabilities — like the probability that a graph Γ_{nN} has exactly k components of size m or that a multigraph Γ_{tnN} has exactly k s-components of size m — are rather intricate. Difficulties for programmers arise especially when they are confronted with alternating series which are composed of many elements. In this case, one can use asymptotic results like those derived in the previous chapter. The formulas from this chapter, thus can be used to derive conditional **asymptotic tests** for testing the null hypothesis of a random clustering of objects from a homogeneous sample.

For example, if we define clusters as s-components of a multigraph $\Gamma(\vec{d}^T)$ and if the number of edges $N = N(\vec{d}^T) = N(n)$ can be assumed of order of magnitude $O(n^{2-1/s}(\log n)^{1/s})$ such that (5-20a) is satisfied, then the value for c can be estimated by inserting t, s, n and $N(\vec{d}^T)$ in (5-20a). From c, the parameter $\lambda = e^{-c}$ of the asymptotical Poisson distribution of the number of s-components or of s-isolated vertices can be calculated immediately. For given error probability α, the null hypothesis of random clusters now can be rejected in favour of a real clustering structure if the number of classes is too large or even if — for very large n — we get more than one proper single-linkage cluster. If N is of order of magnitude satisfying (5-20a) and n is large enough, then we expect only one proper s-component under the null hypothesis because of Theorem 5-9). The critical value $k = k(\alpha)$ of s-isolated vertices, for which the hypothesis of homogeneity can be rejected, can be seen from the tables of the $P(\lambda)$-distribution (with $\lambda = e^{-c}$ and c from (5-20a), see [154], [158]).

On the other hand, let $N(\vec{d}^T)$ in $\Gamma(\vec{d}^T)$ be very small, so that it satisfies (5-25a). Then we can reject the hypothesis of homogeneity if $\Gamma(\vec{d}^T)$ has s-trees with too many vertices or if it has too many s-trees of size 2. The critical number $k = k(\alpha)$ of s-trees with two vertices follows from the asymptotical $P(\nu)$-distribution of the random variable $X_{s,2}^{(T)}$ with $\nu = c/2$. Here, the parameter c follows from inserting t, s, n and $N(\vec{d}^T)$ in (5-25a).

Remark 6-1 While in multigraphs $\Gamma(\vec{d}^T)$ of large size and with a large number of edges, the existence of a second proper s-component already implies inhomogeneities in a data set, the situation is reverse for numbers $N(\vec{d}^T)$ of small order of magnitude. With a random distribution of edges, we expect many small s-components instead of few, but larger ones (see Subsection 4.2.1, where we gave the exact distribution of $Y = n - X_{.1}$ for simple graphs). We can give no rules to decide for given $N(\vec{d}^T)$, which one of the edge sequences (5-20a), (5-22a)–(5-25a) should be accepted as the correct one. However, we recommend to draw n_1, n_2,... objects from the n-element sample, and then to fit the different edge sequences to the respective numbers $N_1(\vec{d}^T)$, $N_2(\vec{d}^T)$,... of the functions of Chapter 5 and to select that edge sequence with the best fit (a procedure similar to that in [132]). It is known that the error probability α of the first kind is not inflicted by this procedure. Nothing is known, however, about the error probability β of the second kind. Yet it should be expected that it takes its minimum for those edge sequences with best fit. •

We also can use the s-projection to derive conditional tests. For example, if the number $N = N(\vec{d}^T) = N(n)$ of edges can be assumed of order of magnitude $O(n^{2-1/s}(\log n)^{1/s})$ such that (5-20a) is satisfied, then as we have seen in Section 5.5, the number $v = v(\vec{d}^T)$ of s-saturated connections in $\Gamma(\vec{d}^T)$ (or of edges in the s-projection $\tilde{\Gamma}_{nv}$) should be of order of magnitude $O(n \log n)$, and Theorem 5-4 can be used to derive test statistics. In this case, we would expect only one nontrivial component and some isolated vertices beside that "giant" component in $\tilde{\Gamma}_{nv}$ if n is large enough. That is, if we can assume that for a sample S and a threshold d, the number v of s-fold connections satisfies the equation $v = \lfloor \frac{1}{2} n (\log n + c + o(1)) \rfloor$. for a constant c, and if too many isolated vertices or too many proper components are in $\tilde{\Gamma}_{n,v(\vec{d}^T)}$ then we can reject the null hypothesis of homogeneous data. The components of that graph then are interpreted as real clusters.

Similarly, other results from Section 5.3 can be used as test statistics for the randomness of the structure of the s-projection $\tilde{\Gamma}_{nv}$ of a graph $\tilde{\Gamma}(d)$ if N is smaller: If N is of order of magnitude $O(n^{2-1/s})$ then the number v of s-saturated connections should be of order of magnitude $O(n)$, and we can use the distribution of the number of cycles or of the degree of a vertex to derive test statistics for testing the structure of a data set. These considerations of course do hold only if the distances are independent for the different blocks: The presence or absence of an edge in one layer of the multigraph must not influence the presence or absence of an edge in other layers. Thus, the blocks of variables should be chosen in a way which guarantees a maximum of independence of the distance measures for different blocks.

In medical research, data sets to be clustered often are not large enough to suppose that the asymptotic results immediately give a good approximation to the finite case values. Especially in those cases, further knowledge — for example, of a good null sequence in (5-16) or in (5-20a) — is of great value (see Section 5.6). This is shown in the following examples from medical research. By these examples, we want to illustrate the application of the classification model which we have developed on the basis of undirected, completely labelled multigraphs in Chapters 3 and 4. We use data from studies of the pharmacokinetics of two drugs: Urapidil (by BYK GULDEN LOMBERG CHEMICALS), an antihypertensivum, and Lidocaine (by ASTRA CHEMICALS), a drug which is generally applied for the treatment and prophylaxis of ventricular tachycardia. Furthermore, we discuss a clinical study already mentioned in Examples 2-1, 2-3 and 3-1 to a broader extend.

All three studies represent a special type of application of cluster analysis: We knew a priori that the samples could be divided into subgroups. For the Urapidil study, we had pharmacokinetic data of normals and of hypertensive patients with normal and impaired renal functions ([161], [193], [411], [412]). For the Lidocaine study, pharmacokinetic data of normals and of persons with renal insufficiency or liver disease have been raised ([103], [173]). In the third study, we wanted to answer the question whether significant differences in blood pressures and heart rates during bicycle exercise between two groups of women could be proved. In this study, the a priori groups are defined by the fact whether a woman had hypertensive disorders during pregnancy or not ([120], see also Subsection 2.2.1 and Subsubsection 2.2.4.4).

Thus for every study, we had a so-called **external criterion**. The motivation to perform cluster analyses on these data was **not** primarily to validate our method on data, for which an a priori knowledge existed (compare the first part of Chapter 4). If these classifications could be repeated with our measurements, then this would imply consequences to the treatments of patients. In the pharmacokinetic studies, it would hint to the fact that patients with renal impairments or liver diseases should be treated with dosages different from those for normals. In the third study, it could prove a pregnancy-induced hypertension to be an indicator of a later-on manifestation of an essential hypertension. Thus, these examples show that there are lots of applications for cluster analysis as a method of inference statistics.

For the remainder of this chapter, we speak of "groups" if we have the the a priori groups of the persons in mind. We speak of "classes" or "clusters", when we denote the relation of these persons to the a posteriori classes — defined as the result of the application of a cluster detecting procedure to the data. Since we cannot suppose that the readers are very familiar with the methodological and mathematical backgrounds of pharmacokinetics, we shall discuss its principles to some extend at the very beginning of the following section.

6.1 Pharmacokinetics of Urapidil in Patients with Normal and Impaired Renal Function

As the effect of a drug correlates directly with its blood concentration and the kinetics, respectively, one may assume that patients with remarkably different distribution volumes and half lives of a pharmacon should be treated with different dosages ([98], [99], [144], [145]). Whether the blood concentration of a drug can be kept in the "therapeutic region" during treatment depends heavily on the elimination half life of a drug and its (fictive) distribution volume. For an optimal therapy therefore, dosage regimen should be adjusted individually to elimination half life and distribution volume of every patient. These variables depend mostly on the renal and liver function (**elimination** and **metabolization**).

Figure 6-1 illustrates the effect of different half lives on concentration-time profiles. For this figure, we simulated a treatment of i.v. bolus injection every 24 h. We supposed equal distribution volumes but different half lives: 96 h in the upper part of Figure 6-1, 6 h in the middle part, and 24 h in the lower part. The concentration in the first curve increases into the toxic range, the second curves has only small peaks reaching into the therapeutic region. The third curve only indicates a good drug regimen.

The antihypertensivum Urapidil is described in [348]. This section is concerned with the question whether there are differences in the pharmacokinetics of Urapidil between patients with normal renal function and those with chronic renal failure. On one side, hypertension can be a hint to a beginning endocrine or renal disease. On the other side, antihypertensive treatment itself could modify renal function. In addition, metabolism and excretion of the antihypertensive compound might be different in patients with chronic renal failure, thereby modifying its pharmacological action. The correct or ideal dosage of Urapidil in hypertensive patients with impaired renal function could be different from those with normal renal function, and — of course — could also differ from normal controls. Thus, the correct or ideal dosage of Urapidil for patients could be completely different from the dose, at which we get a response — a decrease in blood pressure — in normals. In this study, no patients with liver diseases had been included.

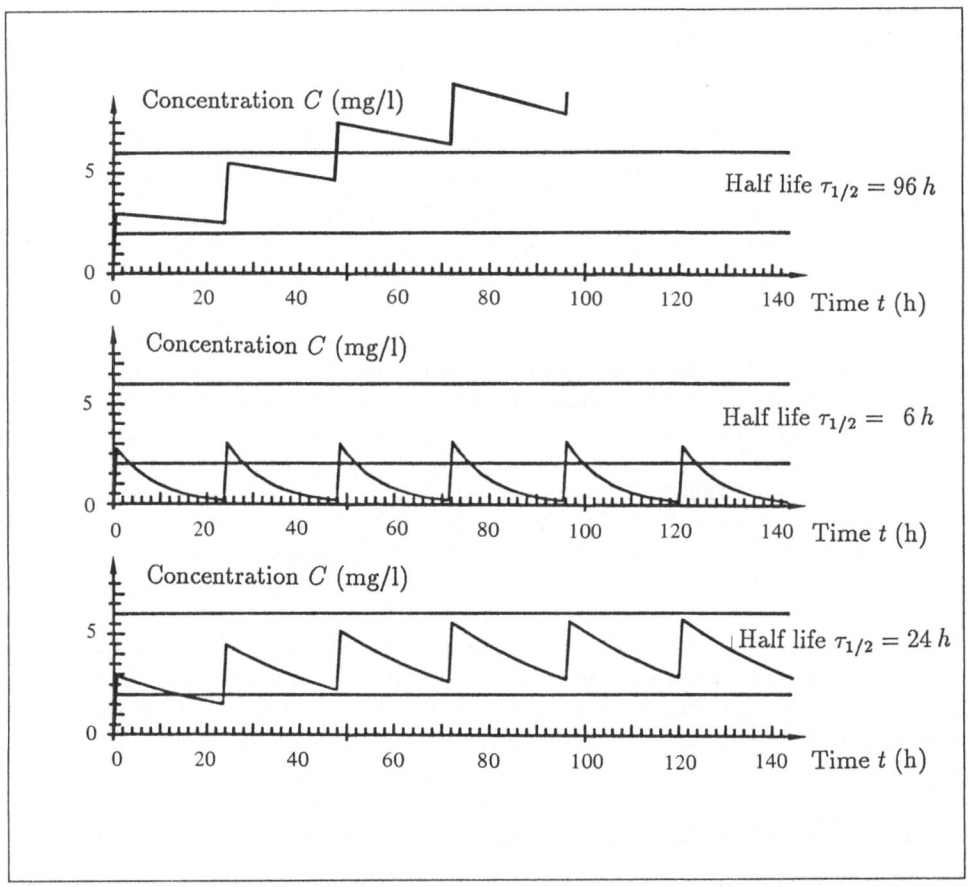

Figure 6-1 *Simulation of a drug therapy for patients with different elimination half lives; the therapeutic blood concentration is marked.*

6.1.1 MATERIAL AND METHODS

Three groups of subjects were included in the study of the pharmacokinetics of Urapidil in men ([161], [411]). Group I consisted of eight normal controls with normal blood pressure and intact renal function; eight patients with essential hypertension and normal glomerular filtration formed group II, and eight patients with various renal disorders resulting in a significant reduction in glomerular filtration rate were included in group III (see Table 6-1 for some anthropometric data). The underlying renal diseases included chronic glomerulonephritis, chronic interstitial nephritis and vascular renal damage. Group I was formed by medical students while groups II and III both consisted of patients only. This is the reason of the age difference between the groups. (The hemodynamics and the effects of the antihypertensive compound in the originally 30 patients with normal and impaired renal function were discussed in [411]. For this part of the study we had to omit two subjects from each of the three groups, because they did not get a rapid bolus injection or there were not enough blood samples for evaluation since some blood samples had been destroyed.)

Table 6-1 *Ratio of females and males and means and standard deviations of age, height and weight in the three groups.*

Group	n	Sex	Age (years)	Height (cm)	Weight (kg)
I	8	6f/2m	24 ± 2.4	169 ± 7.8	56 ± 5.3
II	8	0f/8m	49 ± 5.9	175 ± 9.1	89 ± 14.6
III	8	5f/3m	48 ± 21.8	164 ± 9.2	70 ± 18.1

All subjects were studied on an unrestricted diet. After a supine rest of 45 min, Urapidil (25 mg) was injected intravenously (rapid bolus injection). Blood samples were taken for measurement of the concentration of Urapidil as well as of its main metabolite M1: 5, 10, 15, 30 min and 1, 2, 3, 4, 5 and 6 h after i.v. injection of Urapidil (blood samples 6 h after injection had not always been taken).

Determination of the serum concentrations of Urapidil and its main metabolite M1 was done by high performance liquid chromatography ([298]). The serum concentrations of Urapidil are given in Tables 6-2a–6-2c; the values are rounded to integers. The corresponding concentration-time profiles of Urapidil are shown in Figures 6-2a–6-2c in semilogarithmic scale. Figure 6-2d shows the mean profiles of the three groups.

The serum concentrations of M1 have been used to find an improved compartment model to describe the kinetics of Urapidil. They have not been used for the cluster analysis reported here; thus, they are omitted in this monograph. However, they can be found in [161] and [157] where the improved model is discussed.

6.1.2 Biometrics: Basic Pharmacokinetics of Urapidil

The "2-compartments open model" describes the fate of many drugs in man adequately ([7], [11], [23], [58], [59], [98], [99], [144], [145], [172], [209], [324], [325], [326], [330], [338], [357], [407], [408]). It is based on straightforward assumptions, and its mathematical solutions are consistent with intuitive notions of drug behaviour. The main assumption of this model is to resolve the body into a central compartment and a peripheral compartment (Figure 6-3). These compartments are theoretical spaces; nevertheless, for many drugs the central compartment mainly consists of the serum or blood volume together with the extracellular fluid of highly perfused tissues such as the heart, lungs, liver, kidneys and endocrine glands. Drugs distribute in a few minutes throughout this compartment, and equilibrium is rapidly established and constantly maintained. The peripheral compartment is formed by less perfused tissues. The apparent volumes of both compartments thus depend upon the characteristics of blood flow to each component tissue, on the drug's ability to enter these tissues, and on the drug's affinity for them.

The 2-compartments open model also specifies characteristics of drug passage into and out of the system and drug transfer between the compartments within the system as Figure 6-3 illustrates. In almost all cases, "first-order kinetics" are assumed to describe the exit of drugs from both compartments of the system. Under this assumption the elimination and transfer rates are constants independent of the drug concentration in the compartments.

Table 6-2a *Urapidil (bolus injection): Concentration-time profiles of Urapidil in normals (group I).*

Subject	5	10	15	30	60	120	180	240	300	360
	above: time (min), below: serum concentration (ng/ml)									
ANT	1384	—	923	759	606	474	317	200	145	—
BRU	1230	857	939	516	401	245	111	101	53	—
ROH	849	619	442	308	299	265	160	116	97	—
PAA	491	489	401	289	241	180	88	52	35	—
GRÜ	542	328	302	322	316	285	157	120	120	105
MAY	831	562	472	400	339	235	132	91	76	52
BEG	412	463	374	332	244	191	145	69	66	51
ECK	688	593	540	407	414	321	199	144	105	—

Table 6-2b *Urapidil (bolus injection): Concentration-time profiles of Urapidil in hypertensive patients (group II)*

Subject	5	10	15	30	60	120	180	240	300	360
	above: time (min), below: serum concentration (ng/ml)									
KAL	1788	807	667	462	314	218	144	114	100	87
TOP	646	517	480	413	348	351	348	235	223	—
BLE	666	607	473	351	294	237	116	100	86	—
MEI	1448	909	846	627	629	676	354	213	189	—
SCHA	383	306	324	244	219	170	181	172	142	129
SCHN	471	538	454	424	304	330	214	121	67	—
POH	416	352	336	290	185	184	157	74	45	31
BON	1046	835	747	653	478	439	525	375	227	197

Table 6-2c *Urapidil (bolus injection): Concentration-time profiles of Urapidil in hypertensive patients with chronic renal failure (group III).*

Patient	5	10	15	30	60	120	180	240	300	360
	above: time (min), below: serum concentration (ng/ml)									
JÜN	1295	1009	1019	842	640	415	225	132	53	—
WEB	1533	1140	1136	931	816	703	564	335	259	—
ULL	501	503	510	501	412	305	250	174	114	—
KRO	620	588	507	409	361	345	182	143	63	60
KUH	606	505	478	524	497	371	330	226	165	150
SCHL	632	527	464	388	358	330	283	236	230	187
BED	1257	1109	997	918	639	535	304	290	264	—
WOR	849	611	499	456	340	295	219	98	88	56

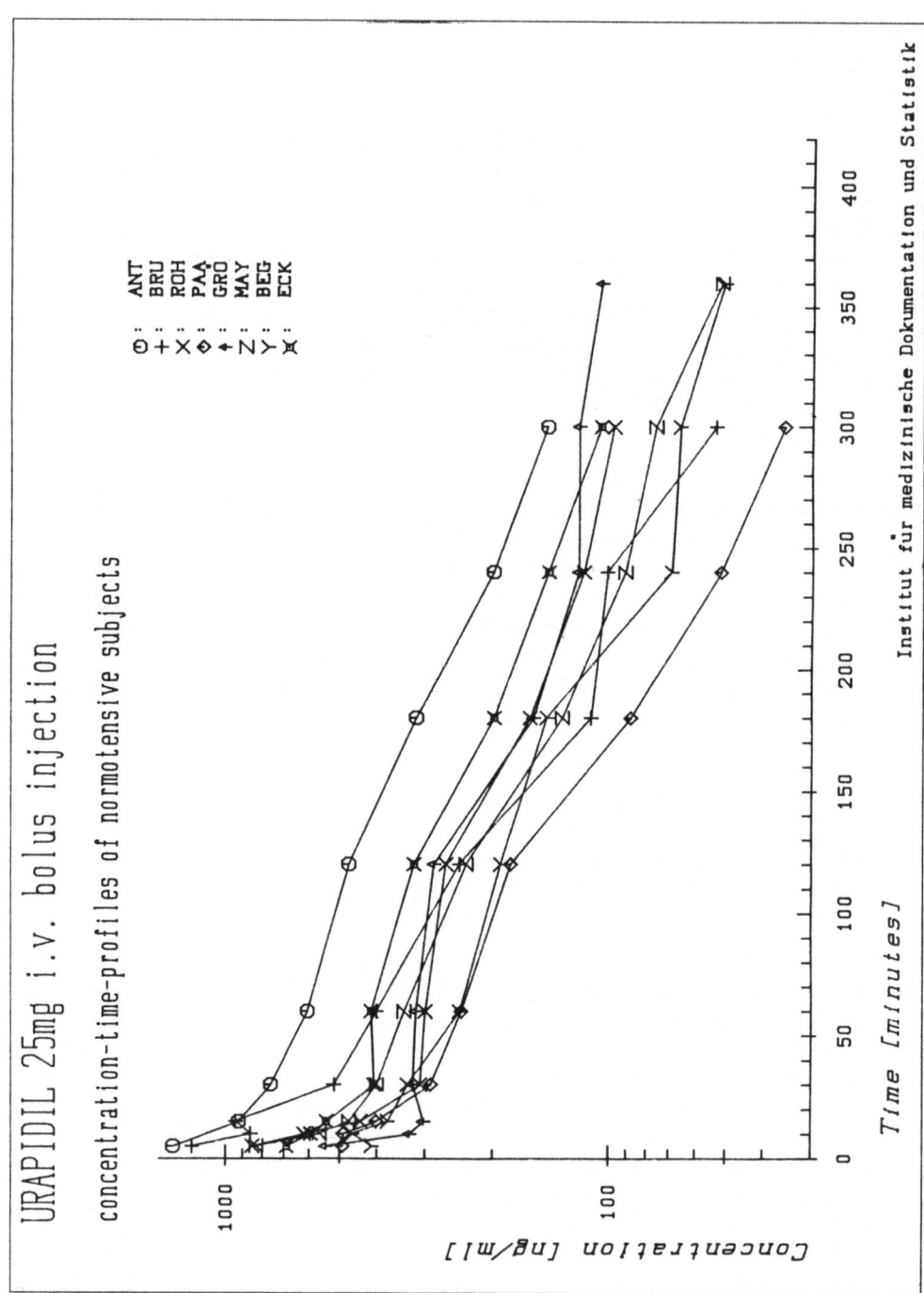

Figure 6-2a *Concentration-time profiles following bolus injection of 25 mg Urapidil in normals (group I).*

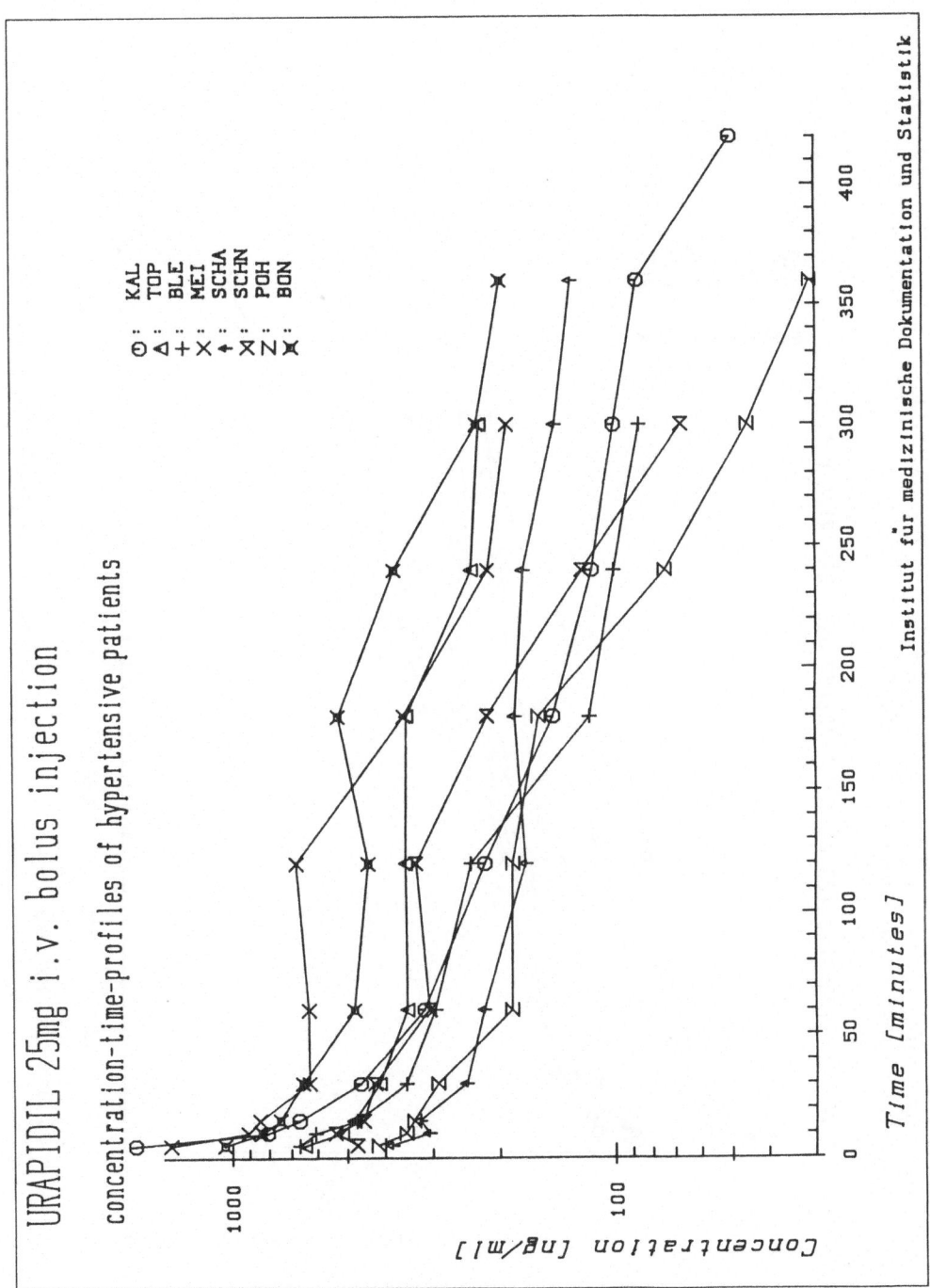

Figure 6-2b *Concentration-time profiles following bolus injection of 25 mg Urapidil in hypertensive patients (group II).*

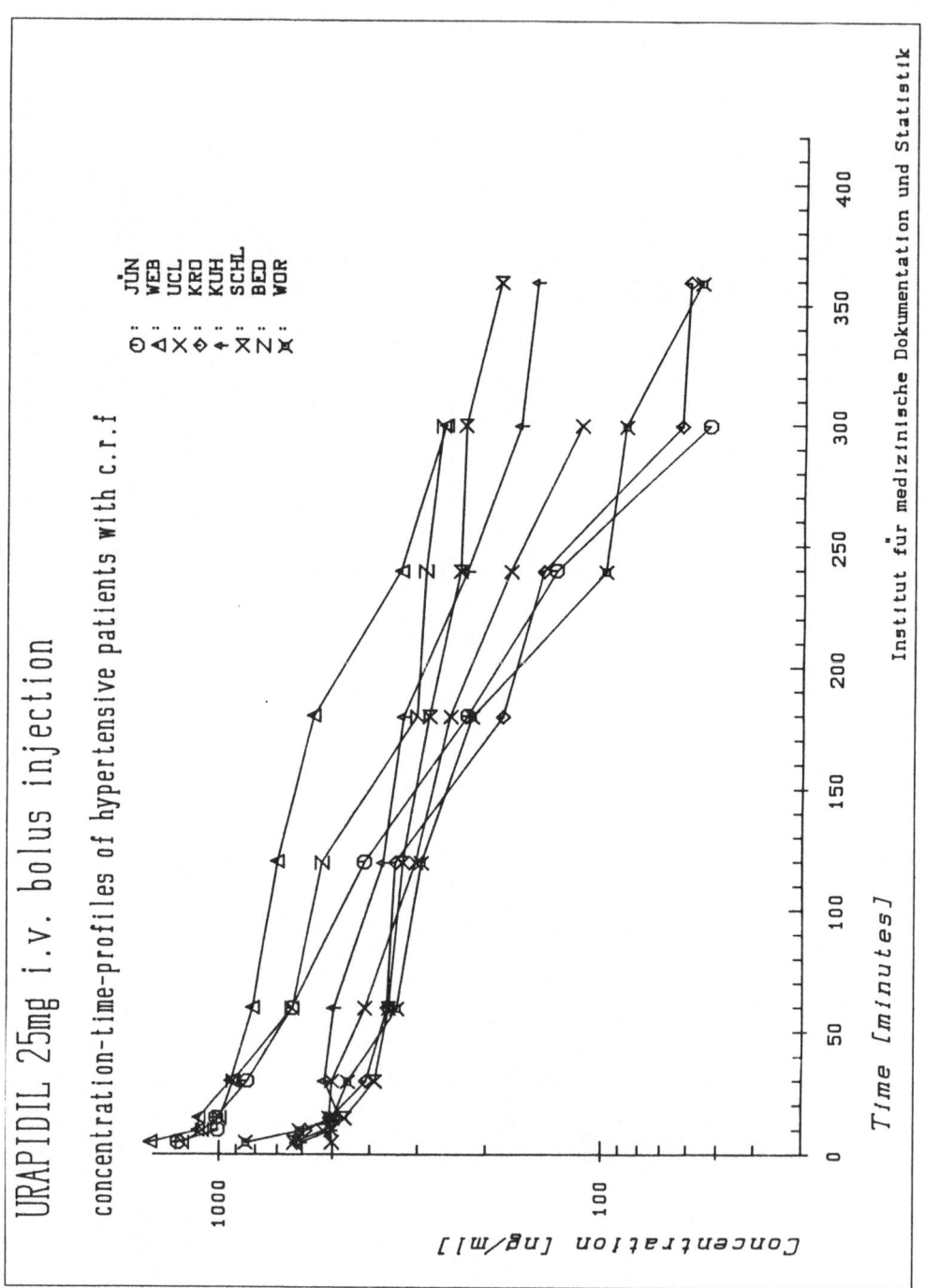

Figure 6-2c *Concentration-time profiles following bolus injection of 25 mg Urapidil in hypertensive patients with chronic renal failure (group III).*

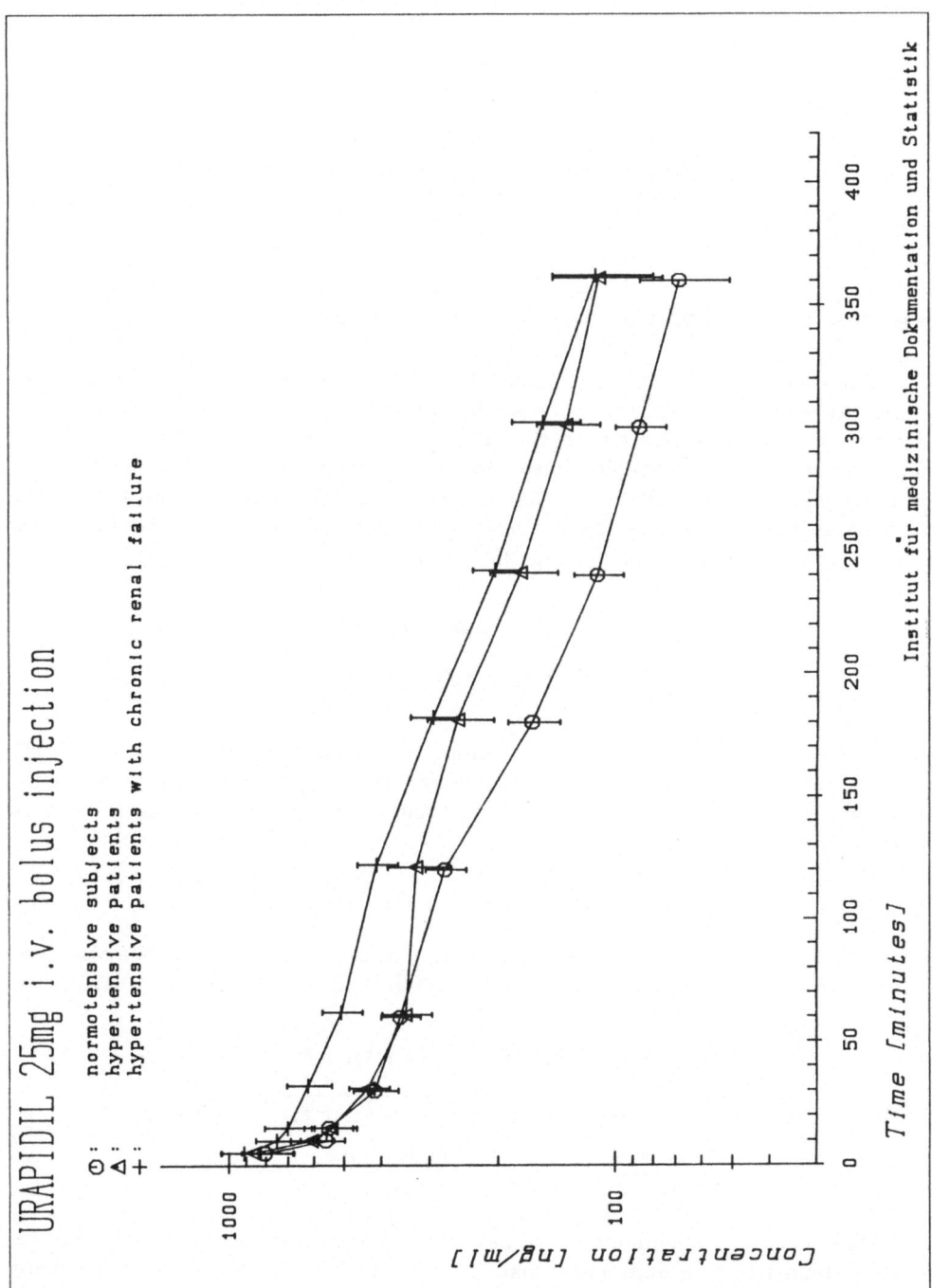

Figure 6-2d *Concentration-time profiles following bolus injection of 25 mg Urapidil: Mean concentration-time curves* $(I = \pm s_{\bar{x}})$.

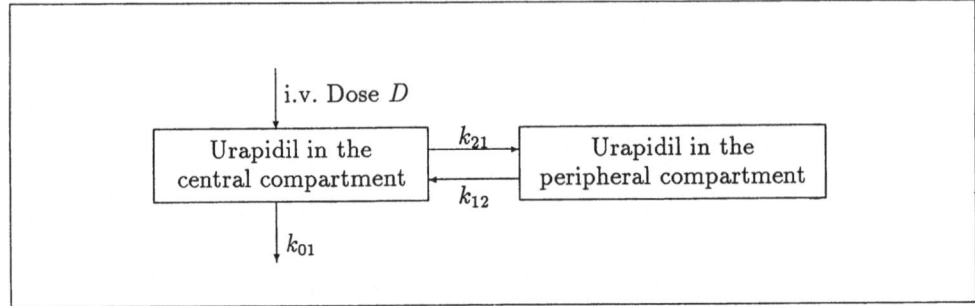

Figure 6-3 *The 2-compartments open model for i.v. bolus injection of Urapidil.*

The simplest method of drug administration for pharmacokinetic analysis is very rapid intravenous administration of a single dose D into the central compartment. Assuming instantaneous distribution throughout the central compartment, we get the following system of two coupled linear differential equations describing the distribution and elimination of the drug, that is, describing the concentration-time profiles $C(t) := x_1(t)$ in the central compartment (blood serum) and $x_2(t)$ in the peripheral compartment ([11], [144], [146], [324], [338], [407], [408]):

(6-1a)
$$\dot{x}_1(t) := \frac{\mathrm{d}x_1(t)}{\mathrm{d}t} = -(k_{01} + k_{21})\, x_1 + k_{12}\, x_2,$$

$$\dot{x}_2(t) := \frac{\mathrm{d}x_2(t)}{\mathrm{d}t} = k_{21}\, x_1 - k_{12}\, x_2.$$

The initial values at the time of application ($t = 0$) are $C(0) = C_0 = x_{10} = D/V_0$ and $x_{20} = 0$ (with V_0 as initial dilution volume). By integration of (6-1a), we get the following well-known relation between time t and drug concentation C in the central compartment:

(6-1b)
$$C(t) = a_{11}\, e^{-\alpha t} + a_{12}\, e^{-\beta t}.$$

The so-called "hybrid" or "macroconstants" a_{11}, α, a_{12}, and β in (6-1b) are composed of the "microconstants" k_{ij} from (6-1) ([11], [144], [146], [324], [338], [407], [408]):

(6-1c)
$$\alpha = \frac{1}{2}\,(k_{01} + k_{21} + k_{12}) + \frac{1}{2}\,\sqrt{(k_{01} + k_{21} + k_{12})^2 - 4k_{01}k_{12}}\,,$$

$$\beta = \frac{1}{2}\,(k_{01} + k_{21} + k_{12}) - \frac{1}{2}\,\sqrt{(k_{01} + k_{21} + k_{12})^2 - 4k_{01}k_{12}}\,,$$

$$a_{11} = \frac{D}{V_0}\frac{k_{12} - \alpha}{\beta - \alpha}\,, \qquad a_{12} = -\frac{D}{V_0}\frac{k_{12} - \alpha}{\beta - \alpha}\,.$$

The slow β- or "elimination" phase, during which drug disappearance from the central compartment is determined mainly by irreversible elimination, is of far greater interest than the so-called α- or "distribution" phase, which represents the rapid process of drug distribution from the central to the peripheral compartment. (The α- or distribution half life of a drug is defined as $\tau_{1/2}(\alpha) = \log 2/\alpha$, while the β- or elimination half life is given by $\tau_{1/2}(\beta) = \log 2/\beta$. The — hypothetical — concentration C_0 at time

Table 6-3a *Urapidil (bolus injection): Pharmacokinetic parameters of normals.*

Subject	AUC (mg·h/l)	β (1/h)	$\tau_{1/2}$ (h)	V_d (l)	V_d/bw (l/kg)	Cl (1/h)	Cl/bw (l/h/kg)
ANT	3.91	0.372	1.86	17.17	0.291	6.39	0.108
BRU	2.77	0.493	1.41	18.28	0.366	9.01	0.180
ROH	3.01	0.307	2.26	26.97	0.428	8.28	0.131
PAA	1.12	0.510	1.36	43.63	0.839	22.25	0.428
GRÖ	1.90	0.239	2.90	54.94	0.901	13.13	0.215
MAY	2.14	0.376	1.84	31.17	0.577	11.66	0.216
BEG	1.37	0.337	2.06	54.10	0.917	18.23	0.309
ECK	2.09	0.355	1.95	33.63	0.686	11.94	0.244

Table 6-3b *Urapidil (bolus injection): Pharmacokinetic parameters of hypertensive patients.*

Subject	AUC (mg·h/l)	β (1/h)	$\tau_{1/2}$ (h)	V_d (l)	V_d/bw (l/kg)	Cl (1/h)	Cl/bw (l/h/kg)
KAL	5.82	0.276	2.51	15.21	0.200	4.29	0.055
TOP	5.00	0.129	5.37	38.77	0.380	5.00	0.049
BLE	2.27	0.332	2.09	33.19	0.369	11.02	0.122
MEI	3.98	0.356	1.95	17.64	0.226	6.28	0.081
SCHA	4.30	0.092	7.53	63.28	0.565	5.82	0.052
SCHN	1.28	0.403	1.72	48.32	0.525	19.47	0.211
POH	1.06	0.397	1.75	59.45	0.646	23.60	0.257
BON	5.40	0.193	3.59	23.99	0.358	4.63	0.069

Table 6-3c *Urapidil (bolus injection): Pharmacokinetic parameters of hypertensive patients with chronic renal failure.*

Subject	AUC (mg·h/l)	β (1/h)	$\tau_{1/2}$ (h)	V_d (l)	V_d/bw (l/kg)	Cl (1/h)	Cl/bw (l/h/kg)
JÜN	2.10	0.613	1.13	19.43	0.249	11.91	0.153
WEB	4.96	0.304	2.28	16.58	0.291	5.04	0.088
ULL	1.62	0.312	2.22	49.58	0.698	15.47	0.218
KRO	1.65	0.410	1.69	36.91	0.499	15.13	0.205
KUH	2.21	0.251	2.76	45.00	0.682	11.30	0.171
SCHL	5.09	0.128	5.42	38.39	0.355	4.91	0.045
BED	5.32	0.238	2.91	19.74	0.420	4.70	0.100
WOR	2.12	0.384	1.81	30.69	0.495	11.78	0.190

Table 6-3d *Urapidil (bolus injection): Pharmacokinetic parameters for the mean concentration-time curves of normals, hypertensive patients and hypertensive patients with renal failure, respectively.*

Group	AUC (mg·h/l)	β (1/h)	$\tau_{1/2}$ (h)	V_d (l)	V_d/bw (l/kg)	Cl (1/h)	Cl/bw (l/h/kg)
I	2.32	0.344	2.01	31.30	0.560	10.77	0.193
II	2.78	0.307	2.26	29.31	0.331	9.00	0.102
III	2.93	0.309	2.24	27.59	0.392	8.53	0.121

$t = 0$ is given by $C_0 = a_{11} + a_{12}$ (both a_{11} and a_{12} must be estimated from the data). From this, the initial dilution volume V_0 can be computed by $V_0 = D/C_0$. Numerically, estimation of C_0 — and thus of V_0, too — is rather sensitive to errors made in taking the measuremants, however. A further parameter, which is more stable than C_0 or V_0, is the distribution volume V_d, defined as the ratio of the clearance Cl and β. (For the 2-compartments open model, V_0 and V_d are different, for the 1-compartment model, they are the same by definition. Details for these definitions can be found in [144], [145], [338].)

As known from previous pharmacokinetic studies, a biphasic decrease of the concentration-time curve is observed for Urapidil with distribution half life of approximately 8–15 min and elimination half life of approximately 2.5 h ([121], [248], [441]). For this part of the study, we therefore assumed that the pharmacokinetics of Urapidil could be adequately described by the 2-compartments open model with first order kinetics. After intravenous administration of Urapidil, the area under the concentration-time curve (AUC) was calculated for every person by curve fit, and the elimination rate constant β of Urapidil was estimated by loglinear regression from the sampling points with time values of at least 1 h after injection. From this, the elimination half life $\tau_{1/2}(\beta)$ was calculated by

$$\tau_{1/2}(\beta) = \log 2/\beta,$$

as mentioned above. The clearance Cl and the distribution volume V_d were individually calculated by

$$Cl = D/AUC, \qquad V_d = Cl/\beta.$$

These values for the individuals of the three groups are given together with those of the distribution volumes per kg body weight (V_d/bw) and clearances per kg body weight (Cl/bw) in Tables 6-3a–6-3c. (The correlation coefficients of the loglinear regressions for estimating the elimination rate constants are 0.9820 ± 0.0064 in group I, 0.9364 ± 0.0131 in group II, and 0.9785 ± 0.0062 in group III, respectively, — $\bar{x} \pm s_{\bar{x}}$ — which indicates a good fit of the β's.) In Table 6-3d, the values of these parameters for the mean concentration-time curves of the three groups are shown.

An analysis of variance with the following linear contrasts for the parameters was carried out:

(a) Normal subjects against all hypertensive patients;
(b) Hypertensive patients without chronic renal failure against patients with chronic renal failure;

(c) All persons without chronic renal failure against hypertensive patients with chronic renal failure.

Comparing normal subjects with the pooled group of all hypertensive patients (contrast (a)), the expectations of only two parameters proved to be statistically different at the 5% level, namely those of the distribution volume per kg body weight and the clearance per kg body weight. We got a p-value between 5% and 10% for the AUC's, again for contrast (a). No statistical differences between the groups II and III could be proved. The differences between the values of the first group and the pooled groups II and III must be interpreted very carefully. One has to keep in mind that this group of normotensive subjects has an average age of 24 years, while in the other groups the averages are 49 years and 48 years, respectively.

6.1.3 CLUSTER ANALYSIS OF THE URAPIDIL DATA

Since renal insufficiency may lead to a slower elimination of the drug and thus to a higher blood level, dosage regimens for patients with normal renal or liver functions cannot applied directly to such patients with impaired renal or liver functions. The same holds for patients with liver diseases because in this case the blood level can be increased by a low metabolization rate. No statistical significant differences in the pharmacokinetic behaviour of Urapidil could be found between hypertensive patients with chronic renal diseases and those without chronic renal diseases using traditional methods of inference statistics. This holds true for the hybrid constants and basic pharmacokinetic parameters like elimination rate constant or distribution volume. Comparing normal subjects with the pooled group of all hypertensive patients, the expectations of only two parameters proved to be statistically different at the 5% level. Standard deviations of the distribution volumes V_d and the elimination half lives $\tau_{1/2}$, however, were rather high within the a priori groups as Figure 6-4 shows. (Triangles represent the values of normal individuals, open circles the hypertensive patients, whereas the other dots represent the data obtained from the hypertensive patients with renal insufficiency.) Thus, by a cluster analysis we hoped to answer the question whether the 24 persons belong to different "real" clusters induced by their distribution volumes V_d and half lives $\tau_{1/2}$.

Especially for the development of better individual dosage regimens, there is need for studies to decide whether patients can be split into different clusters such that dosage regimens could be elaborated for these classes. The development of optimal individual therapies starts with dosage regimens which are based on the pharmacokinetic parameters of such "real clusters" ([220], [327], [329], [354], [355], [356]). Starting from these group-specific regimens, optimal individual dosage regimens can be found easier. (These principles belong methodologically to the topic called **optimal control**, see [118], [176], [177], [221], [249], [293], [336], [399].) Another advantage of starting with some group-specific therapy is that especially the risk of an overdosage can be avoided if the parameters vary extremely between different clusters.

Since we were interested in homogeneous, compact clusters which could provide for hints to potential therapy regimens, we searched for complete-linkage clusters and Ling's k-clusters (with $k \geq 3$ to avoid bridges connecting objects on opposite ends of the sample). We used the L^2- or Euclidean distance. Since both variables (distribution volumes and elimination half lives) should equally influence the classification, we normalised the data before performing the cluster analysis. The cluster analysis then

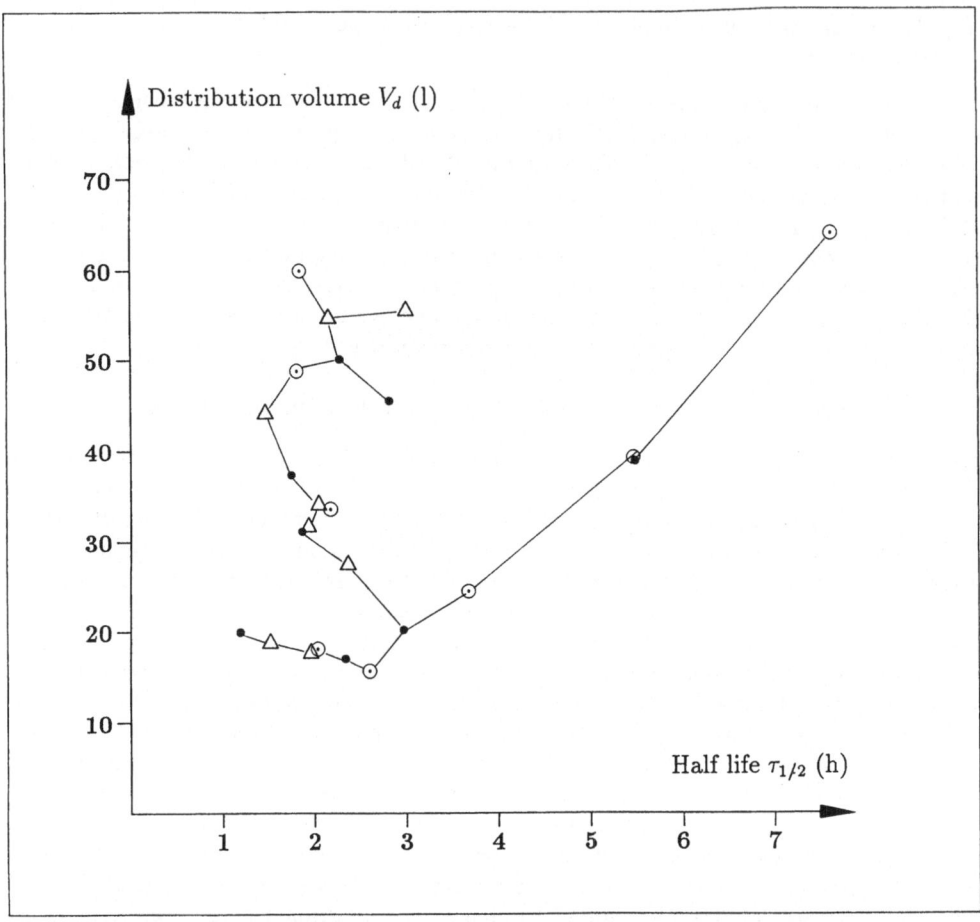

Figure 6-4 *Urapidil (bolus injection): Plot of the elimination half lives and distribution volumes of 24 individuals with minimum spanning tree for the Euclidean distance of the normalised data.*

was performed using the procedure HIERARCHY of D. Wishart's program package CLUSTAN.

In Figure 6-5, the complete-linkage dendrogram is shown for the data of Figure 6-4. For this figure, the participants in the Urapidil study have been labelled with 1–24. Nos. 1–8 indicate normal individuals, while nos. 9–16 denote hypertensive patients, and nos. 17–24 are for those hypertensive patients with renal impairments (see Tables 6-3a–6-3c).

In this dendrogram, it can be seen that two larger clusters are constituted rather early, while a two-element class as well as a one-element class have a rather large isolation index. Thus, to get homogeneous clusters, one should choose this classification into four classes. (The threshold *d* in this case is approximately 2 as Figure 6-5 indicates.) Patient SCHA (no. 13) forms the one-element class. The two-element class consists of the patients TOP and SCHL (nos. 10 and 22). In the third complete-linkage cluster, we find PAA, GRÖ, BEG, SCHN, POH, ULL and KUH (nos. 4, 5, 7, 14, 15, 19

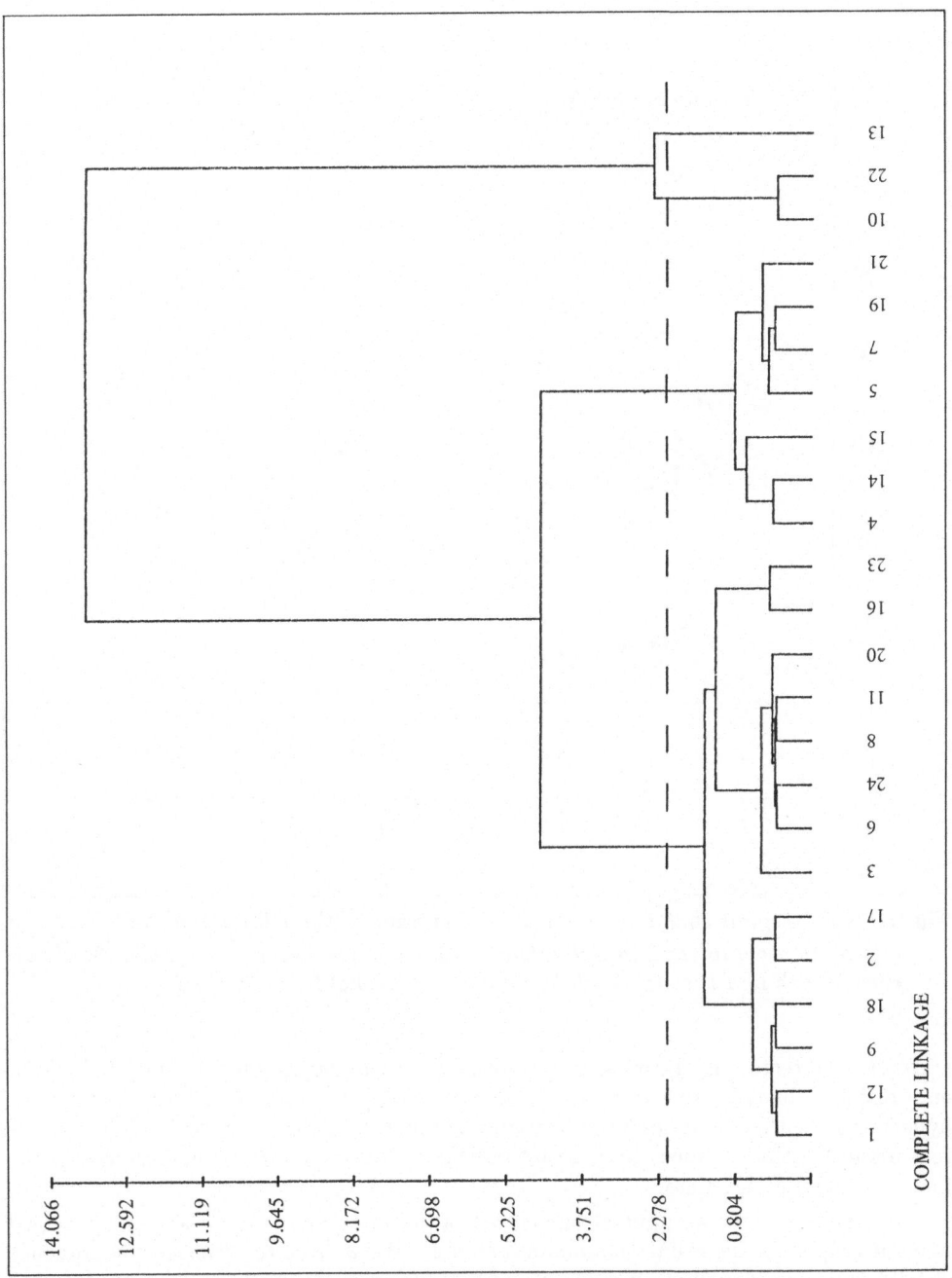

Figure 6-5 *Urapidil (bolus injection): Complete-linkage dendrogram of the 24 individuals from Figure 6-4; the threshold d defining the complete-linkage clusters is marked by a dashed line.*

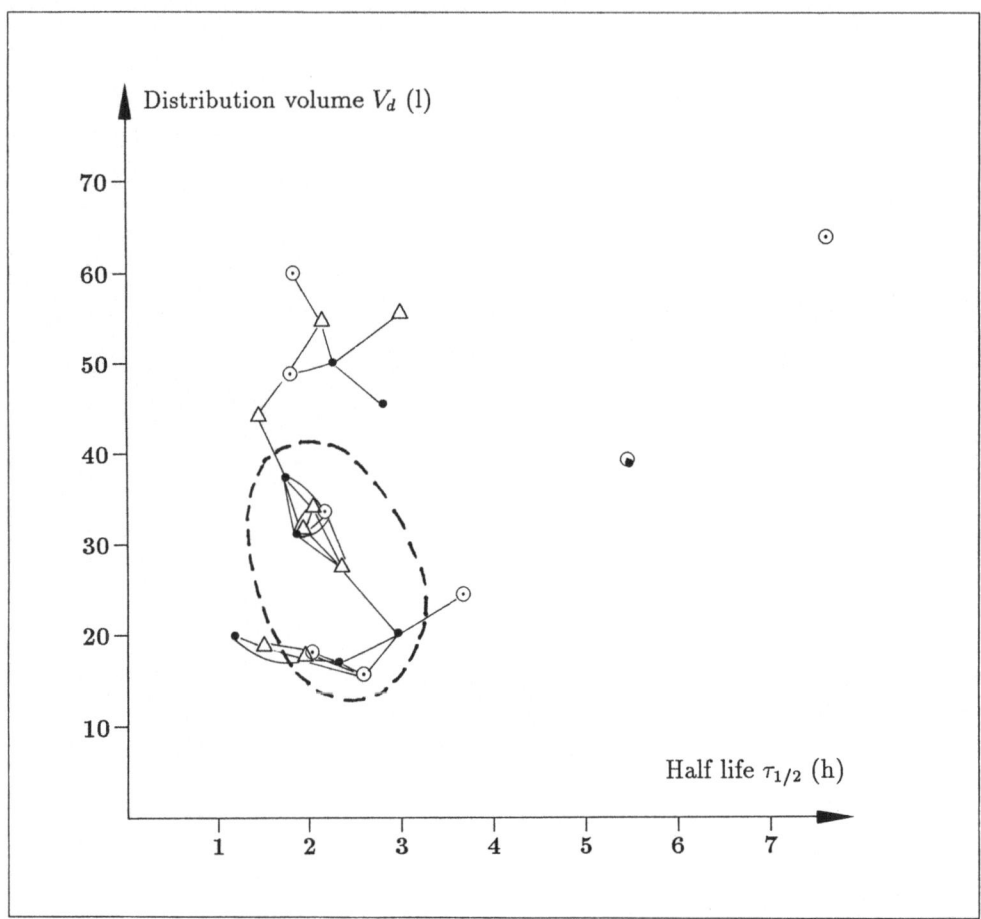

Figure 6-6 *Urapidil (bolus injection): Scattergram of the elimination half lives and distribution volumes of 24 individuals with the 2-projection $\tilde{\Gamma}_{24,37}$; the $(3, \vec{d}^T; 2)$-cluster with half the standard deviations as thresholds is encircled.*

and 21). The remaining 14 persons are gathered in the fourth cluster. (In CLUSTAN, only disjoint clusters are outlined, see Subsubsection 2.2.4.4.) The basic diseases — hypertension either alone or together with chronic renal impairment — have no direct influence, that means, they are no indicators for the clusters on the basis of the pharmacokinetic parameters.

In a second step, we applied our classification method of $(k, \vec{d}^T; s)$-clusters based on multigraphs to the 2-dimensional data $(t = 2)$. We wanted to estimate the influence of both variables, distribution volume and half life simultaneously and were interested in homogeneous clusters. Thus, $s = 2$ and $k = 3$ should be chosen. The degree of compactness of the clusters furthermore depends on the threshold $\vec{d}^T$. As no information about a reasonable medical threshold was available, it was determined statistically. It appeared appropriate to determine patients as "similar" if they differed neither in their half lives of more than half a standard deviation nor in their distribution volumes, that is, $\vec{d}^T = (\frac{1}{2}s_{T_{1/2}}, \frac{1}{2}s_{V_d})$. A total standard deviation appeared to be too much, as we would most likely obtain the whole sample as one single large group. Smaller thresholds would lead to many small groups, providing only for poor hints applicable for the establishment of a therapeutic regimen. (With this choice of $\vec{d}^T$, no normalisation or standardisation of the data is necessary since a threshold for every variable is given.)

For this threshold $\vec{d}^T$, the multigraph $\Gamma_2(\vec{d}^T)$ has a 209 edges in the two layers (out of a total of 552 possible edges). We get one single $(3, \vec{d}^T; 2)$-cluster with ANT, ROH, MAY, ECK, KAL, BLE, MEI, WEB, KRO, SCHL and WOR as members (nos. 1, 3, 6, 8, 9, 11, 12, 18, 20, 22 and 24); no further proper $(3, \vec{d}^T; 2)$-cluster exists besides this one. Also, no direct influence of the basic diseases to the outcome of the cluster analysis is visible for this classification. Figure 6-6 shows the 2-projection $\tilde{\Gamma}_{24,37}$ with the 24 persons as "vertices" and all 37 2-saturated connections (out of a total of 276 ones).

Looking for less compact classes, we kept $s = 2$, we however increased the threshold vector to $\vec{d}^T = (\frac{2}{3}s_{T_{1/2}}, \frac{2}{3}s_{V_d})$ and outlined $(2, \vec{d}^T; 2)$-clusters instead of $(3, \vec{d}^T; 2)$-clusters. Now, we got one single $(2, \vec{d}^T; 2)$-cluster which included all individuals with the exception of TOP, SCHA, BON and SCHL.

Both the number of edges and the number of 2-saturated connections in $\Gamma_{2,24,209}$ are large enough, so that the prepositions of Theorems 5-9 or 5-4 are satisfied. (These numbers perhaps are even larger, indicating that only one completely 2-fold connected multigraph should be expected.) Neither is the probability to get three or more 2-isolated vertices in a random multigraph $\Gamma_{2,24,209}$ below 5% nor is the probability for three or more isolated vertices in a random graph $\Gamma_{24,37}$. Thus, neither of the two multigraphs and their 2-projections indicated a "real" clustering structure, and the clusters may be interpreted as artificial ones (which, nevertheless, may be useful as guidelines for dosage regimens.)

6.2 PHARMACOKINETICS OF LIDOCAINE IN PATIENTS WITH KIDNEY OR LIVER IMPAIRMENTS

Lidocaine had been synthetized in 1943 by Löfgren in Sweden. For some years, it had been used as a local anesthetic. Now, Lidocaine is usually associated with therapeutic control of ventricular arrhythmias after myocard infarction and is generally applied for the treatment and prophylaxis of ventricular tachycardias for instance due to an overdigitalization. Its therapeutic plasma concentrations range from 2–6 μg/ml or 1–5 μg/ml, depending on the author ([24], [171], [299], [338], [422]).

The correct or ideal dosage of Lidocaine needed to keep patients in this therapeutic region is uncertain especially in patients with liver diseases or renal insufficiency, also in patients with problems of metabolization or elimination. As said in the previous section, the effect of the drug correlates directly with the blood concentration and the kinetics respectively. Thus, one may assume that patients with remarkably different distribution volumes and half lives, that is, with different kinetics of Lidocaine, should be treated with different dosages ([65], [401]). In the following, investigations on the pharmacokinetics of Lidocaine in patients with renal diseases and impaired liver functions are reported. The results obtained from these patients are compared with those from normal control persons. The patients with kidney diseases included also persons needing extracorporeal dialysis. Liver diseases included cirrhosis and diffuse parenchyma disease.

Table 6-4 *Means and standard deviations of the anthropometric data of the persons of the five groups; Creat.Cl. denotes the creatinine clearance.*

Group	n	Sex	Age (years)	Height (cm)	Weight (kg)	Creat.Cl. (ml/min)
A	8	5m/3f	60 ± 14	167 ± 7	74 ± 8	8 ± 3
B	7	3m/4f	49 ± 11	167 ± 12	72 ± 10	23 ± 13
C	9	4m/5f	53 ± 9	166 ± 10	61 ± 15	68 ± 22
D	9	6m/3f	48 ± 11	175 ± 12	77 ± 17	10 ± 27
E	10	4m/6f	55 ± 15	165 ± 7	70 ± 8	89 ± 30

6.2.1 MATERIAL AND METHODS

In a study we evaluated for the Department of Internal Medicine II at the University of Köln, 43 persons obtained 100 mg of Lidocaine intravenously. As with the study in the previous section, the aim was to find whether different diseases could result in different kinetics of the drug. Elimination half lives and distribution volumes had been determined for 43 persons from the following five groups:

(A) Eight patients with renal insufficieny, needing extracorporeal dialysis;
(B) seven patients with renal insufficieny, not needing extracorporeal dialysis;
(C) nine patients with liver cirrhosis;
(D) nine patients with diffuse liver parenchyma diseases;
(E) ten normal control persons.

In Table 6-4, the statistics of the most important anthropometric data are reported for the five groups. Statistical dependence between the pharmacokinetics of Lidocaine and the patients' age is reported (increasing effective half lives in elder persons, see [175]); dependence between kinetics and sex could not be proved.

All subjects were studied on unrestricted diet. Lidocaine was applied according to the following rule: A rapid bolus injection of a dose D of 100 mg intravenously was followed by a permanent infusion lasting 1 h. The infusion rate was $v_1 = 2$ mg/min for the first 30 min and $v_2 = 4$ mg/min for the second 30 min. Starting with the bolus injection, over a period of 2 h blood samples were taken for the measurements of the serum concentrations of Lidocaine every ten minutes ([103], [173]). Determination of the serum concentrations was performed using the enzyme-immuno assay of MERCK.

6.2.2 BIOMETRICS: BASIC PHARMACIKINETICS OF LIDOCAINE

In the literature, the 3-compartments open model has been used to describe the pharmacokinetics of Lidocaine as well as the 2-compartments and the 1-compartment model. For the 3-compartments model, the central compartment is the blood, one peripheral compartment consists of the highly perfused tissues such as the heart and liver whereas in the secon peripheral compartment, we find the less perfused tissues ([58]). Equilibrium between the central and the first peripheral compartment is rapidly established since the transfer constant k_{21} is very large compared with the elimnation constant k_{01} from the central compartment (to the environment). Thus, these compartment can

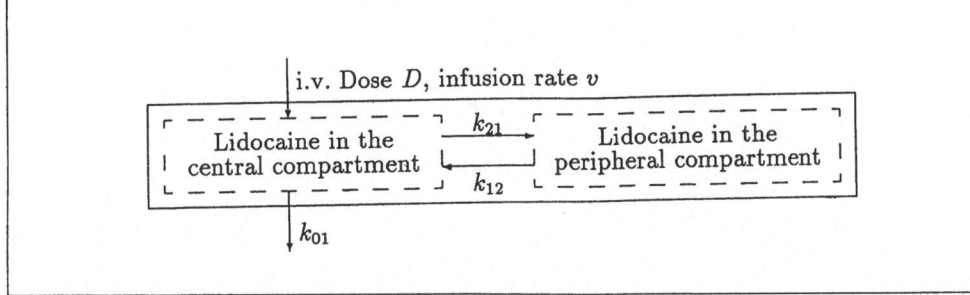

i.v. Dose D, infusion rate v

Lidocaine in the central compartment $\xrightarrow{k_{21}}$ Lidocaine in the peripheral compartment

$\xleftarrow{k_{12}}$

$\downarrow k_{01}$

Figure 6-7 *The 1-compartment (straight lines) and 2-compartments (dashed lines) open model of the Lidocaine kinetics for rapid bolus injection followed by constant infusion.*

be "lumped together", and we get the 2-compartments model of Figure 6-7 ([58], [65], [171], [337], [442]).

Remark 6-2 This lumping of compartments often is necessary for one of two reasons: First, equilibrium is established so rapidly that no measurements can be taken in this early phase ([11]); second, numerically it is rather difficult to solve such systems of "stiff" differential equations ([2]). •

The compartment "blood" is substituted by the fictive central compartment which is composed of the blood together the highly perfused tissues. Naturally, its distribution volume is larger than that of the blood only (fictive distribution volume). The elimination constant k_{01} is composed of the different elimination processes via liver and kidneys as in most 2-compartments models.

An analysis of this 2-compartments model had been performed using the BMDP routine AR ([317]). The α-phase was much shorter than the β-phase (8 min vs. 110 min). Thus with our trial, kinetics could be described very well by the simple 1-compartment open model. (Time intervals between the taking of two blood samples were larger than the α-phase. Thus, for mathematical reasons it was impossible to separate the two phases.) Now, k_{01} describes all elimination and diffusion processes in the body, and we speak of an "effective" elimination constant. The distribution volume V_d is composed of all different volumes. Moreover, by protein bindings it seems to be even larger ([175]).

In a 1-compartment model, when a pharmacon is applied by rapid bolus injection (single dose D) followed by a two-phase permanent i.v. infusion (infusion rate v_1/V_d until time τ_1, then v_2/V_d until time τ_2), kinetics are determined by the differential equations

$$\dot{x}(t) = \frac{v_1}{V_d} - k_{01}\, x(t) \qquad \text{for} \quad 0 \le t < \tau_1,$$

(6-2a)
$$\dot{x}(t) = \frac{v_2}{V_d} - k_{01}\, x(t) \qquad \text{for} \quad \tau_1 \le t < \tau_1,$$

$$\dot{x}(t) = -k_{01}\, x(t) \qquad \text{for} \quad \tau_2 \le t$$

with the initial concentration $C_0 = x(0) = x_0 = D/V_d$ for the concentration of the drug at time $t = 0$ (bolus injection). Here, V_d and V_0 are identical by definition. The value $x(\tau_1)$ at the end of the first infusion is, at the same time, the initial value of the drug's concentration at the start of the second infusion, and $x(\tau_2)$ at the end of this period is the initial value for the phase of sole elimination which is described by the third equation. The solution of these equations is

$$
x(t) = \frac{v_1}{k_{01} V_d} \left(1 - e^{-k_{01} t}\right) + \frac{D}{V_d} e^{-k_{01} t} \qquad \text{for} \quad 0 \leq t \leq \tau_1,
$$

(6-2b)

$$
x(t) = \frac{v_2}{k_{01} V_d} \left(1 - e^{-k_{01}(t-\tau_1)}\right) + x(\tau_1) e^{-k_{01}(t-\tau_1)} \qquad \text{for} \quad \tau_1 \leq t \leq \tau_2,
$$

$$
x(t) = x(\tau_2) e^{-k_{01}(t-\tau_2)} \qquad \text{for} \quad \tau_2 \leq t.
$$

System (6-2a) can easily be integrated ([11], [95], [324], [342]). Again using the subroutine AR from BMDP which allows fitting nonlinear equations to data, we determined the fictive distribution volumes V_d and the effective elimination constants k_{01} by curve fit for every subject. The effective half lives were derived from the elimination constants via

$$
\tau_{1/2} = \frac{\log 2}{k_{01}} \, .
$$

These values are reported for the different groups in Tables 6-5a–6-5e. In Table 6-5f, the estimated statistical parameters for the different groups are given (see also Figure 6-8).

We could find no statistical correlation between these pharmacokinetic parameters V_d and $\tau_{1/2}$ within the five groups. A remarkably large deviation can be seen between the average half lives of Groups C (patients with cirrhosis) and E (normal controls). The same holds for the difference of the average distribution volumes of Groups D (patients with diffuse liver parenchyma afflictions), on one side, and A (patients with renal insufficiency, using extracorporeal dialysis) or E, respectively, on the other side. Figure 6-9 shows the average concentration-time profiles of Lidocaine in the central compartment for the five groups together with the therapeutic range.

From these considerations, and from Figure 6-1, the following two facts can be seen. The larger the initial distribution volume (or dilution volume) V_d in a patient is the higher must be the bolus injection to reach at least the lower border of the therapeutic range of the drug's blood concentration. The larger the effective elimination half life is the lower must be the following infusion rate (or maintenance dose) to prevent the blood concentration from entering the toxic region.

For the kinetics of Lidocaine, a univariate analysis of variance of the elimination half lives, followed by a posteriori tests for linear contrasts between the distinct groups showed that patients with cirrhosis (Group C) had significantly larger half lives of Lidocaine as compared with normal controls ($p < 0.05$). Similarly, the distribution volumes of patients from Group D, in the average, is significantly larger than that of normals and of those patients from Group A.

6.2.3 CLUSTER ANALYSIS OF THE LIDOCAINE DATA

For Lidocaine, the correct or ideal dosage is uncertain in patients with liver diseases or renal insufficiency, also in patients with problems in metabolism or elimination as we mentioned at the beginning of this section. One may assume however that patients with remarkably different distribution volumes and half lives of Lidocaine should be treated with different dosages. Thus, as with the Urapidil study, it was of great interest to uncover cluster structures for the participants of this trial and to look whether these structure correlates with the a priori groups. Elimination half lives and distribution

Table 6-5a *Lidocaine (bolus injection and permanent infusion): Pharmacokinetic parameters of patients with renal insuffiency, extracorporeal dialysis.*

Subject	$\tau_{1/2}$ (min)	V_d (l)	k_{01} (1/min)
1	38.08	15.99	0.018204
2	32.05	33.43	0.021627
3	37.05	42.90	0.018708
4	69.66	42.21	0.009951
5	58.28	59.55	0.011893
6	46.23	57.83	0.014993
7	53.89	98.02	0.012862
8	29.75	26.98	0.023296

Table 6-5b *Lidocaine (bolus injection and permanent infusion): Pharmacokinetic parameters of patients with renal insuffiency, no extracorporeal dialysis.*

Subject	$\tau_{1/2}$ (min)	V_d (l)	k_{01} (1/min)
9	59.45	96.03	0.011659
10	56.68	51.99	0.012229
11	24.30	62.78	0.028528
12	26.22	54.41	0.026440
13	35.18	64.82	0.019705
14	46.07	67.23	0.015044
15	68.14	96.15	0.010172

Table 6-5c *Lidocaine (bolus injection and permanent infusion): Pharmacokinetic parameters of patients with liver cirrhosis.*

Subject	$\tau_{1/2}$ (min)	V_d (l)	k_{01} (1/min)
16	31.67	41.28	0.021886
17	50.91	66.15	0.013614
18	50.23	71.47	0.013799
19	98.87	140.05	0.007011
20	75.22	51.96	0.009215
21	67.20	81.04	0.010315
22	50.91	59.24	0.013616
23	66.30	63.84	0.010455
24	43.61	13.46	0.015893

Table 6-5d *Lidocaine (bolus injection and permanent infusion): Pharmacokinetic parameters of patients with diffuse liver damage.*

Subject	$\tau_{1/2}$ (min)	V_d (l)	k_{01} (1/min)
25	42.13	98.37	0.016454
26	38.62	36.11	0.017947
27	34.66	92.02	0.019959
28	67.48	39.61	0.010272
29	36.46	75.81	0.019013
30	56.06	105.77	0.012365
31	55.71	144.80	0.012443
32	50.25	101.88	0.013793
33	45.54	118.48	0.015222

Table 6-5e *Lidocaine (bolus injection and permanent infusion): Pharmacokinetic parameters of normal control persons.*

Subject	$\tau_{1/2}$ (min)	V_d (l)	k_{01} (1/min)
34	35.11	43.43	0.019741
35	33.03	64.24	0.020983
36	28.03	46.75	0.024725
37	38.96	57.09	0.017790
38	31.74	47.97	0.021835
39	25.26	34.78	0.027437
40	36.12	58.91	0.019191
41	31.39	28.20	0.022085
42	30.85	41.99	0.022469
43	55.89	73.21	0.012402

Table 6-5f *Lidocaine (bolus injection and permanent infusion): Means and standard deviations of the pharmacokinetic parameters of the five groups.*

Group	n	$\tau_{1/2}$ (min)	V_d (l)
A	8	45.7 ± 14.0	47.1 ± 25.3
B	7	45.1 ± 17.1	70.5 ± 18.3
C	9	59.4 ± 19.9	65.4 ± 34.2
D	9	47.4 ± 10.9	90.3 ± 35.2
E	10	$34.6 \pm\ \ 8.4$	49.7 ± 13.7

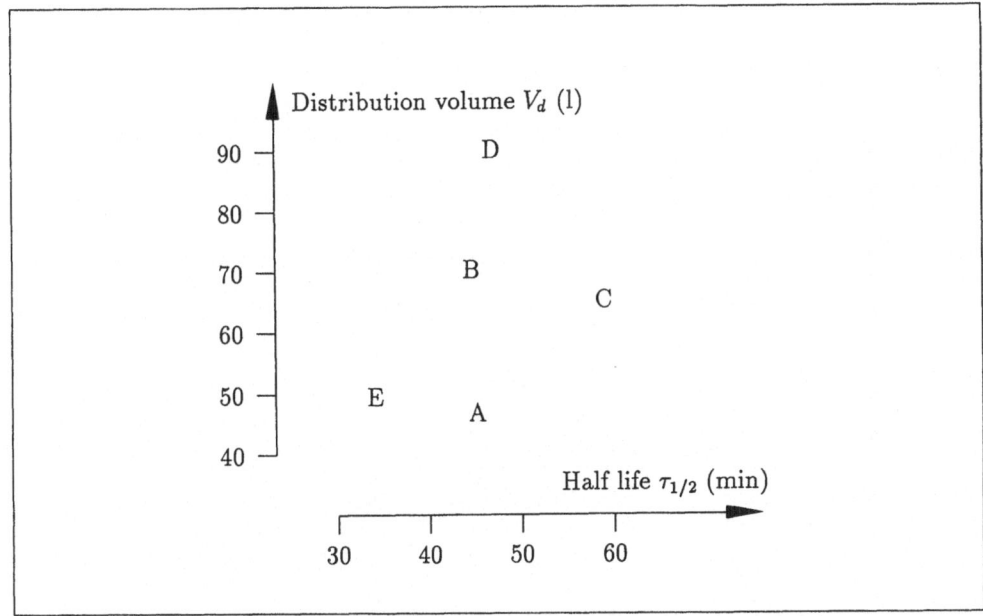

Figure 6-8 *Lidocaine (bolus injection and permanent infusion): Plot of the means $(\overline{\tau}_{1/2}, \overline{V}_d)$ for the five groups.*

volumes of all subjects were determined and used for the classification procedure. By this we got 2-dimensional data, that is, $t = 2$ holds true.

The data are shown in Figure 6-10. Patients with extracorporeal dialyses (Group A) are represented by bullets "•", while the other patients with kidney impairments (Group B) are indicated by small circles "o". Liver patients with cirrhosis (Group C) are represented by "⊗" whereas every "⊙" indicates a patient with liver damage from Group D. Triangles "△" represent the values of normal individuals. It is obvious, that the values of healthy individuals are relatively close to each other. (More tables with data can be found in [103].)

The clusters should be as homogeneous as possible and provide for hints to potential therapy regimens. Thus, we first performed a hierarchical classification using the Euclidean distance for the standardised data and outlined the complete-linkage clusters by CLUSTAN. The dendrogram (for the disjoint complete-linkage clusters at each level) is shown in Figure 6-11.

From the dendrogram of Figure 6-11, one chooses $d = 3$ as threshold to separate clusters. We thus get five a posteriori clusters. The patient, numbered as 19, of Group C makes a one-element class. Patients nos. 7, 9, 15, 21 and 30–33, are gathered in the second complete-linkage cluster. Patients nos. 4, 5, 10, 20, 23, 28 and 43 make the third class. Patients nos. 6, 13, 14, 17, 18, 22, 25, 27, 29, 35, 37 and 40 are all in the fourth class. The remaining 15 persons (nos. 1, 2, 3, 8, 11, 12, 16, 24, 26, 34, 36, 38, 39, 41 and 42) are in the fifth class. A direct influence of the sickness — that is, of the a priori groups — onto the resulting complete-linkage classification cannot be proved. However, it is remarkable that six of the ten controls belong to the same cluster (the last one named above).

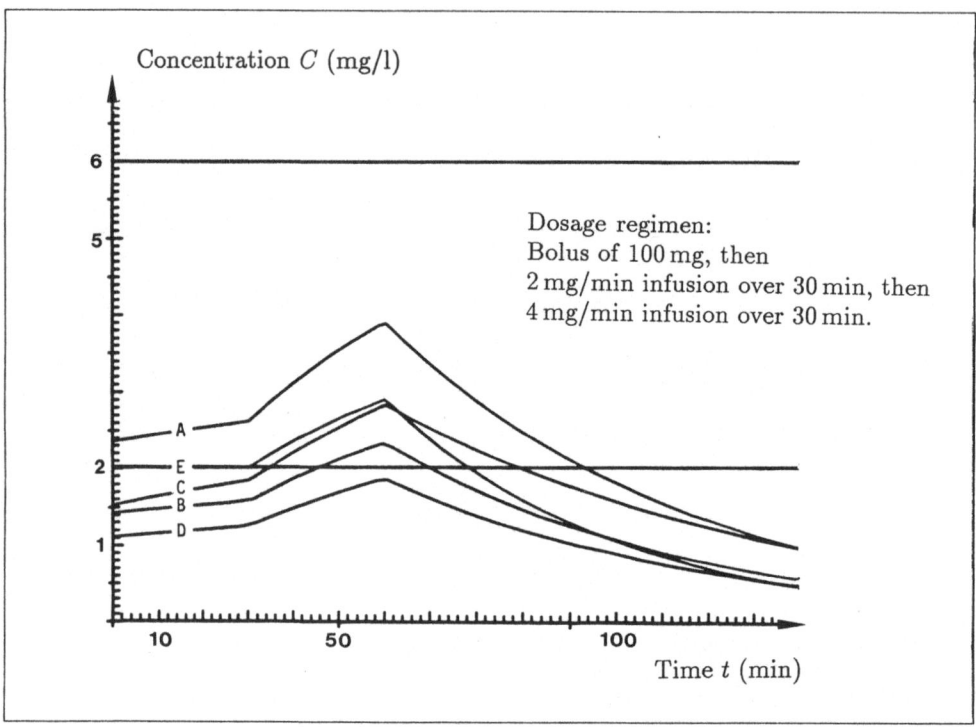

Figure 6-9 *Lidocaine (bolus injection and permanent infusion): Mean concentration-time curves of Lidocaine for the five groups (the therapeutic range is indicated by straight lines).*

The situation is different, when the classification is performed using multigraphs. As with the previous trial, no information about a reasonable medical threshold $\vec{d}^T = (d_1, d_2)$ was available. Thus, again patients were determined as "similar" if they differed in their half lives of not more than half a standard deviation; the same was accepted for their volumes of distribution, that is, $\vec{d}^T = (\frac{1}{2} s_{T_{1/2}}, \frac{1}{2} s_{V_d})$ and $t = s = 2$. As the groups were also supposed to be homogeneous, we had to demand that similarities to at least 3 individuals were to exist, to permit one additional person to belong to this group, that is, we put $k = 3$. (This should avoid bridges connecting otherwise well seperated groups.)

The 2-projection $\tilde{\Gamma}_{43,95}$ of the result of our cluster analysis is shown in Figure 6-12. One ellipsoid $(3, \vec{d}^T; 2)$-cluster exists, composed of 24 persons, nos. 2, 3, 5, 6, 8, 10, 12, 13, 14, 16, 17, 18, 22, 26 and 34–43. All normal individuals, whose kinetic data, in particular, are very homogeneous, belong to this cluster, whereas the patients with the diagnosis diffuse liver damage are widely distributed within the x-y-plane. Only one value of a person of that group belongs to the 3-linkage cluster. This mirrors the fact that this diagnosis is a generic name for a whole series of symptoms; patients with varying diseases are gathered in this group.

In this trial, the outcome of the classification by multigraphs is highly influenced by the a priori groups. This might hint to the fact, that the dosage of Lidocaine should be established according to additional liver or kidney impairment. The result of this

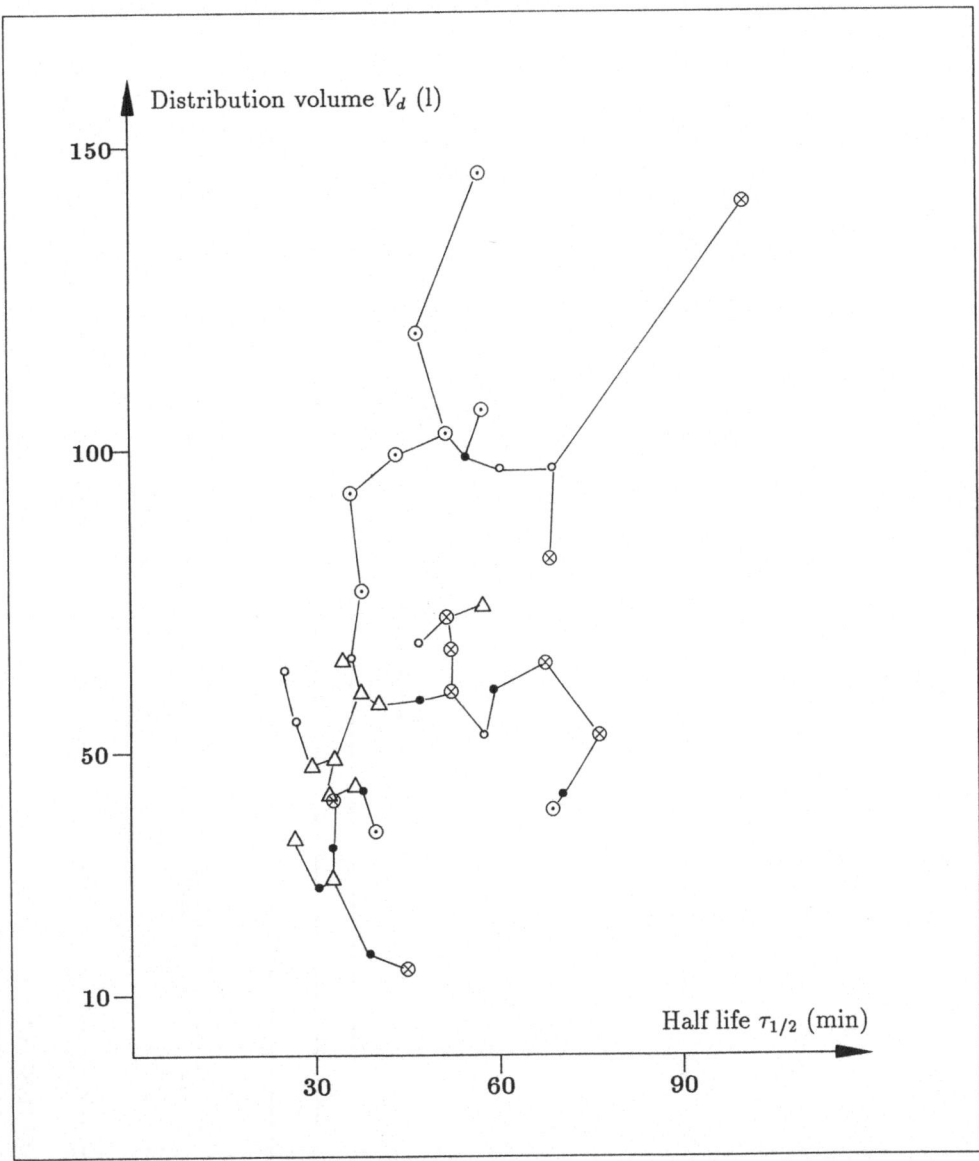

Figure 6-10 *Lidocaine (bolus injection and permanent infusion): Scattergram of the elimination half lives and distribution volumes of 43 subjects for the five groups (with minimum spanning tree of the L^2-distance of the standardised data).*

cluster analysis may be interpreted as follows: No unique dosage regimen can be given for these patients. Patients falling into the 3-linkage cluster due to their distribution volumes and their half lives, should be treated according to a regimen valid for normal individuals. Patients with increased half lives demand a lower maintenance dose, as their elimination is decreased, whereas patients with high distribution volumes need a

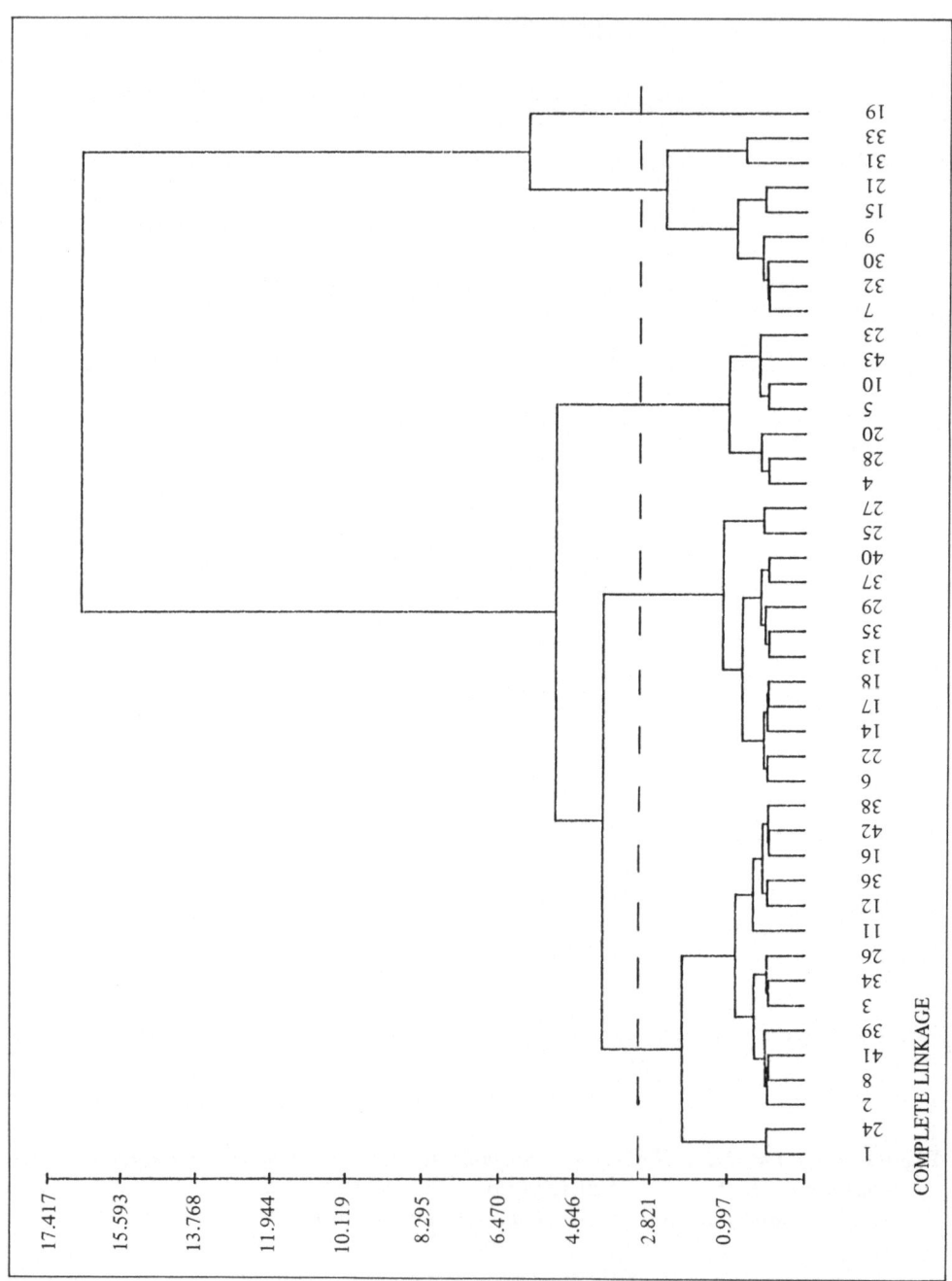

Figure 6-11 *Lidocaine (bolus injection and permanent infusion): Dendrogram of the complete-linkage clusters of the 43 subjects of Figure 6-10; the threshold d for the clustering named in the context is marked by a dashed line.*

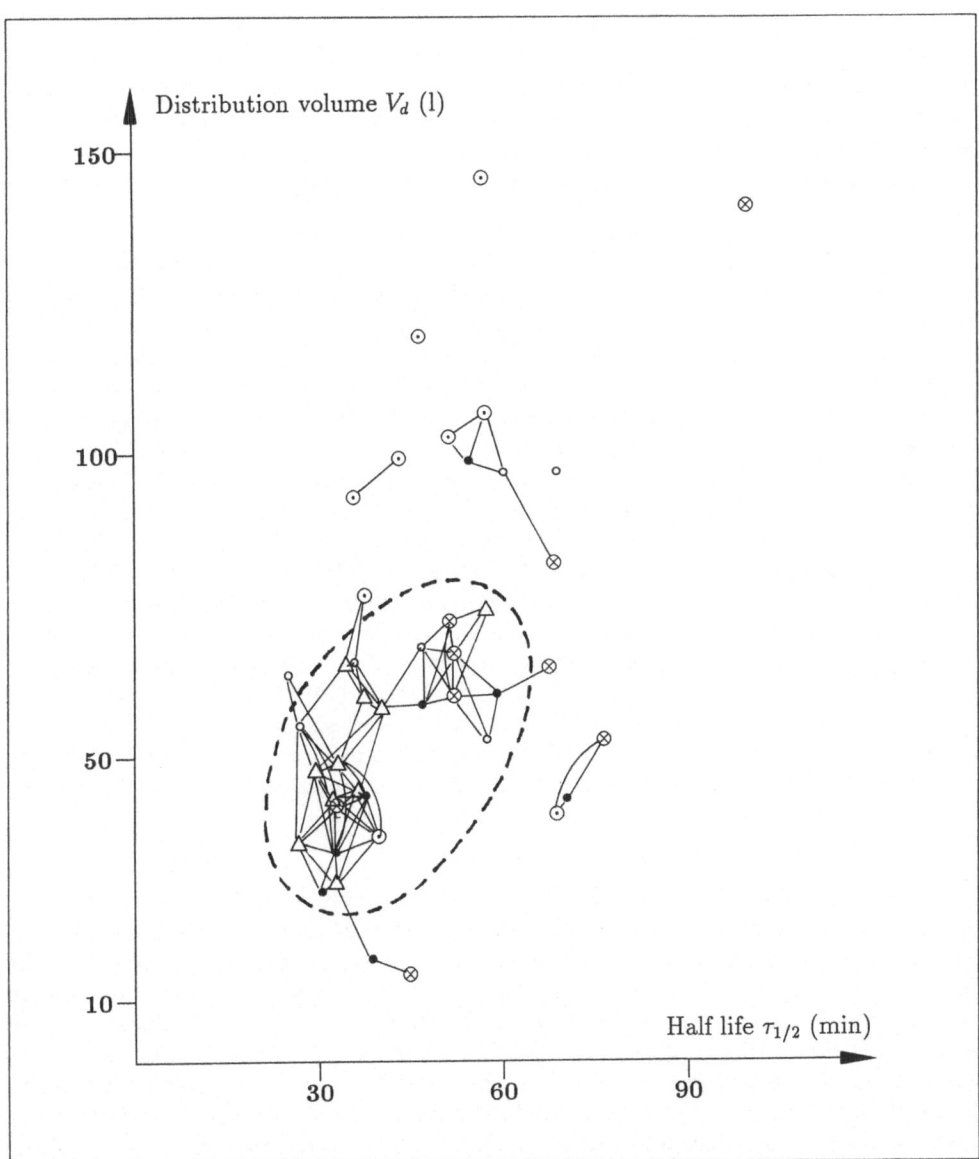

Figure 6-12 *Lidocaine (bolus injection and permanent infusion): Scattergram of the elimination half lives and distribution volumes of 43 subjects with the 2-projection $\tilde{\Gamma}_{43,95}$ as the result of our cluster analysis; the ellipsoid $(3, \vec{d}^T; 2)$-cluster with half the standard deviation as thresholds is encircled.*

higher initial dose, as the initial concentrations of the drug are lower.

If we substitute $k = 3$ by $k = 1$ without changing the threshold $\vec{d}^T$ and again let $s = 2$, now there are four $(1, \vec{d}^T; 2)$-clusters with more than one person and four classes with only one element (2-isolated vertices in the terminology of Chapters 3–5, see Figure 6-12). Persons, numbered 1, 11, 23, 24 and 29, are added to the previous

$(3;2)$-cluster. Subjects, numbered 4, 20 and 28 make a second $(1;2)$-cluster, 7, 9, 21, 30 and 32 a third one, and two patients with diffuse liver diseases, 25 and 27, are in the fourth $(1;2)$-cluster. The remaining four persons (15, 19, 31 and 33), however, remain isolated. This classification also shows that people with liver diseases are fairly well seperated from the other ones. The multigraph $\Gamma_t(\vec{d}^T)$ with the 43 patients as vertices has 533 edges in the two layers for the threshold $\vec{d}^T$ mentioned above (1806 edges are possible).

Despite the fact that the structure of the clusters reflects the groups to a certain extend, the randomness of the clusters should be tested. According to Theorem 5-9 (with $m = 1$), the probability that a random multigraph has k s-isolated edges is asymptotically Poisson distributed with parameter $\lambda = e^{-c}$ if the number of edges satisfies (5-20a), and because of Theorem 5-10 the probability to get one or more s-isolated vertices, tends to 0 if N grows faster than (5-20a). For our multigraph $\Gamma_2(\vec{d}^T)$ with $n = 43$ vertices and $N = 533$ edges, we can assume (5-20a) to be fulfilled. (We get $43^{3/2}(\log 43)^{1/2} = 546.847$; and $N = 533$ is close to this value.) Then we could estimate the values for c and λ by (5-20c), putting $o(1) \equiv 0$. This, however, is acceptable only for $n > 200$ as shown in Section 5.6. For smaller values of n, Formula (5-29) should be preferred. Inserting $t = 2$, $n = 43$, and $N = 533$ into this formula, we get $c = 0.061\,363$. From this, $\lambda = 0.940\,821$ follows as the expected number of 2-isolated vertices under the assumption of drawing edges at random. Hence the probability to get four or more 2-isolated vertices in a multigraph with two layers, 43 vertices and 533 edges is $P_{2,43,533}(X_{2.1} \geq 4) \approx 1 - e^{-\lambda}\left(1 + \lambda + \lambda^2/2 + \lambda^3/6\right) = 0.015\,554$. For a given level of significance of $\alpha = 0.05$, we can reject the null hypothesis that the edges have been drawn at random. This means that we can suppose a real group structure in the data set. Therefore, the clusters may not be regarded as found "at random" but as real ones.

Remark 6-3 With the 2-projection of the original multigraph $\Gamma_{2,43,533}$, we can use the results for simple random graphs in the same way as R.F. Ling does. The 2-projection $\tilde{\Gamma}_{43,95}$ with $n = 43$ vertices has $v = 95$ edges (the 2-connections in the original multigraph). Here we get $\frac{1}{2}43\log 43 = 80.866$; and $v = 95$ is close enough to this value to accept the preposition of Theorem 5-4 as being satisfied (with N substituted by v). Inserting $o(1) \equiv 0$ into (5-16) gives only a poor approximation of the exact probability to get k or more isolated vertices in a random graph for small n. For $n \leq 200$, (5-27c) should be preferred (see Section 5.6). Inserting $n = 43$ vertices, and $v = N_1(n) = 95$ edges into this formula gives estimations for c and $\lambda = e^{-c}$, from which $P_{43,95}(X_{.1} \geq 4) \approx 0.000\,534$ follows as the probability to get four or more isolated vertices in this 2-projection. Theorem 5-4 has been used to compute this probability. •

This example from medicine illustrates that it makes sense to perform a cluster analysis on data which do not split into well separated subsets. We wanted to find a clusters whose elements were very close together such that they could be used to establish rules of thumb for the treatment of patients. In this case, classification procedures are not needed as exploratory methods only (as was their main purpose). They are helpful in verifying a priori group structures and in testing hypotheses on homogeneity.

6.3 PREGNANCY-INDUCED HYPERTENSION

The question for good values for $\vec{d}^T$ as well as for s must be left open to discussion between the biometrician or statistician and the researcher who wants a cluster analysis of his data. Different thresholds obviously yield clusters of different homogeneity. By varying s, the homogeneity of clusters can be controlled, too. The case $s = 1$ allows objects to belong to the same cluster if they are similar in just one block. The case $s = t$, on the other hand assumes that two objects must be similar in all blocks before they will belong to the same group. (In our previous examples, we wanted to find homogeneous clusters, therefore we chose not only $k = 3$ but also $s = t = 2$.) Thus the multigraph model gives a deeper insight into the structure of the data to be clustered. We see exactly in which layers two objects are similar (connected by an edge). Partly, this information is lost when we switch to the s-projection. Here we only count the number of edges connecting two vertices. This is even more informative than calculating a single distance between any pair of objects: In this case we do not know whether two objects are in different clusters since either they differ in all dimensions or they differ significantly in only one dimension and are similar in the remaining $t - 1$ dimensions. This aspect is illustrated by the following example.

As part of a long-term trial about the significance of pregnancy-induced hypertension as a prognostic index for the manifestation of an essential hypertension later on, we wanted to answer the question whether significant differences in blood pressures and heart rates during bicycle exercise between two groups of women could be proved. The first group consisted of normotensive women after a pregnancy without hypertensive disorders. In the second group of normotensive women, all women had hypertensive pregnancy disorders. Upon 21 women measurements of 15 attributes were taken. The data are given in Table 2-1 (see Example 2-1, for more details see [120]). The first nine probands had no hypertensive disorders during their last pregnancy. The last twelve women had pregnancy-induced hypertensions. An exercise-induced hypertension was characterized by a rise in systolic pressure to at least $200\,\text{mm Hg}$ or a rise in diastolic pressure to at least $100\,\text{mm Hg}$ during bicycle exercise. Thus, in nine of the 21 women an exercise-induced hypertension could be proved (O_1, O_2, O_{11}, O_{13}, O_{14}, O_{15}, O_{17}, O_{18} and O_{21} in Table 2-1). In four women, not all 15 measurements had been taken. They have been excluded from the following cluster analysis. (The mean arterial pressures during the second and third trimenon of pregnancy were not available for three women. Height and weight of one woman was missing.) The remaining rows in the data matrix have been renumbered 1–17 as in Example 2-3.

Again, this is a special type of application of cluster analysis. We already knew that the sample can be divided into two groups according to the fact whether a woman had hypertensive disorders during pregnancy or not (external criterion). We wanted to know whether this classification could be repeated with our measurements. This could prove a pregnancy-induced hypertension to be an indicator of a later-on manifestation of an essential hypertension. Figure 6-13 shows the joint distribution of the mean arterial pressure in the second and third trimenon of pregnancy. With CLUSTAN, continuous and qualitative data cannot be used in the same clustering step. Thus for Example 2-3, classifications have been performed using only the continuous variables M_1, M_2, M_5, M_6, M_7, M_{14} and M_{15} in four different linkage procedures. A classification which was induced by the a priori knowledge of a normotensive or hypertensive pregnancy, could not be found (see Figures 2-7a–2-10b).

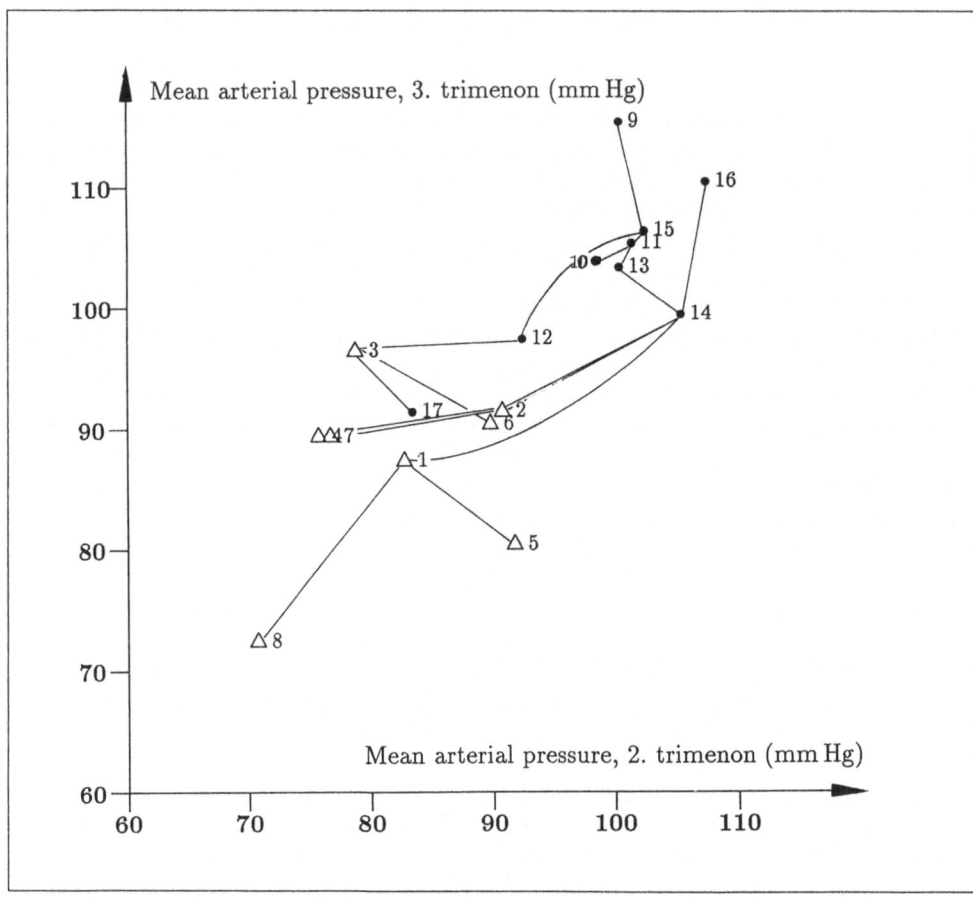

Figure 6-13 *Hypertension Study: Scattergram of mean arterial pressures in the second and third trimenon of pregnancy for 17 women with minimum spanning tree for the normalised data (Euclidean distance).*

We wanted to consider the influence of the family anamnesis (M_3 and M_4) onto the outcome of a cluster analysis. Therefore we used the multigraph model which allows to combine qualitative and quantitative data. For every woman, the Broca index was calculated. It replaced the — possibly correlated — variables "weight" and "height". Together with the age, the Broca index formed the first layer of the multigraph. The items "hypertensive father" and "hypertensive mother" each formed a layer as well as M_7 (total number of pregnancies) and M_8 (reception of an ovulation inhibitor). The mean arterial pressures during the second and third trimenon of pregnancy (M_{14} and M_{15}) were used for the sixth layer. All other variables were excluded from the cluster analysis (M_9–M_{13} were excluded since we wanted to see whether the a posteriori clusters would mirror the a priori groups which were based on the test outcomes during exercise).

All continuous variables — layers 1 and 6 — had been standardised before the classification was performed. Here, the Euclidean distance together with (2-5) was chosen

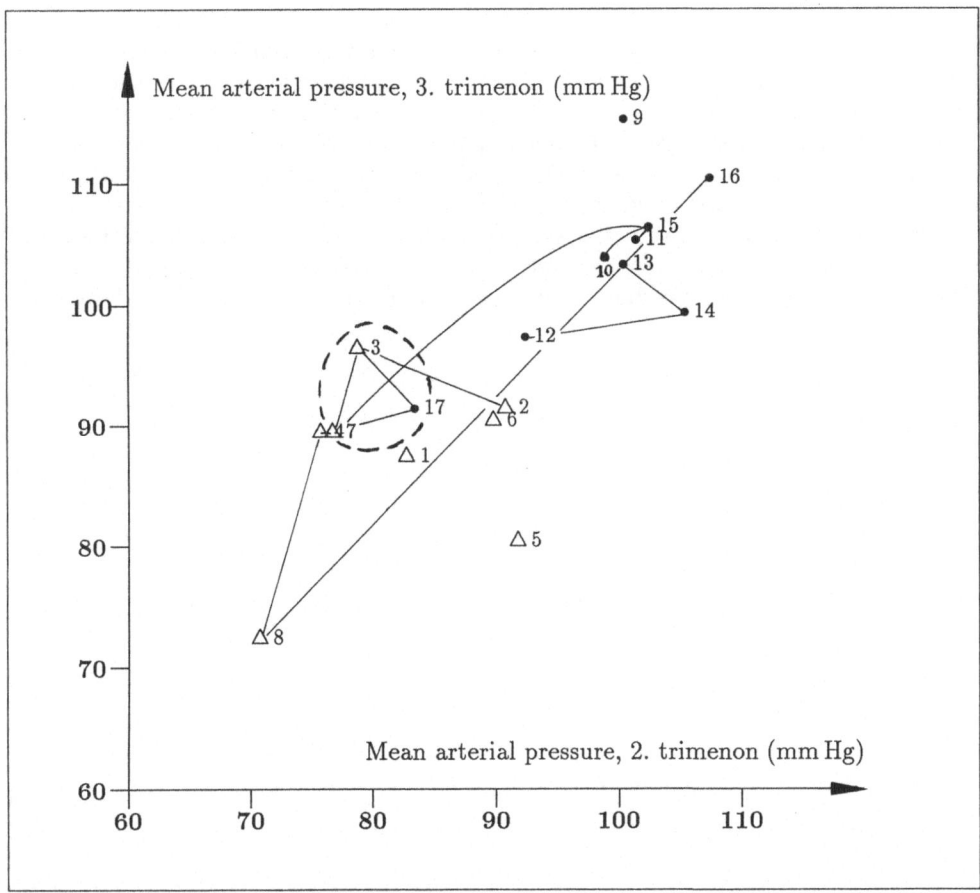

Figure 6-14 *Hypertension Study: Scattergram of mean arterial pressures in the second and third trimenon of pregnancy for 17 women; the 4-saturated connections are pictured as edges; the $(2, \vec{d}^T; 4)$-cluster with 0 and 0.67 as thresholds is outlined.*

as similarity measure. For the layers with binary data, we selected the matching coefficient (2-4a) as similarity measure. For the variable "total number of pregnancies", the similarity between two women was put to 1 if they had the same number of pregnancies; it was put to 0 in all other cases. We chose $d_2 = \cdots = d_5 = 0$ and $d_1 = d_6 = 0.67$ as thresholds (two third of the standard deviations of the continuous variables since they are standardised). Furthermore, $s = 4$ seemed to be appropriate her to avoid too much clusters. By this, we got the 4-projection $\tilde{\Gamma}$ of the multigraph $\Gamma_6(\vec{d}^T)$ of our data as shown in Figure 6-14. This 4-projection has a total of 13 4-saturated connections. It has three 4-isolated vertices (no. 1, 5, 6, 9) and a single proper 4-component or $(1, \vec{d}^T; 4)$-cluster. This $(1, \vec{d}^T; 4)$-cluster contains a $(2, \vec{d}^T; 4)$-cluster (no. 3, 7 and 17).

We performed two other classifications replacing the previous values for d_1 and d_6 by $d_1 = d_6 = 1$ and $d_1 = d_6 = 0.5$, respectively. None of these classifications provided hints that the result was influenced by the fact that the women had an exercise-induced hypertension or not. This also happened, when we increased s to $s = 5$. For this case, choosing $d_1 = d_6 = 0.67$ as previously, the women, numbered 3, 7 and 17, were similar as were 13 and 16. However, for this small number of objects to be classified, the exclusion of four women (about 20% of the sample) could be the reason that no correlations between a priori groups and a posteriori clusters could be found. We may presume this since three of the four women with missing values had pregnancy-induced hypertensive disorders as well as exercise-induced hypertensions later on, while the fourth woman had neither pregnancy-induced nor exercise-induced hypertension.

We want to point out that the woman, numbered 16, is not separated from the rest as long as $s \leq 5$ holds. In Example 2-3, this person had a large isolation index. Now we can see that this was due to one single measurement, namely her bodyweight.

BIBLIOGRAPHY

[1] Abraham, C.: *Evaluation of Clusters on the Basis of Random Graph Theory*. (IBM Research Memo.), IBM Co., Yorktown Height, N.Y., 1962

[2] Aiken, R.C. (Ed.): *Stiff Computation*. Oxford University Press, New York – Oxford 1985

[3] Allen, A., Council, K.A., Sall, J.P.: *SAS User's Guide (Vol. 1: Basics; Vol. 2: Statistics)*. SAS Institute Inc., Cary, NC., 1982

[4] Ameling, W. (Ed.): *First European Simulation Congress ESC83*. Springer-Verlag, Berlin – Heidelberg – New York 1983

[5] Ammon, K.H., Godehardt, E.: Paraphasie und Form der Aphasie: eine Clusteranalyse bei unausgelesenen Patienten. in: Peuser, G. (Ed.): *Brennpunkte der Patholinguistik*. Wilhelm Fink Verlag, München 1978, pp. 35–52

[6] Anderberg, M.R.: *Cluster Analysis for Applications*. Academic Press, New York 1973

[7] Anderson, D.H.: *Compartmental Modeling and Tracer Kinetics*. Springer-Verlag, Berlin – Heidelberg – New York –Tokyo 1983

[8] Anderson, E.: Efficient and Inefficient Methods of Measuring Specific Differences. in: Kempthorne, O. (Ed.): *Statistics and Mathematics in Biology*. Iowa State College Press, Ames 1954, pp. 98–107

[9] Andrews, D.F.: Plots of High-Dimensional Data. *Biometrics* **28** (1972), 125–136

[10] Arminger, G.: *Faktorenanalyse*. B.G. Teubner, Stuttgart 1979

[11] Atkins, G.L.: *Multicompartment Models for Biological Systems*. Methuen & Co., London 1969

[12] Austin, T.L., Fagen, R.E., Penney, W.F., Riordan, J.: The Number of Components in Random Linear Graphs. *Annals of Mathematical Statistics* **30** (1959), 747–754

[13] Baker, F.B.: Stability of Two Hierarchical Grouping Techniques. Case I: Sensitivity to Data Errors. *Journal of the American Statistical Association* **69** (1974), 440–445

[14] Baker, F.B.: Sensitivity of the Complete-Link Clustering Technique to Missing Individuals. *Journal of Educational Statistics* **3** (1978), 233–252

[15] Baker, F.B., Hubert, L.J.: Measuring the Power of Hierarchical Cluster Analysis. *Journal of the American Statistical Association* **70** (1975), 31–38

[16] Baker, F.B., Hubert, L.J.: A Graph-Theoretic Approach to Goodness-of-Fit in Complete-Link Hierarchical Clustering. *Journal of the American Statistical Association* **71** (1976), 870–878

[17] Bandelt, H.-J.: *Graphische Repräsentierung von Merkmalsverteilungen in der biologischen Klassifikation.* (Paper presented at the *9. Jahrestagung der deutschen Gesellschaft für Klassifikation, Karlsruhe 1985.*)

[18] Barbour, A.D.: Poisson Convergence and Random Graphs. *Mathematical Proceedings of the Cambridge Philosophical Society* **92** (1982), 349–359

[19] Barnett, V., Kay, R., Sneath, P.H.A.: A Familiar Statistic in an Unfamiliar Guise — A Problem in Clustering. *The Statistician* **28** (1979), 185–191

[20] Barnett, V., Lewis, T.: *Outliers in Statistical Data.* John Wiley & Son, Chichester – New York 1978

[21] Beale, E.M.L.: Euclidean Cluster Analysis. *Bulletin of the International Statistical Institute* **43** (1969), 92–94

[22] Beineke, L.W., Wilson, R.J. (Eds.): *Selected Topics in Graph Theory (Vol. 1; Vol. 2).* Academic Press, London – New York – San Francisco 1978, 1983

[23] Bellman, R.: *Mathematical Methods in Medicine.* World Scientific Publ. Co., Singapore 1983

[24] Benowitz, N.L., Meister, W.: Clinical Pharmacokinetics of Lignocaine. *Clinical Pharmacokinetics* **3** (1978), 177–201

[25] Benzécri, J.P.: *L'Analyse des Données. (Tome 1: La Taxinomie; Tome 2: L'Analyse des Correspondances).* Dunod, Paris 1976

[26] Bezdek, J.C.: Numerical Taxonomy with Fuzzy Sets. *Journal of Mathematical Biology* **1** (1974), 57–71

[27] Bezdek, J.C.: Cluster Validity with Fuzzy Sets. *Journal of Cybernetics* **3** (1974), 58–73

[28] Bickel, P.J., Breiman, L.: Sums of Functions of Nearest Neighbor Distances, Moment Bounds, Limit Theorems and a Goodness of Fit Test. *Annals of Probability* **11** (1983), 185–214

[29] Binder, D.A.: Bayesian Cluster Analysis. *Biometrika* **65** (1978), 31–38

[30] Binder, D.A.: Comment on "Estimating Mixtures of Normal Distributions and Switching Regressions". *Journal of the American Statistical Association* **73** (1978), 746–747

[31] Blashfield, R.K.: Mixture Model Tests of Cluster Analysis: Accuracy of Four Agglomerative Hierarchical Methods. *Psychological Bulletin* **83** (1976), 377–388

[32] Blashfield, R.K.: Questionnaire on Cluster Analysis Software. *Classification Society Bulletin* **3** (1976), 25–42

[33] Blashfield, R.K., Aldenderfer, M.S.: The Literature on Cluster Analysis. *Multivariate Behavioral Research* **13** (1978), 271–295

[34] Blashfield, R.K., Aldenderfer, M.S., Morey, L.C.: Cluster Analysis Software. in: Krishnaiah, P.R., Kanal, L.N. (Eds.): *Handbook of Statistics (Vol. 2: Classification, Pattern Recognition and Reduction of Dimensionality).* North-Holland, Amsterdam – New York – Oxford 1982, pp. 245–266

[35] Bloedhorn, H., Driever, R., Godehardt, E.: Varianzanalyse bei fehlenden Meßwerten — Eine Übersicht. *Statistical Software Newsletter* **8.2** (1982), 61–72

[36] Blomer, R., Raschewa, C., Thurmayr, R., Thurmayr, R.: A Locally Sensitive Mapping of Multivariate Data onto a Two-dimensional Plane. in: *Medical Data Processing (Proceedings of the Symposium on Medical Data Processing, Toulouse 1976)*. Taylor & Francis, London 1976, pp. 525–529

[37] Bock, H.H.: Statistische Modelle und Bayes'sche Verfahren zur Bestimmung einer unbekannten Klassifikation normalverteilter zufälliger Vektoren. *Metrika* **18** (1972), 120–132

[38] Bock, H.H.: *Automatische Klassifikation. Theoretische und praktische Methoden zur Gruppierung und Strukturierung von Daten (Clusteranalyse)*. Vandenhoeck & Ruprecht, Göttingen 1974

[39] Bock, H.H.: Programme zur Clusteranalyse. *Statistical Software Newsletter* **2.3** (1976), 72–84

[40] Bock, H.H.: On Tests Concerning the Existence of a Classification. in: Institute de Recherche en Informatique et en Automatique (IRIA) (Eds.): *Proceedings of the First International Symposium on Data Analysis and Informatics, Versailles 1977*. IRIA, Le Chesnay 1977, pp. 449–464

[41] Bock, H.H.: Clusteranalyse mit unscharfen Partitionen. in: Bock, H.H. (Ed.): *Klassifikation und Erkenntnis III; Numerische Klassifikation (Studien zur Klassifikation Bd. 6: Proc. 3. Fachtagung der deutschen Gesellschaft für Klassifikation, Königstein 1979)*. Indeks-Verlag, Frankfurt a.M. 1979, pp. 137–163

[42] Bock, H.H.: Clusteranalyse — Überblick und neuere Entwicklungen. *OR Spektrum* **1** (1980), 211–232

[43] Bock, H.H.: Explorative Datenanalyse. in: Victor, N., Lehmacher, W., van Eimeren, W. (Eds.): *Explorative Datenanalyse (Frühjahrstagung der GMDS, München 1980)*. Springer-Verlag, Berlin – Heidelberg – New York 1980, pp. 6–37

[44] Bock, H.H.: Korrespondenzanalyse zur Strukturanalyse und ihre Verwendung zur Clusteranalyse. in: Ihm, P., Dahlberg, I. (Eds.): *Numerische und nicht-numerische Klassifikation zwischen Theorie und Praxis (Studien zur Klassifikation Bd. 10: Proc. 5. Fachtagung der deutschen Gesellschaft für Klassifikation, Hofgeismar 1981)*. Indeks-Verlag, Frankfurt a.M. 1982, pp. 36–53

[45] Bock, H.H.: Statistische Testverfahren im Rahmen der Clusteranalyse. in: Dahlberg, I., Schader, M.R. (Eds.): *Automatisierung in der Klassifikation (Studien zur Klassifikation Bd. 13: Proc. 7. Fachtagung der deutschen Gesellschaft für Klassifikation, Königswinter 1983)*. Indeks-Verlag, Frankfurt a.M. 1983, pp. 161–176

[46] Bock, H.H.: Maßzahlen zum Vergleich von Bäumen, Hierarchien und Sequenzen. in: Bock, H.H. (Ed.): *Anwendungen der Klassifikation: Datenanalyse und numerische Klassifikation (Studien zur Klassifikation Bd. 15: Proc. 8. Fachtagung der deutschen Gesellschaft für Klassifikation, Hofgeismar 1984)*. Indeks Verlag, Frankfurt a.M. 1984, pp. 52–67

[47] Bock, H.H.: Statistical Testing and Evaluation Methods in Cluster Analysis. in: Indian Statistical Institute (Eds.): *Proceedings on the Golden Jubilee Conference in Statistics: Applications and New Directions, Calcutta 1981*. Indian Statistical Institute, Calcutta 1984, pp. 116–146

[48] Bock, H.H.: On Some Significance Tests in Cluster Analysis. *Journal of Classification* **2** (1985), 77–108

[49] Bock, H.H. (Ed.): *Classification and Related Methods of Data Analysis (Proceedings of the First Conference of the International Federation of Classification Societies, Aachen 1987)*. North-Holland, Amsterdam – New York – Oxford – Tokyo 1988, pp. 219–228

[50] Bock, H.H., Ohly, H.P., Bender, D.: *Software zur Clusteranalyse, Netzwerkanalyse und verwandten Verfahren*. Informationszentrum Sozialwissenschaften, Bonn 1983

[51] Bokemeyer, B., Thiele, K.G.: Cluster-Analyse bei 109 Patienten mit systemischem Lupus erythematodes. *Klinische Wochenschrift* **63** (1984), 79–83

[52] Bokemeyer, B., Thiele, K.G.: Cluster-Analyse zur Klassifikation des Lupus erythematodes disseminatus. *Zeitschrift für Rheumatologie* **43** (1984), 272–274

[53] Bollinger, G., Herrmann, A., Möntmann, V.: *BMDP — Statistikprogramme für die Bio-, Human- und Sozialwissenschaften*. Gustav Fischer Verlag, Stuttgart – New York 1983

[54] Bollobás, B.: *Extremal Graph Theory*. Academic Press, London – New York – San Francisco 1978

[55] Bollobás, B.: Vertices of Given Degree in a Random Graph. *Journal of Graph Theory* **6** (1982), 147–155

[56] Bollobás, B.: *Random Graphs*. Academic Press, London – New York – San Francisco 1985

[57] Bollobás, B., Thomason, A.: Random Graphs of Small Order. in: Karoński, M., Ruciński, A. (Eds.): *Random Graphs '83 (Annals of Discrete Mathematics 28: Proceedings of the 1st International Seminar on Random Graphs, Poznań 1983)*. North-Holland, Amsterdam – New York – Oxford 1985, pp. 47–97

[58] Boyes, R.N., Scott, D.B., Jebson, P.J., Godman, M.J., Julian, D.G.: Pharmacokinetics of Lidocaine in Man. *Clinical Pharmacology and Therapeutics* **13** (1971), 105–116

[59] Bozler, G., van Rossum, J.M. (Eds.): *Pharmacokinetics During Drug Development: Data Analysis and Evaluation Techniques*. Gustav Fischer Verlag, Stuttgart – New York 1982

[60] Bruckner, L.A.: On Chernoff Faces. in: Wang, P.C.C. (Ed.): *Graphical Representation of Multivariate Data*. Academic Press, New York 1978, pp. 93–121

[61] Bryant, P.G.: On Testing for Clusters Using the Sample Covariance. *Journal of Multivariate Analysis* **5** (1975), 96–105

[62] Bryant, P.G., Williamson, J.A.: Asymptotic Behavior of Classification Maximum Likelihood Estimates. *Biometrika* **65** (1978), 273–281

[63] Bryant, P.G., Williamson, J.A.: Maximum Likelihood and Classification: A Comparison of Three Approaches. in: Gaul, W., Schader, M. (Eds.): *Classification as a Tool of Research (Proc. 9th Annual Meeting of the Classification Society, Karlsruhe 1985)*. North-Holland, Amsterdam – New York – Oxford – Tokyo 1986, pp. 35–45

[64] Burtin, Yu.D.: On Extreme Metric Characteristics of a Random Graph. II: Limit Distributions. *SIAM Journal on the Theory of Probability and Applications* **20** (1975), 83–101

[65] Bützow, G.H., Dammann, H.G., Pfeiffer, A., Runge, M., Schmoldt, A.: Pharmakokinetik von Lidocain bei Leberschäden. Klinische und tierexperimentelle Untersuchungen. *Intensivmedizin* **17** (1980), 251–255

[66] Cacoullos, T. (Ed.): *Discriminant Analysis and Applications*. Academic Press, New York – London 1973

[67] Calinsky, T., Harabasz, J.: A Dendrite Method for Cluster Analysis. *Communications in Statistics* **3.1** (1974), 1–24

[68] Capobianco, M., Palka, Z.: The Distribution of Popular Persons in a Group. *Social Networks* **5** (1983), 383–393

[69] Cayley, A.: On the Theory of the Analytical Forms Called Trees. *Philosophical Magazine* **13** (1857), 172–176

[70] Cayley, A.: On the Analytical Forms Called Trees. *American Journal of Mathematics* **4** (1881), 266–268

[71] Cayley, A.: A Theorem on Trees. *Quarterly Journal of Pure and Applied Mathematics* **23** (1889), 376–378

[72] Chambers, J.M., Kleiner, B.: Graphical Techniques for Multivariate Data and for Clustering. in: Krishnaiah, P.R., Kanal, L.N. (Eds.): *Handbook of Statistics (Vol. 2: Classification, Pattern Recognition and Reduction of Dimensionality)*. North-Holland, Amsterdam – New York – Oxford 1982, pp. 209–244

[73] Chernoff, H.: Using Faces to Represent Points in k-dimensional Space Graphically. *Journal of the American Statistical Association* **68** (1973), 361–368

[74] Clauß, G., Ebner, H.: *Statistik für Soziologen, Pädagogen, Psychologen und Mediziner (Bd. 1: Grundlagen)*. Verlag Harri Deutsch, Thun – Frankfurt am Main 1982

[75] Cohen, J.E.: *Random Digraphs and Food Webs in Ecology*. (Paper presented at the *2nd International Seminar on Random Graphs, Poznań 1985*.)

[76] Cole, A.J. (Ed.): *Numerical Taxonomy*. Academic Press, New York 1969

[77] Cormack, R.M.: A Review of Classification. *Journal of the Royal Statistical Society Ser. A* **134** (1971), 321–367

[78] Cunningham, K.M., Ogilvie, J.C.: Evaluation of Hierarchical Grouping Techniques: A Preliminary Study. *The Computer Journal* **15** (1972), 209–213

[79] Dahmström, P., Hagnell, M.: *The Formation of Strata Using Cluster Analysis*. (Research Report No. 4), Department of Statistics, Lund (Sweden) 1974

[80] Day, N.E.: Estimating the Components of a Mixture of Normal Distributions. *Biometrika* **56** (1969), 463–474

[81] Day, W.H.E.: Validity of Clusters Formed by Graph-Theoretic Methods. *Mathematical Biosciences* **36** (1977), 299–317

[82] Defay, D.: An Efficient Algorithm for a Complete Link Method. *The Computer Journal* **20** (1977), 364–366

[83] Degens, P.O.: *Clusteranalyse auf topologisch-maßtheoretischer Grundlage*. (Dissertation), Fachbereich Mathematik der Universität, München 1978

[84] Degens, P.O.: Konsistenzforderungen für die automatische Klassifikation. in: Bock, H.H. (Ed.): *Klassifikation und Erkenntnis III; Numerische Klassifikation (Studien zur Klassifikation Bd. 6: Proc. 3. Fachtagung der deutschen Gesellschaft für Klassifikation, Königstein 1979)*. Indeks-Verlag, Frankfurt a.M. 1979, pp. 31–47

[85] Degens, P.O.: Gewichtung bei Clusterverfahren. in: Ihm, P., Dahlberg, I. (Eds.): *Numerische und nicht-numerische Klassifikation zwischen Theorie und Praxis (Studien zur Klassifikation Bd. 10: Proc. 5. Fachtagung der deutschen Gesellschaft für Klassifikation, Hofgeismar 1981)*. Indeks-Verlag, Frankfurt a.M. 1982, pp. 85–93

[86] Degens, P.O.: Hierarchische Clusteranalyse. Approximation und Agglomeration. in: Dahlberg, I., Schader, M.R. (Eds.): *Automatisierung in der Klassifikation (Studien zur Klassifikation Bd. 13: Proc. 7. Fachtagung der deutschen Gesellschaft für Klassifikation, Königswinter 1983)*. Indeks-Verlag, Frankfurt a.M. 1983, pp. 189–202

[87] Degens, P.O.: Phylogenetische Systematik und numerische Taxonomie. in: Bock, H.H. (Ed.): *Anwendungen der Klassifikation: Datenanalyse und numerische Klassifikation (Studien zur Klassifikation Bd. 15: Proc. 8. Fachtagung der deutschen Gesellschaft für Klassifikation, Hofgeismar 1984)*. Indeks Verlag, Frankfurt a.M. 1984, pp. 68–82

[88] Deichsel, G.: *Random walk clustering in großen Datenbeständen.* (Dissertation), Universität Stuttgart 1978

[89] Deichsel, G., Bosch, I.: Automatische Klassifikation von Amöben durch zwei graphentheoretische Verfahren. *EDV in Medizin und Biologie* **3** (1973), 69–73

[90] Deichsel, G., Trampisch, H.J.: *Clusteranalyse und Diskriminanzanalyse.* Gustav Fischer Verlag, Stuttgart – New York 1985

[91] Del Pino, G.E.: On the Asymptotic Distribution of *k*-Spacings with Applications to Goodness-of-Fit Tests. *The Annals of Statistics* **7** (1979), 1058–1065

[92] Dick, N.P., Bowden, D.C.: Maximum Likelihood Estimation for Mixture of Two Normal Distributions. *Biometrics* **29** (1973), 781–790

[93] Dillon, W.R., Goldstein, M.: *Multivariate Analysis — Methods and Applications.* Wiley, New York – Chichester 1984

[94] Dixon, W.J., Brown, M.B., Engelman, L., Frane, J.W., Hill, M.A., Jennrich, R.I., Toporek, J.D.: *BMDP Statistical Software.* University of California Press, Berkeley – Los Angeles – London 1985

[95] Doetsch, G.: *Anleitung zum praktischen Gebrauch der Laplace-Transformation und der Z-Transformation.* Oldenbourg Verlag, München – Wien 1985

[96] Dörfel, H.: Pfadkoeffizienten und Strukturmodelle. *Biometrische Zeitschrift* **14** (1972), 12–26

[97] Dörfel, H.: Schätzen von Pfadkoeffizienten. *Biometrische Zeitschrift* **14** (1972), 209–226

[98] Dost, F.H.: *Der Blutspiegel.* VEB Thieme, Leipzig 1953

[99] Dost, F.H.: *Grundlagen der Pharmakokinetik.* Georg Thieme, Stuttgart 1968

[100] Driever, R.: *Varianzanalyse bei fehlenden Meßwerten.* (Dissertation), Institut für Medizinische Dokumentation und Statistik der Universität, Köln 1985

[101] Dubes, R., Jain, A.K.: Validity Studies in Clustering Methodologies. *Pattern Recognition* **11** (1979), 235–254

[102] Duran, B.S., Odell, P.L.: *Cluster Analysis: A Survey.* Springer-Verlag, Berlin – New York 1974

[103] Dworatzek, R.: *Pharmakokinetik von Lidocain bei Patienten mit Niereninsuffizienz oder Leberparenchymschaden.* (Dissertation), 2. Medizinische Universitätsklinik, Köln 1988

[104] Eberl, W., Hafner, R.: Die asymptotische Verteilung von Koinzidenzen. *Zeitschrift für Wahrscheinlichkeitstheorie und verwandte Gebiete* **18** (1971), 322–332

[105] Eckes, T.: Ein nonparametrischer Test für die Ähnlichkeit zwischen Aufteilungen einer Objektmenge. *Psychologische Beiträge* **24** (1982), 57–84

[106] Eckes, T., Roßbach, H.: *Clusteranalysen.* Kohlhammer, Stuttgart 1980

[107] Edwards, A.W.F., Cavalli-Sforza, L.L.: A Method for Cluster Analysis. *Biometrics* **21** (1965), 362–375

[108] Eigener, M.: *Konstruktion von 2-Stichproben-Tests mit Hilfe clusteranalytischer Methoden.* (Diplomarbeit), Institut für Mathematische Stochastik der Universität, Hamburg 1976

[109] Engelman, L., Hartigan, J.A.: Percentage Points of a Test for Clusters. *Journal of the American Statistical Association* **64** (1969), 1647-1648

[110] Erdős, P., Palka, Z.: Trees in Random Graphs. *Discrete Mathematics* **46** (1983), 145-150

[111] Erdős, P., Rényi, A.: On Random Graphs I. *Publicationes Mathematicae* **6** (1959), 290-297

[112] Erdős, P., Rényi, A.: On the Evolution of Random Graphs. *Publications of the Mathematical Institute of the Hungarian Academy of Sciences* **5** (1960), 17-60

[113] Erdős, P., Rényi, A.: On the Evolution of Random Graphs. *Bulletin of the International Institute of Statistics* **38** (1960), 343-347

[114] Estabrook, G.F.: A Mathematical Model in Graph Theory for Systematic Zoology. *Journal of Theoretical Biology* **12** (1966), 297

[115] Everitt, B.S.: *Cluster Analysis*. Heinemann, London 1974

[116] Everitt, B.S.: *Graphical Techniques for Multivariate Data*. North-Holland, New York 1978

[117] Everitt, B.S.: Unresolved Problems in Cluster Analysis. *Biometrics* **35** (1979), 169-181

[118] Fegley, K.A., Long, V.P. Jr.: Modeling and Analysis in Optimal Therapy. in: Nicolini, C. (Ed.): *Modeling and Analysis in Biomedicine*. World Scientific Publ. Co., Singapore 1984, pp.263-289

[119] Felsenstein, J. (Ed.): *Numerical Taxonomy*. Springer Verlag, Berlin 1983

[120] Feltkamp, H.: *Die Schwangerschaftshypertonie (Untersuchungen zur pathogenetischen Bedeutung vasopressorischer oder vasodepressorischer Regulationssysteme und zur prognostischen Wertigkeit einer späteren Hochdruckmanifestation)*. (Habilitationsschrift), Medizinische Fakultät der Universität, Köln 1985

[121] Fischer, R., Haerlin, R., Steinijans, V., Zech, K., Bruckschen, E.G.: Preliminary Results on the Correlation Between Serum Level and Antihypertensive Effect of Urapidil (Ebrantil). *Methods and Findings in Experimental Clinical Pharmacology* **3** (Suppl. 1, 1981), 89S-93S

[122] Fisher, L., van Ness, J.W.: Admissible Clustering Procedures. *Biometrika* **58** (1971), 91-104

[123] Fisher, R.A.: The Use of Multiple Measurements in Taxonomic Problems. *Annals of Eugenics* **7** (1936), 179-188

[124] Fisher, R.A.: The Precision of Discriminant Functions. *Annals of Eugenics* **10** (1940), 422-429

[125] Fisher, W.D.: *Clustering and Aggregation in Economics*. John Hopkins, Baltimore 1969

[126] Fleiss, J.L., Lawlor, W., Platman, S.R., Fieve, R.R.: On the Use of Inverted Factor Analysis for Generating Typologies. *Journal of Abnormal Psychology* **77** (1971), 127-132

[127] Fleiss, J.L., Zubin, J.: On the Methods and Theory of Clustering. *Multivariate Behavioural Research* **4** (1969), 235-250

[128] Florek, K., Łukaszewicz, J., Perkal, J., Steinhaus, H., Zubrzycki, S.: Sur la Liaison et la Division des Points d'un Ensemble Fini. *Colloquia Mathematicae* **2** (1951), 282-285

[129] Florek, K., Łukaszewicz, J., Perkal, J., Steinhaus, H., Zubrzycki, S.: Taksonomia Wroclawska. *Przeglad Antropologiczny (Poznań)* **17** (1951), 193-207

[130] Flury, B., Riedwyl, H.: *Angewandte multivariate Statistik*. Gustav Fischer Verlag, Stuttgart – New York 1983

[131] Forst, H.T.: *Zur Klassifizierung von Städten nach wirtschafts- und sozialstatistischen Merkmalen*. Physica-Verlag, Würzburg 1974

[132] Frank, O.: Estimation of the Number of Connected Components in a Graph by Using a Sampled Subgraph. *Scandinavian Journal of Statistics* **5** (1978), 177–188

[133] Frank, O., Harary, F.: Cluster Inference by Using Transitivity Indices in Empirical Graphs. *Journal of the American Statistical Association* **77** (1982), 835–840

[134] Friedman, H.P., Rubin, J.: On Some Invariant Criteria for Grouping Data. *Journal of the American Statistical Association* **62** (1967), 1159–1178

[135] Friedman, J.H., Rafsky, L.C.: Graph-Theoretic Measures of Multivariate Association and Prediction. *The Annals of Statistics* **11.2** (1983), 377–391

[136] Fukunaga, K., Hosteller, L.D.: Optimization of k-Nearest Neighbor Density Estimates. *IEEE Transactions on Information Theory* **IT-19.5** (1973), 320–326

[137] Fukunaga, K., Narendra, P.M.: A Branch and Bound Algorithm for Computing k-Nearest Neighbors. *IEEE Transactions on Computers (Corresp.)* **C-24.7** (1975), 750–753

[138] Gabriel, K.R., Odoroff, C.L.: Use of Three-Dimensional Biplots for Diagnosis of Models. in: Gaul, W., Schader, M. (Eds.): *Classification as a Tool of Research (Proc. 9th Annual Meeting of the Classification Society, Karlsruhe 1985)*. North-Holland, Amsterdam – New York – Oxford – Tokyo 1986, pp. 153–159

[139] Gaul, W., Schader, M. (Eds.): *Classification as a Tool of Research (Proc. 9th Annual Meeting of the Classification Society, Karlsruhe 1985)*. North-Holland, Amsterdam – New York – Oxford – Tokyo 1986

[140] Giacomelli, F., Wiener, J., Kruskal, J.B., von Pomeranz, J., Loud, A.V.: Subpopulations of Blood Lymphocytes Demonstrated by Quantitative Cytochemistry. *Journal of Histochemistry and Cytochemistry* **19** (1971), 426–433

[141] Gilbert, E.N.: Enumeration of Labelled Graphs. *Canadian Journal of Mathematics* **8** (1956), 405–411

[142] Gilbert, E.N.: Random Graphs. *Annals of Mathematical Statistics* **30** (1959), 1141–1144

[143] Gitman, I., Levine, M.D.: An Algorithm for Detecting Unimodal Fuzzy Sets and Its Application as a Clustering Technique. *IEEE Transactions on Computers* **C-19.7** (1970), 583–593

[144] Gladtke, E., von Hattingberg, H.M.: *Pharmakokinetik*. Springer-Verlag, Berlin – Heidelberg – New York 1977

[145] Gladtke, E., Heimann, G. (Eds.): *Pharmacokinetics*. Gustav Fischer Verlag, Stuttgart – New York 1980

[146] Glaser, E.: *Pharmakokinetik — Grundlagen, lineare Modelle, Rechenverfahren, Auswertemethoden*. pmi Verlag, Frankfurt 1985

[147] Gnanadesikan, R.: *Methods for Statistical Data Analysis of Multivariate Observations*. John Wiley & Son, New York 1977

[148] Godehardt, E.: *Heuristische und wahrscheinlichkeitstheoretische Ansätze zur Cluster Analysis*. (Diplomarbeit), Institut für Mathematische Statistik und Dokumentation der Universität, Düsseldorf 1974

[149] Godehardt, E.: *Eine Erweiterung der Sätze vom Erdős-Rényi-Typ auf ungerichtete, vollständig indizierte Zufallsmultigraphen*. (Dissertation), Institut für Mathematische Statistik und Dokumentation der Universität, Düsseldorf 1980

[150] Godehardt, E.: An Extension of the Theorems of Erdős-Rényi-Type to Random Multigraphs. in: Bereanu, B., Grigorescu, S., Iosifescu, M., Postelnicu, T. (Eds.): *Proceedings of the Sixth Conference on Probability Theory, Braşov 1979*. Editura Academiei RSR, Bucureşti 1981, pp. 417–425

[151] Godehardt, E.: *Anwendungen der Graphentheorie auf die Modellierung komplexer biologischer Systeme*. (Paper presented at the *28. Biometrischen Kolloquium, Aachen 1982*.)

[152] Godehardt, E.: Zufallsmultigraphen: Ein mathematisches Modell als Hilfsmittel zum Auffinden von Strukturen. in: Ihm, P., Dahlberg, I. (Eds.): *Numerische und nicht-numerische Klassifikation zwischen Theorie und Praxis (Studien zur Klassifikation Bd. 10: Proc. 5. Fachtagung der deutschen Gesellschaft für Klassifikation, Hofgeismar 1981)*. Indeks-Verlag, Frankfurt a.M. 1982, pp. 103–112

[153] Godehardt, E.: Limit Theorems for Undirected, Completely Labelled Random Multigraphs: Extensions of the Theorems of P. Erdős and A. Rényi. in: Révész, P. (Ed.): *Limit Theorems in Probability and Statistics (Vol. I, Vol. II). (Colloquia Mathematica Societatis János Bolyai 36: Proceedings of the 2nd Colloquium on Limit Theorems in Probability and Statistics, Veszprém 1982)*. North-Holland, Amsterdam – New York – Oxford – Tokyo 1984, pp. 499–518 (Vol. I)

[154] Godehardt, E.: Random Multigraphs: A Mathematical Model for Cluster Analysis. in: Dudewicz, E.J., Plachky, D., Sen, P.K. (Eds.): *Selected Papers presented at the 16th European Meeting of Statisticians, Marburg, September 3–7, 1984 (Statistics & Decisions, Supplement Issue No. 2)*. R. Oldenbourg Verlag, München 1985, pp. 315–319

[155] Godehardt, E.: *Applications of the Theory of Random Multigraphs in Cluster Analysis*. (Paper presented at the *2nd International Seminar on Random Graphs, Poznań 1985*.)

[156] Godehardt, E.: Theorems of Erdős-Rényi-Type Applied to Random Graphs of Small Order. *Notes from New York Graph Theory Day XII* **12** (1986), 29–36

[157] Godehardt, E.: *Explorative mathematische Modelle in der Medizin: Nichtlineare Regression und Numerische Klassifikation*. (Habilitationsschrift), Institut für Medizinische Dokumentation und Statistik der Universität, Köln 1986

[158] E. Godehardt and H. Herrmann, Multigraphs as a Tool for Numerical Classification. in: Bock, H.H. (Ed.): *Classification and Related Methods of Data Analysis (Proceedings of the First Conference of the International Federation of Classification Societies, Aachen 1987)*. North-Holland, Amsterdam – New York – Oxford – Tokyo 1988, pp. 219–228

[159] Godehardt, E., Richter, O.: Application of Graph Theory in Modelling of Biological Systems. in: Ameling, W. (Ed.): *First European Simulation Congress ESC83*. Springer-Verlag, Berlin – Heidelberg – New York 1983, pp. 580–587

[160] Godehardt, E., Steinebach, J.: On a Lemma of P. Erdős and A. Rényi about Random Graphs. *Publicationes Mathematicae* **28.3–4** (1981), 271–273

[161] Godehardt, E., Wambach, G., Heitz, W., Steinijans, V., Haerlin, R., Kaufmann, W.: Pharmacokinetics of Urapidil in Patients with Normal and Impaired Renal Function. in: Amery, A. (Ed.): *Treatment of Hypertension with Urapidil: Preclinical and Clinical Update*. Royal Society of Medicine Services, London – New York 1986, pp. 71–86

[162] Gordon, A.D., Henderson, J.J.: An Algorithm for Euclidean Sum of Squares Classification. *Biometrics* **33** (1977), 355–362

[163] Gower, J.C.: Some Distance Properties of Latent Root and Vector Methods Used in Multivariate Analysis. *Biometrika* **53** (1966), 325–338

[164] Gower, J.C.: A Comparison of Some Methods of Cluster Analysis. *Biometrics* **23** (1967), 623–637

[165] Gower, J.C.: Multivariate Analysis and Multidimensional Geometry. *The Statistician* **17** (1967), 13–28

[166] Gower, J.C.: A General Coefficient of Similarity and Some of Its Properties. *Biometrics* **27** (1971), 857–871

[167] Gower, J.C., Ross, G.J.S.: Minimum Spanning Trees and Single Linkage Cluster Analysis. *Journal of The Royal Statistical Society Ser. C: Applied Statistics* **18** (1969), 54–64

[168] Gower, J.S., Banfield, C.F.: Goodness-of-Fit Criteria in Cluster Analysis and Their Empirical Distributions. in: Corsten, L.C.A., Postelnicu, T. (Eds.): *Proceedings of the 8th International Biometric Conference, Constanta 1974*. Academiei RSR, Bucureşti 1975

[169] Gray, J.B., Ling, R.F.: *K*-Clustering as a Detection Tool for Influential Subsets in Regression. *Technometrics* **26.4** (1984), 305–330

[170] Gray, P.M.D., Murray, A.M., Young, N.A.: Wright's Formulae for the Nunber of Connected Sparsely Edged Graphs. *Journal of Graph Theory* **1** (1977), 331–334

[171] Greenblatt, D.J., Bolognini, V., Koch-Weser, J., Harmatz, J.S.: Pharmacokinetic Approach to the Clinical Use of Lidocaine Intravenously. *Journal of the American Medical Association* **236** (1976), 273–277

[172] Greenblatt, D.J., Koch-Weser, J.: Clinical Pharmacokinetics, I. *The New England Journal of Medicine* **290** (1975), 702–705

[173] Griebenow, R., Godehardt, E., Saborowski, F., Dworatzek, R., Köhler, A., Evers, J.: Eliminationshalbwertzeit von Lidocain bei Patienten mit eingeschränkter Nieren- und Leberfunktion. *Herzmedizin* **6.3** (1983), Abstract 48 (Kurzfassungen der Vorträge der *141. Tagung Rheinisch-Westfälischen Gesellschaft für Innere Medizin, Düsseldorf 1983*)

[174] Grimmett, G.: Random Graphs. in: Beineke, L.W., Wilson, R.J. (Eds.): *Selected Topics in Graph Theory (Vol. 2)*. Academic Press, London – New York – San Francisco 1983, pp. 201–235

[175] Grossman, S.H., Davis, D., Kitchell, B.B., Shand, D.G., Routledge, P.A.: Diazepam and Lidocaine Plasma Protein Binding in Renal Disease. *Clinical Pharmacology and Therapeutics* **31** (1982), 350–357

[176] Hacisalihzade, S.S.: Optimierung der Medikamenten-Dosierung mit Hilfe der Simulation am Beispiel des Parkinsonismus. in: Möller, D.P.F. (Ed.): *Erwin-Riesch Arbeitstagung: Systemanalyse biologischer Prozesse (1. Ebernburger Gespräch)*. Springer-Verlag, Berlin – Heidelberg – New York – Tokyo 1984, pp. 177–182

[177] Hamza, M.H., Xu-Shi, J.: An Algorithm for Adaptive Control. *International Journal of Systems Sciences* **14** (1983), 317–324

[178] Hansen, P., Delattre, M.: Complete-Link Cluster Analysis by Graph Coloring. *Journal of the American Statistical Association* **73** (1978), 397–403

[179] Hansert, E.: Ein Modell zur Analyse von Merkmals-Clustern bei Alternativmerkmalen. in: Lange, H.J., Wagner, G. (Eds.): *Computergestützte ärztliche Diagnostik*. Schattauer Verlag, Stuttgart 1978, pp. 187–196

[180] Harary, F.: *Graphentheorie*. R. Oldenbourg Verlag, München – Wien 1974

[181] Harman, H.H.: *Modern Factor Analysis*. Chicago University Press, Chicago 1976

[182] Hartigan, J.A.: Representation of Similarity Matrices by Trees. *Journal of the American Statistical Association* **62** (1967), 1140–1158

[183] Hartigan, J.A.: *Clustering Algorithms*. Wiley-Interscience, New York 1975

[184] Hartigan, J.A.: Distribution Problems in Clustering. in: van Ryzin, J. (Ed.): *Classification and Clustering*. Academic Press, New York 1977, pp. 45–71

[185] Hartigan, J.A.: Asymptotic Distributions for Clustering Criteria. *The Annals of Statistics* **6** (1978), 117–131

[186] Hartigan, J.A.: Consistency of Single Linkage for High Density Clusters. *Journal of the American Statistical Association* **76** (1981), 388–394

[187] Hartigan, J.A.: Statistical Theory in Clustering. *Journal of Classification* **2** (1985), 63–76

[188] Hartigan, J.A., Hartigan, P.M.: The Dip Test of Multimodality. *The Annals of Statistics* (1984, contributed)

[189] Hartigan, J.A., Wong, M.A.: A k-Means Clustering Algorithm (Algorithm AS 136). *Journal of The Royal Statistical Society Ser. C: Applied Statistics* **28** (1979), 100–108

[190] Hartung, J., Elpelt, B.: *Multivariate Statistik*. R. Oldenbourg Verlag, München – Wien 1984

[191] Hässig, K.: *Graphentheoretische Methoden des Operations Research*. B.G. Teubner, Stuttgart 1979

[192] Hein, O.: *Graphentheorie für Anwender*. Bibliographisches Institut, Mannheim 1977

[193] Heitz, W.: *Untersuchungen zur Pharmakodynamik und -kinetik des Antihypertensivums Urapidil*. (Dissertation), 2. Medizinische Universitätsklinik, Köln 1987

[194] Henze, N.: The Limit Distribution for Maxima of Weighted r-th Nearest Neighbor Distances. *Journal of Applied Probability* **19** (1982), 334–354

[195] Henze, N.: An Asymptotic Result on the Maximum Nearest Neighbor Distance between Independent Random Vectors with an Application for Testing Goodness-of-Fit in $\mathcal{R}^p$ on Spheres. *Metrika* **30** (1984), 245–260

[196] Hirschfeld, H.D.: A Connection between Correlation and Contingency. *Proceedings of the Cambridge Philosophical Society* **31** (1935), 520–524

[197] Holmes, J.M.C.: A Comparison of Numerical Taxonomic Techniques Using Measurements on the Genera Gammarus and Marinogammarus (Amphipoda). *Biological Journal of the Linnean Society* **7** (1975), 183–214

[198] Hosmer, D.W.: A Comparison of Iterative Maximum Likelihood Estimates of the Parameters of a Mixture of Two Normal Distributions under Three Different Types of Sample. *Biometrics* **29** (1973), 761–770

[199] Hotelling, H.: Relations between Two Sets of Variables. *Biometrika* **28** (1936), 129–149

[200] Hubert, L.J.: Monotone Invariant Clustering Procedures. *Psychometrika* **38** (1973), 47–62

[201] Hubert, L.J.: Approximate Evaluation Techniques for the Single-Link and Complete-Link Hierarchical Clustering Procedures. *Journal of the American Statistical Association* **69** (1974), 698–704

[202] Hubert, L.J., Baker, F.B.: Data Analysis by Single-Link and Complete-Link Hierarchical Clustering. *Journal of Educational Statistics* **1** (1976), 87–111

[203] Hubert, L.J., Baker, F.B.: The Comparison and Fitting of Given Classification Schemes. *Journal of Mathematical Psychology* **16** (1977), 233–253

[204] Hubert, L.J., Baker, F.B.: An Empirical Comparison of Baseline Models for Goodness-of-Fit in r-Diameter Hierarchical Clustering. in: van Ryzin, J. (Ed.): *Classification and Clustering*. Academic Press, New York 1977, pp. 131–153

[205] Hymes, D. (Ed.): *The Use of Computers in Anthropology*. Mouton & Co., London – The Hague – Paris 1965

[206] Ihm, P.: Automatic Classification in Anthropology. in: Hymes, D. (Ed.): *The Use of Computers in Anthropology*. Mouton & Co., London – The Hague – Paris 1965, pp. 357–376

[207] Ihm, P., Himmelmann, G.W., Hinz, U., Fürsch, H.: Taxonometrische Untersuchungen an Epilachna-Stichproben aus Zentralafrika. *Biometrische Zeitschrift* **9** (1967), 159–179

[208] Ihm, P., Trautner, R., Wolf, H.: Lineare algebraische Methoden in der numerischen Taxonomie. *Biometrische Zeitschrift* **13** (1971), 161–202

[209] Ingram, D., Bloch, R. (Eds.): *Mathematical Methods in Medicine (Part 1: Statistical and Analytical Techniques)*. John Wiley and Sons, Chichester – New York 1984

[210] Jambu, M., Lebeaux, M.-O.: *Cluster Analysis and Data Analysis*. North-Holland, Amsterdam 1983

[211] Janowitz, M.F.: Monotone Equivariant Cluster Methods. *SIAM Journal of Applied Mathematics* **37.1** (1979), 148–165

[212] Jardine, C.J., Jardine, N., Sibson, R.: The Structure and Construction of Taxonomic Hierarchies. *Mathematical Biosciences* **1** (1967), 173–179

[213] Jardine, N., Sibson, R.: The Construction of Hierarchic and Non-hierarchic Classifications. *The Computer Journal* **11** (1968), 117–184

[214] Jardine, N., Sibson, R.: A Model for Taxonomy. *Mathematical Biosciences* **2** (1968), 465–482

[215] Jardine, N., Sibson, R.: Choice of Methods for Automatic Classification. *The Computer Journal* **14** (1971), 404–406

[216] Jardine, N., Sibson, R.: *Mathematical Taxonomy*. Wiley, New York 1971

[217] Jarvis, R.A., Patrick, E.A.: Clustering Using a Similarity Measure Based on Shared Near Neighbours. *IEEE Transactions on Computers* **C-22** (1973), 1025–1034

[218] Johnson, S.C.: Hierarchical Clustering Schemes. *Psychometrika* **32** (1967), 241–254

[219] Jost, A.: *Zur Nosologie des Diabetes mellitus mittels Clusteranalyse*. (Dissertation), Psychosomatische Abteilung der Universität, Köln 1985

[220] Jusko, W.J.: Concepts for Population Pharmacokinetics of Theophylline. in: Gladtke, E., Heimann, G. (Eds.): *Pharmacokinetics*. Gustav Fischer Verlag, Stuttgart – New York 1980, pp. 181–190

[221] Kalaba, R., Spingarn, K.: *Control, Identification, and Input Optimization*. Plenum Press, New York – London 1982

[222] Karoński, M.: On a Definition of Cluster and Pseudocluster for Multivariate Normal Populations. *Bulletin of the International Statistical Institute* **45** (1973), 593–598

[223] Karoński, M.: A Review of Random Graphs. *Journal of Graph Theory* **6** (1982), 349–389

[224] Karoński, M.: *Random Graphs for Cluster Analysis*. (Paper presented at the *9. Jahrestagung der deutschen Gesellschaft für Klassifikation, Karlsruhe 1985*.)

[225] Karoński, M., Palka, Z.: On the Size of a Maximal Induced Tree in a Random Graph. *Mathematica Slovaca* **30.2** (1980), 151–155

[226] Karoński, M., Palka, Z.: Addendum and Erratum to the Paper "On the Size of a Maximal Induced Tree in a Random Graph". *Mathematica Slovaca* **31.1** (1981), 107–108

[227] Kelly, F.P., Ripley, B.D.: A Note on Strauss's Model for Clustering. *Biometrika* **63** (1976), 357–360

[228] Kel'mans, A.K.: Asymptotic Formulas for the Probability of k-Connectedness of Random Graphs. *SIAM Journal on the Theory of Probability and Applications* **17** (1972/73), 243–254

[229] Kendall, M.G.: Cluster Analysis. in: Watanabe, S. (Ed.): *Frontiers in Pattern Recognition*. Academic Press, New York – London 1972, pp. 291–309

[230] Kendall, M.G.: The Basic Problems of Cluster Analysis. in: Cacoullos, T. (Ed.): *Discriminant Analysis and Applications*. Academic Press, New York – London 1973, pp. 179–191

[231] Kennedy, J.W.: The Random-Graph Like State of Matter. in: Heller, S.R., Potenzone, R. (Eds.): *Computer Applications in Chemistry*. Elsevier Science Publishers B.V., Amsterdam 1983, pp. 151–178

[232] Kohlsche, A.J.: Entscheidungstheoretische Clusteranalyse-Modelle: Neuentwicklungen und Anwendungen. in: Bock, H.H. (Ed.): *Anwendungen der Klassifikation: Datenanalyse und numerische Klassifikation (Studien zur Klassifikation Bd. 15: Proc. 8. Fachtagung der deutschen Gesellschaft für Klassifikation, Hofgeismar 1984)*. Indeks Verlag, Frankfurt a.M. 1984, pp. 92–99

[233] Kohlsche, A.J.: *Entscheidungstheoretische Clusteranalysemodelle: Eine Bestandsaufnahme von Theorie und Programmen*. (Paper presented at the 9. Jahrestagung der deutschen Gesellschaft für Klassifikation, Karlsruhe 1985.)

[234] Komlós, J., Sulyok, M., Szemerédi, E.: Second Largest Component in a Random Graph. *Studia Scientiarum Mathematicarum Hungarica* **15** (1980), 391–395

[235] Koontz, W.L., Fukunaga, K.: A Nonparametric Valley-Seeking Technique for Cluster Analysis. *IEEE Transactions on Computers* **C-21.2** (1972), 171–178

[236] Koontz, W.L., Fukunaga, K.: Asymptotic Analysis of a Nonparametric Clustering Technique. *IEEE Transactions on Computers* **C-21.9** (1972), 967–974

[237] Koontz, W.L., Narendra, P.M., Fukunaga, K.: A Graph-Theoretic Approach to Nonparametric Cluster Analysis. *IEEE Transactions on Computers* **C-25.9** (1976), 936–944

[238] Krauth, J.: *Grundlagen der Mathematischen Statistik für Bio-Wissenschaftler*. Verlag Anton Hein, Meisenheim 1975

[239] Krauth, J.: Sequentielle Rangklassifikationsverfahren. in: Ihm, P., Dahlberg, I. (Eds.): *Numerische und nicht-numerische Klassifikation zwischen Theorie und Praxis (Studien zur Klassifikation Bd. 10: Proc. 5. Fachtagung der deutschen Gesellschaft für Klassifikation, Hofgeismar 1981)*. Indeks-Verlag, Frankfurt a.M. 1982, pp. 58–66

[240] Krauth, J.: Evaluation von Verfahren der automatischen Klassifikation. in: Dahlberg, I., Schader, M.R. (Eds.): *Automatisierung in der Klassifikation (Studien zur Klassifikation Bd. 13: Proc. 7. Fachtagung der deutschen Gesellschaft für Klassifikation, Königswinter 1983)*. Indeks-Verlag, Frankfurt a.M. 1983, pp. 203–212

[241] Krauth, J.: Classifikation Procedures for Ordered Categorical Data. in: Gaul, W., Schader, M. (Eds.): *Classification as a Tool of Research (Proc. 9th Annual Meeting of the Classification Society, Karlsruhe 1985)*. North-Holland, Amsterdam – New York – Oxford – Tokyo 1986, pp. 249–255

[242] Krauth, J., Lienert, G.A.: *KFA. Die Konfigurationsfrequenzanalyse und ihre Anwendung in Psychologie und Medizin.* Verlag Karl Alber, Freiburg – München 1973

[243] Krishnaiah, P.R., Kanal, L.N. (Eds.): *Handbook of Statistics (Vol. 2: Classification, Pattern Recognition and Reduction of Dimensionality)*. North-Holland, Amsterdam – New York – Oxford 1982

[244] Krolak-Schwerdt, S.: A Graph Theoretic Allocation Criterion for Single Linkage. in: Degens, P.O., Hermes, H.-J., Opitz, O. (Eds.): *Die Klassifikation und ihr Umfeld (Studien zur Klassifikation Bd. 17: Proc. 10. Fachtagung der deutschen Gesellschaft für Klassifikation, Münster 1986)*. Indeks-Verlag, Frankfurt a.M. 1986, pp. 202–210

[245] Kruskal, J.B.: Multidimensional Scaling by Optimizing Goodness-of-Fit to a Nonmetric Hypothesis. *Psychometrika* **29** (1964), 1–27

[246] Kruskal, J.B.: Nonmetric Multidimensional Scaling: A Numerical Method. *Psychometrika* **29** (1964), 115–129

[247] Kuiper, F.K., Fisher, L.: A Monte Carlo Comparison of Six Clustering Procedures. *Biometrics* **31** (1975), 777–783

[248] Kukovetz, W.R. und Mitarb.: Humankinetik und Metabolitenmuster von Urapidil. *Arzneimittel-Forschung (Drug Research)* **27** (1977), 2406–2411

[249] Kusuoka, H., Kodama, S., Maeda, H.: Optimal Control in Compartmental Systems and its Application to Drug Administration. *Mathematical Biosciences* **53** (1981), 59–77

[250] Lachenbruch, P.A.: *Discriminant Analysis.* Hafner Press, New York 1975

[251] Lance, G.N., Williams, W.T.: A General Theory of Classificatory Sorting Strategies. I. Hierarchical Systems. *The Computer Journal* **9** (1967), 373–380

[252] Langenmayr, A., Späth, H.: Cluster-Analyse neurotischer Symptome bei Kindern und Jugendlichen. *Zeitschrift für Klinische Psychologie* **6** (1977), 83–99

[253] Larsen, R.J., Holmes, C.L., Heath, C.W. Jr.: A Statistical Test for Measuring Unimodal Clustering: A Description of the Test and of Its Application to Cases of Acute Leukemia in Metropolitan Atlanta, Georgia. *Biometrics* **29** (1973), 301–309

[254] Laue, R.: *Elemente der Graphentheorie und ihre Anwendung in den biologischen Wissenschaften.* Vieweg, Braunschweig 1971

[255] Lebart, L., Morineau, A., Fénelon, J.-P.: *Statistische Datenanalyse: Methoden und Programme.* Verlag Harri Deutsch, Thun – Frankfurt am Main 1984

[256] Lee, K.L.: Multivariate Tests for Clusters *Journal of the American Statistical Association* **74** (1979), 708–714

[257] Lennington, R.K., Flake, R.H.: Statistical Evaluation of a Family of Clustering Methods. in: Estabrook, G.F. (Ed.): *Proceedings of the 8th International Conference on Numerical Taxonomy.* Freeman, San Francisco 1975, pp. 1–37

[258] Lerman, I.C.: *Les Bases de la Classification Automatique.* Gauthier-Villars, Paris 1970

[259] Ling, R.F.: *Cluster Analysis.* (Technical Report No. 18 – Contract NONR-609(52)), Department of Statistics, Yale University 1971

[260] Ling, R.F.: On the Theory and Construction of k-Clusters. *The Computer Journal* **15** (1972), 326–332

[261] Ling, R.F.: A Probability Theory of Cluster Analysis. *Journal of the American Statistical Association* **68** (1973), 159–164

[262] Ling, R.F.: A Computer Generated Aid for Cluster Analysis. *Communications of the ACM* **16.6** (1973), 355–361

[263] Ling, R.F.: The Expected Number of Components in Random Linear Graphs. *The Annals of Probability* **1.5** (1973), 876–881

[264] Ling, R.F.: An Exact Probability Distribution on the Connectivity of Random Graphs. *Journal of Mathematical Psychology* **12** (1975), 90–98

[265] Ling, R.F., Killough, G.G.: Probability Tables for Cluster Analysis Based on a Theory of Random Graphs. *Journal of the American Statistical Association* **71** (1976), 293–300

[266] Mahalanobis, P.C.: On the Generalized Distance in Statistics. *Proceedings of the National Institute of Sciences of India* **2** (1936), 49–55

[267] Maronna, R., Jacovkis, P.M.: Multivariate Clustering Procedures with Variable Metrics. *Biometrics* **30** (1974), 499–505

[268] Marriott, F.H.C.: Practical Problems in a Method of Cluster Analysis. *Biometrics* **27** (1971), 501–514

[269] Marriott, F.H.C.: Optimization Methods of Cluster Analysis. *Biometrika* **69** (1982), 417–421

[270] Matula, D.W.: k-Komponents, Clusters and Slicings in Graphs. *SIAM Journal of Applied Mathematics* **22** (1972), 459–480

[271] Matula, D.W.: *The Largest Clique Size in a Random Graph.* (Technical Report), Department of Computer Science, Southern Methodist University, Dallas 1976

[272] Matula, D.W.: Graph Theoretic Techniques for Cluster Analysis Algorithms. in: van Ryzin, J. (Ed.): *Classification and Clustering.* Academic Press, New York 1977, pp. 95–129

[273] Matula, D.W.: k-Blocks and Ultablocks in Graphs. *Journal of Combinatorial Theory* **B-24** (1978), 1–13

[274] Matula, D.W.: Divisive vs. Agglomerative Average Linkage Hierarchical Clustering. in: Gaul, W., Schader, M. (Eds.): *Classification as a Tool of Research (Proc. 9th Annual Meeting of the Classification Society, Karlsruhe 1985).* North-Holland, Amsterdam – New York – Oxford – Tokyo 1986, pp. 289–301

[275] McQuitty, L.L.: Elementary Linkage Analysis for Isolating Orthogonal and Oblique Types and Typal Relevancies. *Educational and Psychological Measurement* **17** (1957), 207–229

[276] McRae, D.J.: MICKA, A FORTRAN IV Iterative k-Means Cluster Analysis Program. *Behavioural Science* **16** (1971), 423–424

[277] Meisel, W.S.: *Computer-Oriented Approaches to Pattern Recognition.* Academic Press, New York 1972

[278] Milligan, G.W.: An Examination of the Effect of Six Types of Error Perturbation on Fifteen Clustering Algorithms. *Psychometrika* **45** (1980), 325–342

[279] Milligan, G.W.: A Monte Carlo Study of Thirty Internal Criterion Measures for Cluster Analysis. *Psychometrika* **46** (1981), 187–199

[280] Milligan, G.W.: A Review of Monte Carlo Tests of Cluster Analysis. *Multivariate Behavioral Research* **16** (1981), 379–401

[281] Mojena, R.: Hierarchical Grouping Methods and Stopping Rules: An Evaluation. *The Computer Journal* **20** (1977), 359–363

[282] Möller, D.P.F. (Ed.): *Erwin-Riesch Arbeitstagung: Systemanalyse biologischer Prozesse (1. Ebernburger Gespräch).* Springer-Verlag, Berlin – Heidelberg – New York – Tokyo 1984

[283] Molliere, J.-L.: What's the Real Number of Clusters? in: Gaul, W., Schader, M. (Eds.): *Classification as a Tool of Research (Proc. 9th Annual Meeting of the Classification Society, Karlsruhe 1985).* North-Holland, Amsterdam – New York – Oxford – Tokyo 1986, pp. 311–320

[284] Morineau, A., Lebart, L.: Specific Clustering Algorithms for Large Data Sets and Implementation in SPAD Software. in: Gaul, W., Schader, M. (Eds.): *Classification as a Tool of Research (Proc. 9th Annual Meeting of the Classification Society, Karlsruhe 1985).* North-Holland, Amsterdam – New York – Oxford – Tokyo 1986, pp. 321–329

[285] Müller, G.: *Anwendung von cluster- und diskriminanzanalytischen Verfahren zwecks automatischer Funktionsprüfung des Herzens.* (Paper presented at the *9. Jahrestagung der deutschen Gesellschaft für Klassifikation, Karlsruhe 1985.*)

[286] Murtagh, F.: A Very Fast, Exact Nearest Neighbour Algorithm for Use in Information Retrieval. *Information Technology: Research and Development* **1** (1982), 275–283

[287] Murtagh, F.: Expected-Time Complexity Results for Hierarchic Clustering Algorithms Which Use Cluster Centres. *Information Processing Letters* **16** (1983), 237 241

[288] Murtagh, F.: A Probability Theory of Hierarchic Clustering Using Random Dendrograms. *Journal of Statistical Computation and Simulation* **18** (1983), 145–157

[289] Murtagh, F.: A Survey of Recent Advances in Hierarchical Clustering Algorithms. *The Computer Journal* **26.4** (1983), 354–359

[290] Murtagh, F.: Counting Dendrograms: A Survey. *Discrete Applied Mathematics* **7** (1984), 191–199

[291] Murtagh, F.: A Review of Fast Techniques for Nearest Neighbour Searching. in: International Association for Statistical Computing (Eds.): *COMPSTAT 1984.* Physica Verlag, Wien 1984, pp. 143–147

[292] Myers, B.R.: On Spanning Trees, Weighted Compositions, Fibonacci Numbers, and Resistor Networks. *SIAM Review* **17.3** (1975), 465–474

[293] Nathanson, M.H., Hillman, R.S., Georgakis, C.: Towards an Optimal Drug-Delivery Regimen for Methotrexate Chemotherapy. *Applied Mathematics and Computation* **12** (1983), 99–117

[294] Naus, J.I., Rabinowitz, L.: The Expectation and Variance of the Number of Components in Random Linear Graphs. *The Annals of Probability* **3.1** (1975), 159–161

[295] Needham, R.M.: Automatic Classification: Models and Problems. in: Medical Research Council (Eds.): *Mathematics and Computer Science in Biology and Medicine.* H.M.S.O., London 1965, pp. 111–114

[296] Nicolini, C. (Ed.): *Modeling and Analysis in Biomedicine.* World Scientific Publ. Co., Singapore 1984

[297] Nie, N.H., Hull, C.H., Jenkins, I.G., Steinbrenner, K., Bent, D.H.: *SPSS-X — Statistical Package for the Social Sciences.* McGraw-Hill, New York – San Francisco 1984

[298] Nieder, M., Dilger, C., Haerlin, R.: Quantitation of Urapidil and its Metabolites in Human Serum by High Performance Liquid Chromatography. *Journal of High Resolution Chromatography & Chromatography Communications* **8** (1985), 224–229

[299] Ochs, H.R., Greenblatt, D.J., Bodem, G.: Clinical Pharmacokinetics of Some Antiarrhythmic Drugs. *Herz* **4** (1979), 330–343

[300] Olson, E.C., Miller, R.L.: A Mathematical Model Applied to a Study of the Evolution of Species. *Evolution* **5** (1951), 256–338

[301] Opitz, O. (Ed.): *Numerische Taxonomie in der Marktforschung.* Franz Vahlen, München 1978

[302] Palásti, I.: On the Strong Connectedness of Directed Random Graphs. *Studia Scientiarum Mathematicarum Hungarica* **1** (1966), 205–214

[303] Palka, Z.: Isolated Trees in a Random Graph. *Zastosowania Matematyki (Applicationes Mathematicae)* **17.2** (1982), 309–316

[304] Palka, Z.: On the Number of Given Degree in a Random Graph. *Journal of Graph Theory* **8** (1984), 167–170

[305] Paykel, E.S., Rassaby, E.: Classification of Suicide Attempters by Cluster Analysis. *British Journal of Psychiatry* **133** (1978), 45–52

[306] Pearson, K.: Contributions to the Mathematical Theory of Evolution. 1. Dissection of Frequency Curves. *Philosophical Transactions of the Royal Society Ser. A* **185** (1894), 71–110

[307] Pearson, K.: On Lines and Planes of Closest Fit to Systems of Points in Space. *Philosophical Magazine* **2** (1901), 559–572

[308] Pearson, K.: On the Coefficient of Racial Likeness. *Biometrika* **13** (1926), 105–117

[309] Pearson, K.: Note on Standardisation of Method of Using the Coefficient of Racial Likeness. *Biometrika* **20** (1928), 376–378

[310] Peitgen, H.-O., Richter, P.H.: *The Beauty of Fractals.* Springer, Berlin – Heidelberg – New York – Tokyo 1986

[311] Penrose, L.S.: Distance, Size, and Shape. *Annals of Eugenics* **18** (1954), 337–343

[312] Perkal, J.: Une Méthode Taxonomique et ses Applications aux Sciences Naturelles. *Colloquia Mathematicae* **2** (1951), 319

[313] Perruchet, C.: Significance Tests for Clusters: Overview and Comments. in: Felsenstein, J. (Ed.): *Numerical Taxonomy.* Springer Verlag, Berlin 1983, pp. 199–208

[314] Pollard, D.: Strong Consistency of *k*-Means Clustering. *The Annals of Statistics* **9** (1981), 135–140

[315] Pollard, D.: A Central Limit Theorem for *k*-Means Clustering. *Annals of Probability* **10** (1982), 919–926

[316] Popper, K.L.: *Logik der Forschung.* J.C.B. Mohr (Paul Siebeck), Tübingen 1934, 1982

[317] Ralston, M.L., Jennrich, R.I., Sampson, P.F., Uno, F.K.: *Fitting Pharmacokinetic Models with BMDPAR.* (BMDP Technical Report No. 58), BMDP Statistical Software, Department of Biomathematics, University of California, Los Angeles 1979

[318] Ramesh, A., Ball, M.O., Colbourn, C.J.: Bounds for All-Terminal Reliability in Planar Networks. in: Karoński, M., Palka, Z. (Eds.): *Random Graphs '85 (Annals of Discrete Mathematics 33: Proceedings of the 2nd International Seminar on Random Graphs, Poznań 1985).* North-Holland, Amsterdam – New York – Oxford 1987, pp. 261–273

[319] Rand, W.M.: Objective Criteria for the Evaluation of Clustering Methods. *Journal of the Amarican Statistical Association* **66** (1971), 846–850

[320] Rao, C.R.: The Utilization of Multiple Measurements in Problems of Biological Classification. *Journal of the Royal Statistical Society Ser. B* **10** (1948), 159–203

[321] Rao, C.R.: Cluster Analysis Applied to a Study of Race Mixture in Human Populations. in: van Ryzin, J. (Ed.): *Classification and Clustering*. Academic Press, New York 1977, pp. 175–198

[322] Read, R.C. (Ed.): *Graph Theory and Computing*. Academic Press, New York 1972

[323] Rényi, A.: On Connected Graphs. I. *Publications of the Mathematical Institute of the Hungarian Academy of Sciences* **4** (1959), 385–388

[324] Rescigno, A., Segre, G.: *Drug and Tracer Kinetics*. Blaisdell, Waltham, Ma. 1966

[325] Richter, O.: *Mathematische Modelle für die klinische Forschung: Enzymatische und pharmakokinetische Prozesse*. Springer-Verlag, Berlin Heidelberg New York 1982

[326] Richter, O.: Pharmakokinetische Grundlagen. in: Kuemmerle, H.-P. (Ed.): *Basiswissen Klinische Pharmakologie*. Hippokrates Verlag, Stuttgart 1984, pp. 75–106

[327] Richter, O., Reinhardt, D.: Methods for Evaluating Optimal Dosage Regimens and Their Application to Theophylline. *International Journal of Clinical Pharmacology, Therapy and Toxicology* **20** (1982), 564–575

[328] Riddell, R.J., Uhlenbeck, G.E.: On the Virial Development of the Equation of State of Monoatomic Gases. *Journal of Chemical Physics* **21** (1953), 2056–2064

[329] Riegelman, S., Sheiner, L.B., Beal, S.L.: Population Based Approach to Pharmacokinetics and Bioavailability Studies in Patients. in: Gladtke, E., Heimann, G. (Eds.): *Pharmacokinetics*. Gustav Fischer Verlag, Stuttgart – New York 1980, pp. 83–96

[330] Riggs, D.S.: *The Mathematical Approach to Physiological Problems*. The Williams and Wilkins Company, Baltimore 1963

[331] Rohlf, F.J.: Hierarchical Clustering Using the Minimun Spanning Tree. *The Computer Journal* **16** (1973), 93–95

[332] Rohlf, F.J.: Graphs Implied by the Jardine-Sibson Overlapping Clustering Methods, B_k. *Journal of the American Statistical Association* **69** (1974), 705–710

[333] Rohlf, F.J.: A New Approach to the Computation of the Jardine-Sibson B_k Clusters. *The Computer Journal* **18** (1975), 164–168

[334] Rohlf, F.J.: Single-Link Clustering Algorithms. in: Krishnaiah, P.R., Kanal, L.N. (Eds.): *Handbook of Statistics (Vol. 2: Classification, Pattern Recognition and Reduction of Dimensionality*. North-Holland, Amsterdam – New York – Oxford 1982, pp. 267–284

[335] Röhr, M., Lohse, H., Ludwig, R.: *Statistik für Soziologen, Pädagogen, Psychologen und Mediziner (Bd. 2: Statistische Verfahren)*. Verlag Harri Deutsch, Thun – Frankfurt am Main 1982

[336] Rosen, R.: On Control and Optimal Control in Biodynamic Systems. *Bulletin of Mathematical Biology* **42** (1980), 889–897

[337] Rowland, M., Thomson, P.D., Guichard, A., Melmon, K.L.: Disposition Kinetics of Lidocaine in Normal Subjects. *Annals of the New York Academy of Sciences* **179** (1971), 383–398

[338] Rowland, M., Tozer, T.N.: *Clinical Pharmacokinetics*. Lea & Febiger, Philadelphia 1980

[339] Roy, B.: An Algorithm for a General Constrained Set Covering Problem. in: Read, R.C. (Ed.): *Graph Theory and Computing*. Academic Press, New York 1972, pp. 267–283

[340] Ruspini, E.: Numerical Methods for Fuzzy Clustering. *Information Science* **2** (1970), 319–350

[341] van Ryzin, J. (Ed.): *Classification and Clustering*. Academic Press, New York 1977

[342] Sanchez, D.A., Allen, R.C. Jr., Kyner, W.T.: *Differential Equations — An Introduction*. Addison-Wesley, Reading, Ma. 1983

[343] Schäffer, J.B.: *Beschreibung und Benutzeranleitung des Wishartschen Clusteranalyse-Pakets (CLUSTAN)*. (GSF-Bericht MD295), Gesellschaft für Strahlen- und Umweltschutz mbH, München 1979

[344] Schilling, M.F.: Goodness of Fit Testing in $\mathcal{R}^m$ Based on the Weighted Empirical Distribution of Certain Nearest Neighbor Statistics. *The Annals of Statistics* **11** (1983), 1–12

[345] Schilling, M.F.: An Infinite-Dimensional Approximation for Nearest Neighbor Goodness of Fit. *The Annals of Statistics* **11** (1983), 13–24

[346] Schmidt, B.: Systemanalyse und Modellaufbau. in: Möller, D.P.F. (Ed.): *Erwin-Riesch Arbeitstagung: Systemanalyse biologischer Prozesse (1. Ebernburger Gespräch)*. Springer-Verlag, Berlin – Heidelberg – New York – Tokyo 1984, pp. 16–28

[347] Schneider, B.: Die Logik der Modellbildung. in: Möller, D.P.F. (Ed.): *Erwin-Riesch Arbeitstagung: Systemanalyse biologischer Prozesse (1. Ebernburger Gespräch)*. Springer-Verlag, Berlin – Heidelberg – New York – Tokyo 1984, pp. 3–15

[348]] Schoetensack, W., Bruckschen, E.G., Zech, K.: Urapidil. in: Scriabine, A. (Ed.): *New Drugs Annual: Cardiovascular Drugs*. Raven Press, New York 1983, pp. 19–48

[349] Schubö, W., Uehlinger, H.-M.: *SPSS-X. Handbuch der Programmversion 2*. Gustav Fischer Verlag, Stuttgart – New York 1984

[350] Schultz, J.V., Hubert, L.J.: Data Analysis and the Connectivity of Random Graphs. *Journal of Mathematical Psychology* **10** (1973), 421–428

[351] Schultz, J.V., Hubert, L.J.: An Empirical Evaluation of an Approximate Result in Random Graph Theory. *British Journal of Mathematical and Statistical Psychology* **28** (1975), 103–111

[352] Scott, A.J., Knott, M.: A Cluster Analysis Method for Grouping Means in the Analysis of Variance. *Biometrics* **30** (1974), 507–512

[353] Scott, A.J., Symons, M.J.: Clustering Methods Based on Likelihood Ratio Criteria. *Biometrics* **27** (1971), 387–388

[354] Sheiner, L.B., Beal, S.L.: Evaluation of Methods for Estimating Population Pharmacokinetic Parameters. I. Michaelis-Menten Model: Routine Clinical Pharmacokinetic Data. *Journal of Pharmacokinetics and Biopharmaceutics* **8** (1980), 553

[355] Sheiner, L.B., Beal, S.L.: Evaluation of Methods for Estimating Population Pharmacokinetic Parameters. II. Biexponential Model and Experimental Pharmacokinetic Data. *Journal of Pharmacokinetics and Biopharmaceutics* **9** (1981), 635

[356] Sheiner, L.B., Beal, S.L.: Estimation of Pooled Pharmacokinetic Parameters Describing Populations. in: Endrenyi, L. (Ed.): *Kinetic Data Analysis. Design and Analysis of Enzyme and Pharmacokinetic Experiments*. Plenum Press, New York – London 1981, pp. 271–284

[357] Sheppard, C.W.: The Theory of the Study of Transfers Within a Multi-Compartment System Using Isotopic Tracers. *Journal of Applied Physics* **19** (1948), 70

[358] Shier, D.R.: Testing for Homogeneity Using Minimum Spanning Trees. *The UMAP Journal* **3** (1982), 273–283

[359] Sibson, R.: A Model for Taxonomy II. *Mathematical Biosciences* **6** (1970), 405–430

[360] Sibson, R.: Some Observations on a Paper by Lance and Williams. *The Computer Journal* **14** (1971), 156–157

[361] Sibson, R.: SLINK: An Optimally Efficient Algorithm for the Single-Link Cluster Method. *The Computer Journal* **16** (1973), 30–34

[362] Skarabis, H.: *Mathematische Grundlagen und praktische Aspekte der Diskrimination und Klassifikation*. Physica-Verlag, Würzburg 1970

[363] Sneath, P.H.A.: Some Thoughts on Bacterial Classifications. *Journal of General Microbiology* **17** (1957), 184–200

[364] Sneath, P.H.A.: The Application of Computers to Taxonomy. *Journal of General Microbiology* **17** (1957), 201–206

[365] Sneath, P.H.A.: Recent Developments in Theoretical and Quantitative Taxonomy. *Systematic Zoology* **10** (1961), 118–139

[366] Sneath, P.H.A.: A Comparison of Different Clustering Methods as Applied to Randomly-Spaced Points. *Classification Society Bulletin* **1** (1966), 2–18

[367] Sneath, P.H.A.: The Future Outline of Bacterial Classification. *Classification Society Bulletin* **3** (1968), 28–45

[368] Sneath, P.H.A.: Evaluation of Clustering Methods. in: Cole, A.J. (Ed.): *Numerical Taxonomy*. Academic Press, New York 1969, pp. 257–272

[369] Sneath, P.H.A.: A Method for Testing the Distinctness of Clusters: A Test of the Disjunction of Two Clusters in Euclidean Space as Measured by their Overlap. *Journal of the International Association of Mathematical Geology* **9** (1977), 123–143

[370] Sneath, P.H.A.: Cluster Significance Tests and Their Relation to Measures of Overlap. in: Institute de Recherche en Informatique et en Automatique (IRIA) (Eds.): *Proceedings of the First International Symposium on Data Analysis and Informatics, Versailles 1977*. IRIA, Le Chesnay 1977, pp. 15–36

[371] Sneath, P.H.A.: A Significance Test for Clusters in UPGMA Phenograms Obtained from Squared Euclidean Distances. *Classification Society Bulletin* **4** (1977), 2–14

[372] Sneath, P.H.A.: Basic Program for a Significance Test for 2 Clusters in Euclidean Space as Measured by Their Overlap. *Computers and Geosciences* **5** (1979), 143–155

[373] Sneath, P.H.A., Sokal, R.R.: *Principles of Numerical Taxonomy*. W.H. Freeman Comp., San Francisco 1973

[374] Sodeur, W.: *Empirische Verfahren zur Klassifikation*. B.G. Teubner, Stuttgart 1974

[375] Sokal, R.R.: Distance as a Measure of Taxonomic Similarity. *Systematic Zoology* **10** (1961), 70–79

[376] Sokal, R.R.: Clustering and Classifikation: Background and Current Directions. in: van Ryzin, J. (Ed.): *Classification and Clustering*. Academic Press, New York 1977, pp. 1–15

[377] Sokal, R.R., Michener, C.D.: A Statistical Method for Evaluating Systematic Relationships. *University of Kansas Science Bulletin* **38** (1958), 1409–1438

[378] Sokal, R.R., Michener, C.D.: The Effects of Different Numerical Techniques on the Phenetic Classification of Bees of the Hoplitis Complex (Megachilidas). *Proceedings of the Linnean Society (London)* **178** (1967), 59

[379] Sokal, R.R., Rohlf, F.J.: The Comparison of Dendrograms by Objective Methods. *Taxonomy* **11** (1962), 33–40

[380] Sokal, R.R., Sneath, P.H.A.: *Principles of Numerical Taxonomy.* W.H. Freeman Comp., San Francisco 1961

[381] Sonntag, I.: Random Networks of Catalytic Biochemical Reactions. *Biometrical Journal* **26.7** (1984), 790–807

[382] Sonntag, I.: Applications of the Percolation Theory to Random Networks of Biochemical Reactions. *Biometrical Journal* **26.7** (1984), 809–813

[383] Sørensen, T.: A Method of Establishing Groups of Equal Amplitude in Plant Sociology Based on Similarity of Species Content and its Application to Analyses of the Vegetation on Danish Commons. *Biologiske Skrifter* **5** (1948), 1–34

[384] Sparks, D.N.: Euclidean Cluster Analysis (Algorithm AS 58). *Journal of the Royal Statistical Society Ser. C (Applied Statistics)* **22** (1973), 126–130

[385] Späth, H.: *Cluster-Analyse-Algorithmen zur Objektklassifizierung und Datenreduktion.* R. Oldenbourg Verlag, München – Wien 1977

[386] Späth, H. (Ed.): *Fallstudien Cluster-Analyse.* R. Oldenbourg Verlag, München – Wien 1977

[387] Späth, H. (Ed.): *Ausgewählte OR-Software in FORTRAN.* R. Oldenbourg Verlag, München – Wien 1979

[388] Späth, H.: *Cluster-Formation und -Analyse.* R. Oldenbourg Verlag, München – Wien 1983

[389] Spearman, C.: General Intelligence, Objectively Determined and Measured. *American Journal of Psychology* **15** (1904), 201–293

[390] Spouge, J.L.: Random Graph Problems in Polymer Chemistry. in: Karoński, M., Ruciński, A. (Eds.): *Random Graphs '83 (Annals of Discrete Mathematics 28: Proceedings of the 1st International Seminar on Random Graphs, Poznań 1983).* North-Holland, Amsterdam – New York – Oxford 1985, pp. 251–262

[391] Stahl, H.: Cluster Analysis of Large Data Sets. in: Gaul, W., Schader, M. (Eds.): *Classification as a Tool of Research (Proc. 9th Annual Meeting of the Classification Society, Karlsruhe 1985).* North-Holland, Amsterdam – New York – Oxford – Tokyo 1986, pp. 423–430

[392] Stanfel, L.E.: Applications of Clustering Theory to Cancer Mortality Data. *Computers and Biomedical Research* **19** (1986), 117–141

[393] Steinhausen, D.: *Vergleich von Clusteranalyse-Prozeduren in den Statistikpaketen BMDP, SAS, SPSS-X.* (Paper presented at the *9. Jahrestagung der deutschen Gesellschaft für Klassifikation, Karlsruhe 1985.*)

[394] Steinhausen, D., Langer, K.: *Clusteranalyse — Einführung in Methoden und Verfahren der automatischen Klassifikation.* Walter de Gruyter, Berlin 1977

[395] Stepanov, V.E.: On the Probability of Connectedness of the Random Graph $G_m(t)$. *SIAM Journal on the Theory of Probability and Applications* **15** (1970/71), 55–67

[396] Stolarski, P.: *On the Application of Nonparametric Density Estimators to Classifikation.* (Paper presented at the *9. Jahrestagung der deutschen Gesellschaft für Klassifikation, Karlsruhe 1985.*)

[397] Strauss, D.J.: A Model for Clustering. *Biometrika* **62** (1975), 467–475

[398] Strauss, D.J.: The Evaluation of Clustering Techniques. *Trabajos de Estadistica y de Investigacion Operativa* **28** (1977), 167–182

[399] Swan, G.W.: *Applications of Optimal Control Theory in Biomedicine.* Marcel Dekker, New York – Basel 1984

[400] Szczotka, F.A.: On a Method of Ordering and Clustering of Objects. *Zastosowania Matematyki (Applicationes Mathematicae)* **13** (1972), 23–34

[401] Thomson, P.T., Melmon, K.L., Richardson, J.A., Cohn, K., Steinbrunn, W., Cudihee, R., Rowland, M.: Lidocaine Pharmacokinetics in Advanced Heart Failure, Liver Disease, and Renal Failure in Humans. *Annals of Internal Medicine* **78** (1973), 499–508

[402] Thurstone, L.L.: *Multiple Factor Analysis.* Chicago University Press, Chicago 1947

[403] Torgerson, W.S.: Multidimensional Scaling of Similarity. *Psychometrika* **30** (1965), 379–393

[404] Überla, K.: *Faktorenanalyse.* Springer-Verlag, Berlin – Heidelberg – New York 1971

[405] Uhlenbeck, G.E.: Successive Approximation Methods in Classical Statistical Mechanics. *Physica* **26** (1960), 17–27

[406] Victor, N., Lehmacher, W., van Eimeren, W. (Eds.): *Explorative Datenanalyse (Frühjahrstagung der GMDS, München 1980).* Springer-Verlag, Berlin – Heidelberg – New York 1980

[407] Wagner, J.G.: *Fundamentals of Clinical Pharmacokinetics.* Drug Intelligence, Hamilton, Ill. 1975

[408] Wagner, J.G.: Linear Pharmacokinetic Equations Allowing Direct Calculation of Many Needed Pharmacokinetic Parameters from the Coefficients and Exponents of Polyexponential Equations Which Have Been Fitted to the Data. *Journal of Pharmacokinetics and Biopharmaceutics* **4** (1976), 443–467

[409] Wallenstein, S.R., Naus, J.I.: Probabilities for a k-th Nearest Neighbor Problem on the Line. *Annals of Probability* **1** (1973), 188–190

[410] Wallenstein, S.R., Naus, J.I.: Probabilities of the Size of Largest Clusters and Smallest Intevrals. *Journal of the American Statistical Association* **69** (1974), 690–697

[411] Wambach, G., Godehardt, E., Lang, R., Heitz, W., Meurer, K.-A., Kaufmann, W.: Pharmacodynamics of Urapidil in Essential Hypetrension and in Chronic Renal Failure. in: Amery, A. (Ed.): *Treatment of Hypertension with Urapidil: Preclinical and Clinical Update.* Royal Society of Medicine Services, London – New York 1986, pp. 63–69

[412] Wambach, G., Meurer, K.-A.: Urapidil — Profil eines neuen Antihypertensivums. *Münchner medizinische Wochenschrift* **126** (1984), 1345–1348

[413] Wang, P.C.C. (Ed.): *Graphical Representation of Multivariate Data.* Academic Press, New York 1978

[414] Ward, J.H.: Hierarchical Grouping to Optimize an Objective Function. *Journal of the American Statistical Association* **58** (1963), 236–244

[415] Watanabe, S. (Ed.): *Frontiers in Pattern Recognition.* Academic Press, New York – London 1972

[416] Weber, E.: *Einführung in die Faktorenanalyse.* VEB Gustav Fischer Verlag, Jena 1974

[417] Weiss, L.: A Test of Goodness of Fit Based on the Largest Sample Spacing. *Journal of the Society of Industrial and Applied Mathematics (SIAM Journal)* **8** (1960), 295–299

[418] Westcott, M.: Asymptotic Results for Poisson Cluster Processes. *Advances in Applied Probability* **6** (1974), 227–233

[419] White, R.F., Lewinson, T.M.: Probabilistic Clustering for Attributes of Mixed Type with Biopharmaceutical Applications. *Journal of the American Statistical Association* **72** (1977), 271–277

[420] Whittle, P.: Random Graphs and Polymerisation Prozesses. in: Karoński, M., Ruciński, A. (Eds.): *Random Graphs '83 (Annals of Discrete Mathematics 28: Proceedings of the 1st International Seminar on Random Graphs, Poznań 1983).* North-Holland, Amsterdam – New York – Oxford 1985, pp. 337–348

[421] Williams, W.T., Lance, G.N., Dale, M.B., Clifford, H.T.: Controversy Concerning the Criteria for Taxonometric Strategies. *The Computer Journal* **14** (1971), 162–165

[422] Winter, M.E.: *Basic Clinical Pharmacokinetics.* Applied Therapeutics Inc., Spocane, Wa. 1980

[423] Wirth, M., Estabrook, G.F., Rogers, D.J.: A Graph Thoery Model for Systematic Biology, with an Example for the Oncidiinae (Orchdaceae). *Systematic Zoology* **15** (1966), 59–69

[424] Wishart, D.: Mode Analysis: A Generalization of Nearest Neighbour Which Reduces Chaining Effects. in: Cole, A.J. (Ed.): *Numerical Taxonomy.* Academic Press, New York 1969, pp. 282–308

[425] Wishart, D.: A Numerical Classification Method for Deriving Natural Classes. *Nature* **221** (1969), 97–98

[426] Wishart, D.: An Algorithm for Hierarchical Classification. *Biometrics* **25** (1969), 165–170

[427] Wishart, D.: Treatment of Missing Values in Cluster Analysis. in: Corsten, L.C., Hermans, J. (Eds.): *COMPSTAT 1978.* Physica-Verlag, Wien 1978, pp. 281–287

[428] Wishart, D.: *CLUSTAN-Benutzerhandbuch.* Gustav Fischer Verlag, Stuttgart – New York 1984

[429] Wishart, D.: Hierarchical Cluster Analysis with Messy Data. in: Gaul, W., Schader, M. (Eds.): *Classification as a Tool of Research (Proc. 9th Annual Meeting of the Classification Society, Karlsruhe 1985).* North-Holland, Amsterdam – New York – Oxford – Tokyo 1986, pp. 453–460

[430] Wolfe, J.H.: Pattern Clustering by Multivariate Mixture Analysis. *Multivariate Behavioural Research* **5** (1970), 329–350

[431] Wolfe, J.H.: *A Monte Carlo Study of the Sampling Distribution of the Likelihood Ratio for Mixture of Multinormal Distribution.* (Technical Bulletin STB 72-2), U.S. Naval Personnel and Training Research Laboratory, San Diego 1981

[432] Wong, M.A.: A Hybrid Clustering Algorithm for Identifying High Density Clusters. *Journal of the American Statistical Association* **77** (1982), 841–847

[433] Wong, M.A., Lane, T.: A kth Nearest Neighbor Clustering Procedure. *Journal of the Royal Statistical Society Ser. B* **45** (1983), 362–368

[434] Wright, E.M.: The Number of Connected Sparsely Edged Graphs. *Journal of Graph Theory* **1** (1977), 317–330

[435] Wright, E.M.: The Number of Connected Sparsely Edged Graphs. II. Smoth Graphs and Blocks. *Journal of Graph Theory* **2** (1978), 299–305

[436] Wright, E.M.: The Number of Connected Sparsely Edged Graphs. III. Asymptotic Results. *Journal of Graph Theory* **4** (1980), 393–407

[437] Wright, W.E.: An Axiomatic Specifikation of Euclidean Cluster Analysis. *The Computer Journal* **17** (1974), 355–364

[438] Young, F.W., Edds, T., Kent, D., Kuhfeld, W.F.: Interactive Hypergraphics for Data Analysis. in: Gaul, W., Schader, M. (Eds.): *Classification as a Tool of Research (Proc. 9th Annual Meeting of the Classification Society, Karlsruhe 1985)*. North-Holland, Amsterdam – New York – Oxford – Tokyo 1986, pp. 461–470

[439] Zahn, C.T.: Graph-Theoretical Methods for Detecting and Describing Gestalt Clusters. *IEEE Transactions on Computers* **C-20** (1971), 68–87

[440] Zajicek, G., Maayan, Ch., Rosenmann, E.: The Application of Cluster Analysis to Glomerular Histopathology. *Computers in Biomedical Research* **10** (1977), 471–481

[441] Zech, K., Sturm, E., Steinijans, V.: Pharmakokinetik und Metabolismus von Urapidil bei Tier und Mensch. in: Kaufmann, W., Bruckschen, E.G. (Eds.): *Urapidil: Darstellung einer neuen antihypertensiven Substanz*. Excerpta Medica, Amsterdam – Genf – Princeton – Tokio 1982, pp. 50–64

[442] Zito, R.A., Reid, P.R., Longstreth, J.A.: Variability of Early Lidocaine Levels in Patients. *American Heart Journal* **94** (1977), 292–296

[443] Zupan, J.: *Clustering of Large Data Sets*. Research Studies Press, Letchworth 1982